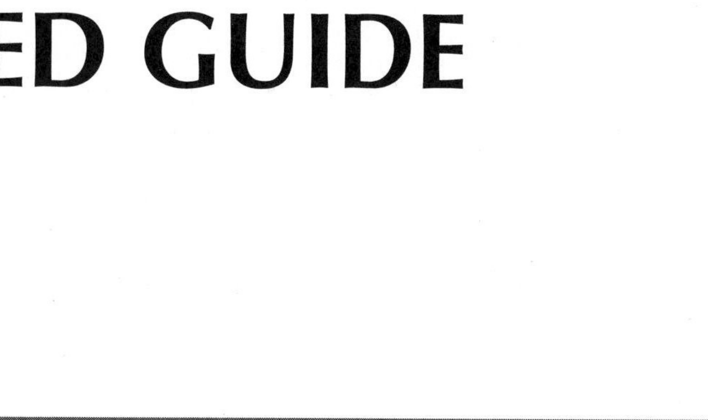

AN ILLUSTRATED GUIDE TO PRUNING

AN ILLUSTRATED GUIDE TO PRUNING

Second Edition

EDWARD F. GILMAN

DELMAR
THOMSON LEARNING™

Australia Canada Mexico Singapore Spain United Kingdom United States

An Illustrated Guide to Pruning, Second Edition
by Edward F. Gilman

Business Unit Director:
Susan L. Simpfenderfer

Executive Editor:
Marlene McHugh Pratt

Acquisitions Editor:
Zina M. Lawrence

Developmental Editor:
Andrea Edwards

Editorial Assistant:
Elizabeth Gallagher

Executive Production Manager:
Wendy A. Troeger

Production Manager:
Carolyn Miller

Production Coordinator:
Matthew J. Williams

Executive Marketing Manager:
Donna J. Lewis

Channel Manager:
Nigar Hale

Cover Design:
Dutton and Sherman Design

Printed in the United States
2 3 4 5 XXX 05 04 03 02

For more information contact Delmar, 3 Columbia Circle, PO Box 15015, Albany, NY 12212-5015.

Or find us on the World Wide Web at
www.thomsonlearning.com or www.Agriscience.Delmar.com

Library of Congress Cataloging-in-Publication Data

Gilman, Edward F.
An illustrated guide to pruning / Edward F. Gilman—2nd ed.
p. cm.
Rev. ed. of: Trees for urban and suburban landscapes. 1997.
Includes bibliographical references (p.).
ISBN: 0-7668-2271-0
1. Ornamental trees—Pruning. 2. Trees in cities. I. Gilman, Edward F. Trees for urban and suburban landscapes. II. Title.

SB435.76 .G54 2001
635.9'77—dc21 2001055805

NOTICE TO THE READER

TABLE OF CONTENTS

PREFACE

One of the most rapidly expanding, yet untapped, areas of plant care in landscapes is pruning technology. This newfound interest is due to increased number of plants installed in landscapes and a new awareness by the general public and by professional aborists and horticulturists that proper pruning can influence risk, plant quality, growth rate, plant health, longevity, and aesthetics. Though there are many basic texts written about pruning, there is a need for one that is extremely well illustrated that also includes more advanced techniques and protocol based on the latest research.

The basic and advanced techniques presented in hundreds of illustrations, photographs, and charts will bring a new technical understanding to students and professionals alike. Nursery operators, arborists, and landscape maintenance professionals will find this an indispensable guide for training and sales. In addition to all of the common techniques, instructors will find that this text presents many new ideas and technology that are just beginning to emerge in the green industry.

Foremost in the text is an examination of which stems and branches to remove from trees and shrubs to effect change in their architecture. This is the essence of pruning and the cornerstone of preventive aboriculture. Special care has been taken to present highly technical topics such as nursery pruning and structural tree pruning in the landscape in a comprehensive yet understandable manner. Less focus is placed on orchard fruit tree production and conifer pruning because there are fewer pruning issues on these plants. Detailed illustrations are used throughout to present all major topics. Many readers will find that they can gain tremendous insight simply by reviewing the illustrations and charts.

Many changes and additions have been made to the second edition of *An Illustrated Guide to Pruning*. Most changes suggested by current users of the Guide were incorporated. Eight new chapters were added. A chapter on shrub pruning completes the treatment of all woody landscape plants. Two new chapters present the objectives and protocol for growing trees in a production wholesale nursery. Two new chapters examine the structural pruning of shade trees in established landscapes. One new chapter takes a close look at industry standards and guides the reader through the process of writing pruning specifications for landscape trees. The differences between pruning shade trees and pruning small ornamental and fruit trees are made clearer. Through his extensive travel around the world, the author presents a global approach to pruning by including examples from temperate and tropical climates. The second edition of *An Illustrated Guide to Pruning* is a much more complete pruning book.

ADDED FEATURES

- Objectives and Key Words are now listed at the start of every chapter to make it easier to focus on important topics. Three new sections, "Check Your Knowledge," "Challenge Questions," and "Suggested Exercises" were added at the end of every chapter to help put into practice those key points presented in each unit.
- Dozens of illustrations have been added to make learning even easier.
- Tree structure and biology chapters introduce all the terms and principles needed to understand why pruning should be performed in the prescribed manner.
- Nursery production pruning chapters emphasize the need for beginning with quality stock and show how to develop protocols for pruning virtually any tree.
- Landscape structural pruning chapters devote many illustrations to understanding why pruning is required for safe, sustainable landscapes and how to accomplish this.

ENHANCED CONTENT

- Glossary entries have been doubled to nearly 200 terms.
- Established mature tree pruning has been expanded to include restoration pruning on abused or damaged trees.
- The specifications section has been expanded with more examples so the reader can better understand how to develop good pruning specifications.
- Dozens of illustrations have been added and modified to expand understanding of important terms and techniques.
- This book provides extensive coverage of the skills needed to conduct all the pruning types for trees and shrubs in urban and suburban landscapes.

ABOUT THE AUTHOR

The author began teaching pruning in 1986 since arriving at the University of Florida. He received a Bachelors of Science in Forestry and a Ph.D. in Plant Pathology from Rutgers University. Before joining the faculty at the University of Florida, he worked in the tree care and landscape industry for four years. He continues to conduct tree care seminars and workshops on many continents and received the 1999 author citation award from the International Society of Arboriculture (ISA) for his writing in tree care. The American Horticulture Society named him to the G. B. Gunlogson award in 2001 for creative development of horticultural software. He is a past president of the Florida chapter ISA and urban forestry council and continues to serve on numerous national and regional professional committees.

ACKNOWLEDGMENTS

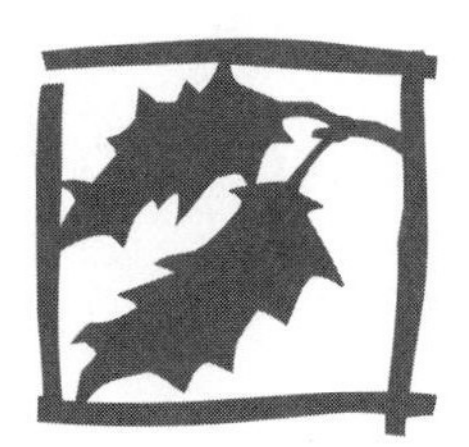

Special thanks go to the following people for taking the time and effort to critically review this manuscript: Dane Buell, Dr. Kim Coder, Dr. Jason Grabosky, C. Way Hoyt, Geoffrey Kempter, Michael Marshall, Brent Reeves, Dr. Dennis P. Ryan III, Dr. Tom Smiley, and Dr. Kevin Smith.

Delmar and the author also wish to express their thanks to the content reviewers. Their input and expertise added greatly to this new edition.

Leslie A. Blackburn
Boise State University
Boise, Idaho

Glenn Herold
Illinois Central College
East Peoria, Illinois

Philip Gibson
Gwinnett Technical College
Lawrenceville, Georgia

I also want to thank the many nurseries such as Select Trees, Skinners Nursery, and Marshall Tree Farm who continue to allow me to learn with them.

DEDICATION

To my three girls, Betsy, Samantha, and Megan

CHAPTER 1

INTRODUCTION

OBJECTIVES

1) Decide if and why trees need to be pruned.
2) Present a protocol for inspecting, evaluating, and pruning trees.
3) Introduce the objectives and strategies of pruning trees.
4) List the faults in trees that are correctable and not correctable with pruning.

KEY WORDS

Border tree
Defects
Good structure
Mature tree
Medium-aged trees
Poor structure
Preventive tree care
Pruning cycles
Pruning objectives
Root problems
Root pruning
Structure
Tree evaluation
Tree inspection

INTRODUCTION

Historically, pruning has been performed in response to the short-term desires of people, with not enough thought given to how pruning affects tree and shrub structure and health. This philosophy continues today. These two objectives are not mutually exclusive and can both be met after the elements of tree biology, growth, and development have been understood. This book examines the basic elements (the science) before presenting the principles of pruning and training (the art). A quality pruning program incorporates the science into the art to deliver trees and shrubs that are healthy, strong, and aesthetically pleasing.

Proper pruning of young trees by removing or shortening live branches has a significant impact on their future growth. Second to placing a tree in a proper location, planting it correctly, and protecting the root system from injury, pruning probably has the biggest impact on longevity. Pruned trees are likely to live longer and be stronger and healthier than trees that are not pruned. Planting properly trained trees with **good structure** makes it easier for caretakers in the landscape to complete the job begun in the nursery of developing and maintaining structurally sound trees.

Unfortunately, many trees are not trained or pruned properly in the nursery or landscape. Nursery operators, horticulturists, landscape managers, and arborists have a responsibility to their customers to learn how to prune young and **medium-aged trees** in order to minimize the need for expensive corrective measures later. Homeowners also should have a basic understanding in order to evaluate the pruning needs in their yards. Pruning young trees to prevent problems later is a simple process and is inexpensive compared to treating problems later. I used to think it was easy to learn how to prune, but I have come to realize it is not easy for a portion of the population to learn new pruning practices. Some people can "see" what needs to be done after learning some basic principles. Others need much more detail and many examples before they grasp the concepts. Keep this in mind when you teach pruning.

Methods and techniques are now available for **preventive tree care** and management. The communities that have implemented this type of management have found that trees are better able to withstand stresses such as ice, snow, and other storm damage. Pruning live branches, especially in the upper portion of the canopy, is a big part of preventive tree care, and preventive strategies are described in detail in this book. A preventive program places trees under the care of a professional who develops a plan that should include a pruning program designed to meet specific objectives. Once objectives are established each tree or group of trees is placed on a **pruning cycle.** Specific **pruning objectives** should be defined and met with each pruning.

Certain pruning practices, such as flush cuts, and removing large branches and large codominant stems are well known to initiate decay inside the trunk and branches. This book provides guidelines for developing and maintaining sound trunk and branch structure by removing small-diameter live branches. *Live branch pruning in the upper and middle portion of the tree is the most important procedure on young and medium-aged trees.*

Few pruning books published in the past have included enough meaningful drawings and photographs to be useful to me or to my students. This book was written around the drawings and photographs. They serve as the main teaching tool because you cannot learn pruning by reading words. The only way to learn how to prune and train trees is to learn the basic techniques then prune them and watch how they respond over the next several months or years. The concepts presented in the drawings will provide enough information to allow you to begin pruning trees quickly, correctly, and more efficiently. They are not based on theory, but are founded in research and the experience of many individuals in the landscape, nursery, and arboriculture professions. As you read portions of this book, go out and prune some trees. Try something presented in these pages that you have not done before. To teach yourself a new technique and build confidence, you can first try it on a small number of trees. Then incorporate the techniques into your operation.

OBJECTIVES OF PRUNING

The main objectives of pruning trees are to create and maintain strong **structure** by guiding the tree's architecture, and to produce a healthy tree with a functional and pleasing form by removing the smallest possible amount of living tissue at any one pruning. This means that pruning should begin early in the life of a tree, and should be performed at regular intervals when a tree is young. This prevents lateral branches and stems from growing too fast and spoiling good structure. It may take twenty-five to thirty years of regular pruning to develop good trunk and branch structure. Each time, live branches should be removed to direct growth to more desirable tree parts.

 **Tip:** *Create good, strong structure by removing live branches to direct growth.*

In many ways, pruning corrects **defects** (Tables 1-1 and 1-2). The job of the pruner becomes recognizing the defects and deciding on the treatment. Young trees should be pruned regularly to develop a strong structure so that they can live many years without creating a safety hazard. Older trees are pruned to maintain good structure and to minimize **poor structure** and other conditions in the tree that could place people or property at risk. All defects in the tree may not be correctable with pruning (Table 1-3).

TABLE 1-1. Defects in young and medium-aged trees wholly or partially correctable with pruning.

Multiple leaders	Dead branches
Flat tops	Small diameter circling roots
Forks	Clustered branches
Branch unions with included bark	Unwanted flower or fruit production
Rubbing branches	Topped trees
Pest-infested branches	Lion's-tailed trees
Deformed branches	Root loss
Trees spaced too closely in the nursery	Some foliage or twig diseases
Low, codominant stems	Branchless trunks
Long branch stubs	Side branches taller than the leader
Fast growing branches on young trees	Water sprouts
Dense canopy	Double leaders

TABLE 1-2. Defects in medium-aged and mature trees partially correctable with pruning.

Defects
Branch unions with included bark
Branches with cracks
Dead branches
Branches blocking a view
Branches hanging too low over a building
Branches rubbing each other
Long branches with most weight at the ends
Dense canopy shading turf below
Some vascular wilt diseases (DED)
Trees topped through sapwood only
Lion's tailed trees
Storm-damaged
Leaning trees
Heavy branches with large diameter compared to the trunk
Multiple branches at same point on trunk
Unbalanced canopy
Circling roots

TABLE 1-3. Defects in mature trees difficult to correct with pruning.

Defects
Root loss, as during construction
Decayed roots near trunk
Topped trees if cuts were made through heartwood
Codominant stems
Hollowed or decayed trunks
Heavy top
Crack in main branch or trunk

Regardless of tree age, you should usually remove branches that are dead, broken, split, dying, diseased, or touching other branches. *Never prune if you do not know why the tree needs pruning,* and do not indiscriminately remove branches with live foliage because this stresses the tree. Light pruning causes only slight stress, but heavy pruning can introduce severe stress because of the large amount of stored energy removed with the branches. Overpruned trees also lose too much foliage resulting in a reduction of photosynthesis. A loss in photosynthetic capacity can reduce the tree's ability to defend itself against insects, disease, and mechanical injury. Root growth can also be slowed, initiating a downward spiral in vigor.

Trees, such as many of the elms and oaks, that mature at a height greater than about 35 feet should be trained so that they develop branches spaced along one dominant trunk. This form has proven more resistant to storm damage than trees with multiple trunks (Wood et al., 2001). Branches should be considerably smaller in diameter than the trunk (preferably less than half). This helps secure them to the tree and helps the main trunk dominate the tree. This is accomplished by removing live branches from fast-growing limbs to slow their growth. Small maturing, ornamental trees can be trained with one or several trunks depending on the landscape situation and wishes of the property owner. Fruit trees are pruned to create strong architecture and generate healthy fruit that is easy to pick.

Healthy trees with roots damaged or removed during construction of a building or parking lot are sometimes pruned with the intention of aiding survival. The least vigorous branches are pruned from the outside of the canopy, removing no more than about 15 to 20 percent of the live foliage. There is one research report supporting this and many arborists believe it improves tree survival. There is some evidence that dieback on medium-aged trees can be minimized with this treatment. Of course the key to tree recovery is soil moisture management and providing loose soil suitable for rapid root growth beyond the point of root injury. Live branches *should not* be removed from trees in poor health; they need abundant photosynthetic capacity to help them regain health.

Tip: *It is better to remove a small amount of live foliage often than a lot all at once.*

Root pruning has been practiced for centuries to prepare trees for transplanting. Trees that are root-pruned inside of what will become the root ball prior to transplanting appear to undergo less shock at transplanting than those that are not root-pruned (Gilman, 1997). An integral part of the art of bonsai is root pruning, which slows the growth rate and helps keep the tree very small. On construction sites, roots of existing trees are pruned, occasionally, with mechanical root-pruning devices and hand saws, before construction begins, with the hope of preparing the tree for the impact of construction. If loose soil with good drainage and adequate soil moisture is provided beyond the point of root pruning, new roots will often generate quickly, helping the tree to recover. As a part of reconstruction efforts, the roots are pruned when they raise sidewalks and curbing. This could cause tree instability and tree failure if done too close to the trunk. Roots circling a nursery container should be cut at planting to prevent subsequent root girdling of the trunk. Roots growing close to the trunk in a circling fashion should also be cut, especially if they are small. Although a detailed discussion of this process is beyond the scope of this book, it is important to point out that root cuts should be made cleanly with a sharp pruning tool. Tearing roots with heavy equipment or cutting with dull shovels is not recommended. New roots often grow from just behind a clean cut, whereas beyond a jagged tear, they will die back for several inches to as much as several feet (Figure 1-1).

PRUNING STRATEGIES

Appropriate pruning strategies depend on tree age (Table 1-4). Live branches should be regularly shortened and removed on nursery trees in order to accomplish objectives. Leave as much foliage as possible on nursery trees in order to speed growth. *Many growers remove too many lower branches too soon resulting in slow growth and weak trunks.*

Live branches usually need to be removed from young and medium-aged landscape trees to accomplish desired objectives. There is no other way to establish a dominant leader or space branches apart. This is the best method for establishing good structure by directing future growth. On the other hand, few live branches should be removed from mature trees because they need as much foliage as possible to remain healthy. Mature tree pruning concentrates on reducing conditions in the tree that could place

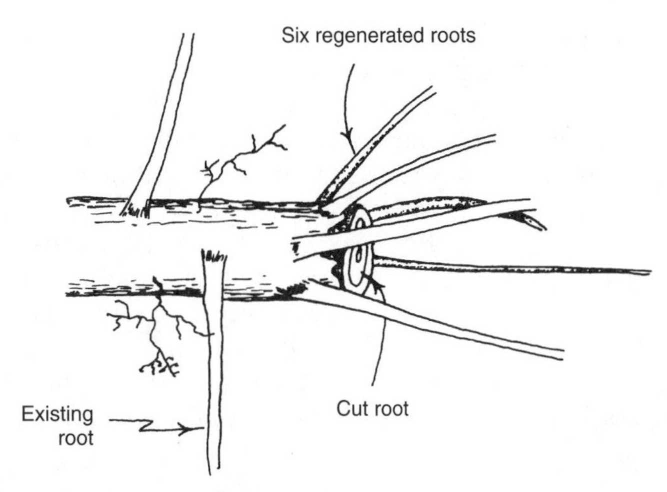

FIGURE 1-1. New roots originate from just behind the cut root. Existing roots are often stimulated and their growth rate increases in response to root-pruning.

people or property at risk by removing dead branches, and reducing length on very long branches or those that are poorly attached.

INSPECT THE TREE BEFORE CLIMBING

Trees should be inspected carefully before they are climbed or pruned to check for potentially unsafe conditions (Table 1-5). **Tree inspection** is a careful process of checking the trees for defects that could lead to tree failure while climbing. These conditions could result in a climber getting injured or killed while in the tree. Most climbers' deaths occur when trees break or fall over with the climber in it. **Root problems** are often the cause of this tree failure so check for these problems carefully. Root problems to look for include roots circling close to the trunk, cut roots (especially close to the trunk), decayed roots, lack of trunk flare, deep planting, and mulch piled close on the trunk.

Tip: *Inspect the tree before climbing to check for unsafe conditions.*

Trunk and branch decay and cracks can also cause failure of tree parts when climbers enter the tree. Dead branches could become dislodged. One of the sprouts on a previously topped tree could fail if a climber secures a safety rope to it. Trees left to stand alone following removal of surrounding trees could be unstable and fail from the added weight of a climber. Last but not least, be sure climbing equipment conforms to the national safety standard, American National Standards Institute (ANSI) Z133.1, prior to entering the tree.

EVALUATE THE TREE BEFORE PRUNING

After inspecting the tree to determine if it is safe to climb, step away and walk around it at a little distance before you begin to prune. This process of **tree evaluation** is crucial to providing quality tree care. If the tree is part of a landscape, take special note to

TABLE 1-4. Strategies for pruning shade trees of different ages in approximate order of importance.

YOUNG TREES IN THE NURSERY:
Eliminate circling roots with specially-designed containers and root pruning
Establish strong structure by developing one dominant trunk
Shorten or remove aggressive low branches
Leave shortened lower branches on trunk until about 12 months (warm climates) to 24 months (cooler climates) before marketing
Space main branches along trunk by shortening others
Establish a pleasing form and full canopy
Eliminate touching branches
YOUNG TREES IN THE LANDSCAPE:
Eliminate circling roots on the soil surface by cutting them
Establish strong structure by developing and maintaining one dominant trunk
Shorten aggressive low branches
Space main branches along trunk by shortening others
Eliminate touching branches
MEDIUM-AGED TREES:
Cut girdling roots and other roots circling close to the trunk
Maintain or establish one dominant trunk by reducing length of others
Shorten branches below lowest permanent limb
Shorten aggressive low branches that will be in the way later
Prevent stems on low branches from growing up into the permanent canopy
Space main branches 18 to 36 inches apart by shortening others
Reduce length of over-extended branches
Remove dead branches
Thin edge of canopy
Eliminate touching branches
MATURE TREES:
Remove dead branches
Minimize potential hazards by reducing length of over-extended limbs
Thin branches from the edge of the canopy to reduce wind resistance
Remove as little live tissue as possible to accomplish objectives

look at it from the angle from which people are most likely to view it. Ask yourself what functions the tree is providing the property. Will pruning maintain or enhance those functions? Determine why you are pruning this tree. It is to:

- develop or improve the architecture or structure of the trunk and branches.
- remove structural defects or restore storm-damaged branches.
- reduce risk in the landscape and improve tree health by removing dead branches.
- reduce weight at the ends of long or defective branches to minimize the potential for branch breakage.
- clear branches from a walkway, street, or structure.
- alter the form.

TABLE 1-5. Inspect trees for root, trunk and other problems before climbing them.

ROOT PROBLEMS	
Circling roots	Soil cracking close to trunk
Roots absent from one side	Root plate separating from surrounding soil
Severed roots	Flat lower trunk
Mushrooms (Root rot)	Tree planted too deeply
Deflected roots	No root flare at soil line
Shallow root system	Cracks in roots
TRUNK PROBLEMS	
Fungus (mushroom) conks on the trunk	
Large branches previously removed	
Open trunk and branch hollows	
Decayed trunk	
Cracks and cavities in branches and trunk	
Abnormal taper or bulges in trunk	
SOME OTHER PROBLEMS	
Dead branches hanging in tree	
Included bark in branch unions	
Previous topped tree	
Former forest-grown tree now in the open	
Stinging insects	
Electrical hazards	
Severe lean, especially if it appears to have occurred recently	

- enhance flowering.
- eliminate fruit.
- reduce size.
- slow growth rate.
- thicken or thin the canopy.
- remove diseased tissue.
- direct future growth.
- shape the canopy.

After deciding why pruning is needed and determining the customer's expectations and budget, visualize the tree you are about to prune as it will appear ten, twenty, thirty years from now. Then choose the branches that should be pruned to provide the desired effect and remove them in the correct manner. Table 1-6 summarizes the general protocol for pruning.

Determine if the tree can tolerate removal of live branches. Very old trees, trees in poor health or on the decline, trees with large sections of the bark missing, and recently transplanted trees may best be left unpruned, as this provides the most leaf tissue and can help them regain their health. This happens because growth regulators and carbohydrates produced in the leaves help initiate and maintain the root growth needed to help these trees recover. Removing live branches slows recovery by reducing root

TABLE 1-6. Protocol for pruning.

Step 1: Step away from the tree
Step 2: Determine why the tree needs pruning; i.e. what are your objectives?
Step 3: Visualize the tree five to thirty years from now
Step 4: Select branches and stems to be removed or shortened
Step 5: Remove or shorten branches or stems properly
Step 6: Notify the customer of any potentially hazardous conditions in the trees

growth. Instead of pruning live branches from a historic tree with special importance to a community, consider other alternatives such as cabling or bracing to help keep the tree healthy. Consider moving wires and other utilities instead of reducing a large tree to provide clearance.

LEGAL OBLIGATIONS

Be sure to check with neighbors if you will be pruning trees located on the border of a property. Obtaining written permission to prune a **border tree** is a good way to help prevent misunderstandings. In many states, a tree located on an adjacent property cannot be pruned by you in a manner that devalues the tree. For example, simply pruning a tree back to the edge of the property is inappropriate and could leave you with a liability. The same goes for root pruning. You may be under legal obligation to follow the latest industry standards such as ANSI A300 and ANSI Z133.1 (see Chapter 15) when conducting tree care services. Be sure to check these standards before proceeding.

If you are an arborist pruning landscape trees, you may have a legal obligation to notify the customer if you see conditions in trees that could place people or property at risk. For example, if during your inspection, you notice that a branch is so poorly secured to the trunk that it could split from the tree and fall on the house, you might be legally required to notify the potential customer in writing (keep a copy).

PRUNING SEVERITY

Pruning live branches causes injury to a tree and elicits a wound response. For this reason, it is best to develop a management plan designed to remove as little live foliage as possible each time the tree is pruned, while meeting the pruning objective. But remember, pruning live branches is *essential* for developing good trunk and branch structure. Trees and individual limbs that are pruned lightly or not at all experience the most growth. Total tree growth will be greater if more branches are left on the tree, but individual shoots will be shorter. Removing several small branches rather than one or two big branches takes more time but results in a tree that sprouts less, looks more natural, and appears less like it was pruned. However, you should not interpret this principle as a license to ignore a branch that should be removed or pruned to improve structure.

When young nursery trees are pruned for the first time at two years or older, a large portion of the canopy may have to be removed to encourage proper trunk and branch development. Vigorous sprouts often emerge from undesirable positions on a tree in response to severe pruning, whereas light annual or twice annual pruning helps maintain a nice-looking tree form and minimizes sprouting. Generally, if more than ¼ (medium-aged trees) to ⅓ (young trees) of the foliage must be removed, you have waited too long to prune. Whereas at one pruning you could remove branches holding about half the foliage on a young, nursery-sized tree, sprouting would be minimized and the tree stressed less if you removed half of the undesirable branches now and waited for between several months and a year to remove the others.

Mature trees with well-develop structure need minimal pruning of their live branches. Thinning, cleaning the canopy, or other procedures described in the later chapters may be desirable in some instances, but you should remove no more than about 10 to 15 percent of the live foliage. As trees grow older, you should decrease the amount of living tissue you remove, but increase the removal of dead branches. Live branches should be removed from mature trees only with good reason. Overpruning large trees can be extremely damaging, resulting in sunscald, decay, excessive sprouting, excess loss of the photosynthetic area, root decline, and even death. The negative effects on the root system from overpruning are often underestimated because the roots are below ground and often forgotten.

Pruning shoots and branches slows or stops root growth while the tree attempts to replace removed branches.

CHECK YOUR KNOWLEDGE

1) One good way to prune a healthy medium-aged tree that has sustained root damage from construction equipment would be to:

   a. remove interior foliage shaded by the outside branches.
   b. remove the least vigorous branches from the outside of the canopy.
   c. never prune live branches from a root-damaged tree.
   d. remove live branches in proportion to estimated root loss.

2) Which of the following faults or tree defects are partially correctable in vigorous mature trees?

   a. decayed roots near the trunk
   b. codominant stems
   c. dense canopy shading turf below
   d. cracks in large limbs and branches

3) Which of the following would NOT be appropriate reasons for pruning?

   a. improve trunk and branch architecture
   b. reduce weight at ends of long branches
   c. eliminate fruit production
   d. reduce the size of the canopy near power lines
   e. all of the above are appropriate

4) Which of the following would be LEAST appropriate in young trees?
   a. remove dead branches only
   b. remove live branches to help develop good, strong structure
   c. remove live branches from the end of an aggressive, low limb
   d. shorten some branches to space main limbs along the trunk

5) Which of the following would be LEAST appropriate in mature trees?
   a. remove dead branches only
   b. thin some live branches from the edge of the canopy to reduce wind resistance
   c. remove live branches from the end of a long, low limb
   d. reduce the canopy to increase light to turf below

6) What is one of the leading causes of climber injury and death when working in trees?
   a. tree falls over with climber in it due to root problems
   b. tree splits at point of included bark
   c. decayed area on a branch breaks
   d. safety rope fails

Answers: b, c, e, a, d, a

CHALLENGE QUESTIONS

1) List five faults or defects in mature trees that are at least partially correctable with pruning.
2) What is the main difference in strategy between pruning mature trees and pruning medium-aged trees?

SUGGESTED EXERCISES

1) Walk in a nearby landscape and determine what faults are evident on several trees of different ages.
2) Divide into two or more groups of about five people. Select as many trees as there are groups. Have each group evaluate and inspect one tree. Have each group prepare a written and oral report to the entire class in the field by the tree. Have the rest of the class critique the presentation and the report.

CHAPTER 2

PLANT SELECTION AND MANAGEMENT: THE BEST WAY TO MINIMIZE PRUNING NEEDS

OBJECTIVES

1) Determine which trees to save on a construction site to minimize pruning needs later.
2) Distinguish between good-quality and poor-quality nursery trees.
3) Show how landscape design and management practices can influence pruning needs.
4) Contrast the advantages and disadvantages of planting trees with different canopy forms.
5) Show the possible consequences of not pruning decurrent trees.

KEY WORDS

Codominant stems
Cultivars
Decurrent growth
Dominant trunk
Excurrent growth
Fail
Good structure (form)
Included bark
Large wound
Live crown ratio
Multiple trunks
Poor form
Quality nursery tree
Single-leadered tree
Species selection
Tree habit
Weak branch attachments

GOOD URBAN DESIGN

The best methods for reducing the need for pruning in urban landscapes include designing the site appropriately and choosing the correct tree or shrub for the location (Table 2-1 and Gilman, 1997). Good site design incorporates enough soil space to allow for adequate root growth. Good design also reduces or prevents conflicts between trees and urban structures. Well-designed landscapes allow trees and shrubs to grow without the need for drastic pruning measures to keep trees in bounds. Trees and shrubs that grow to a large size can be a maintenance problem requiring frequent pruning if placed in a location better suited for a smaller plant.

SPECIES SELECTION

The pruning requirement at a landscape site can be minimized through proper **species selection.** Unfortunately, trees are frequently placed in the landscape according to their current size and shape, and not the size they will attain in thirty to fifty years. The practice of failing to locate the tree according to the mature size anticipated at that site often guarantees that the tree will need regular pruning to keep it from growing into wires or into some other structure. Roots may need to be pruned if the tree was planted too close to a sidewalk or driveway. In many cases, these practices are destructive to the tree. Thus, it is critical to determine what size and shape of tree are best suited for the planting site before selecting the species and planting it. Pruning cannot gracefully control tree shape and height to the degree that many people think.

Some trees require less pruning than others to form a strong structure. Baldcypress, sweet gum, black gum, fir, spruce, pine, and other conical-shaped trees with an **excurrent growth** habit usually require less structural pruning than trees with a **decurrent growth** habit (Figure 2-1). Many decurrent trees need regular training in the first twenty-five years of their life in order to develop good structure with a dominant trunk. However, some decurrent trees need only a moderate amount of pruning (Appendix 1).

Some tree **cultivars** have been selected and propagated by nursery operators and researchers for their good structural development. Planting these cultivars (instead of the species) can help reduce the amount of pruning needed to develop high-quality trees (Figure 2-2). For example, the 'Bradford' Callery pear is a very popular tree cultivar for landscapes, but it can break apart in ice, snow, or wind storms. There is no dominant trunk, and all the major branches grow upright, originating from nearly the same

TABLE 2-1. How to reduce the need for pruning.

Design urban space appropriately to allow for adequate root growth
Plant good quality nursery stock
Place security lights away from trees or under the canopies
Place trees away from overhead wires
Locate trees away from buildings
Place trees that grow to be large only in large soil spaces

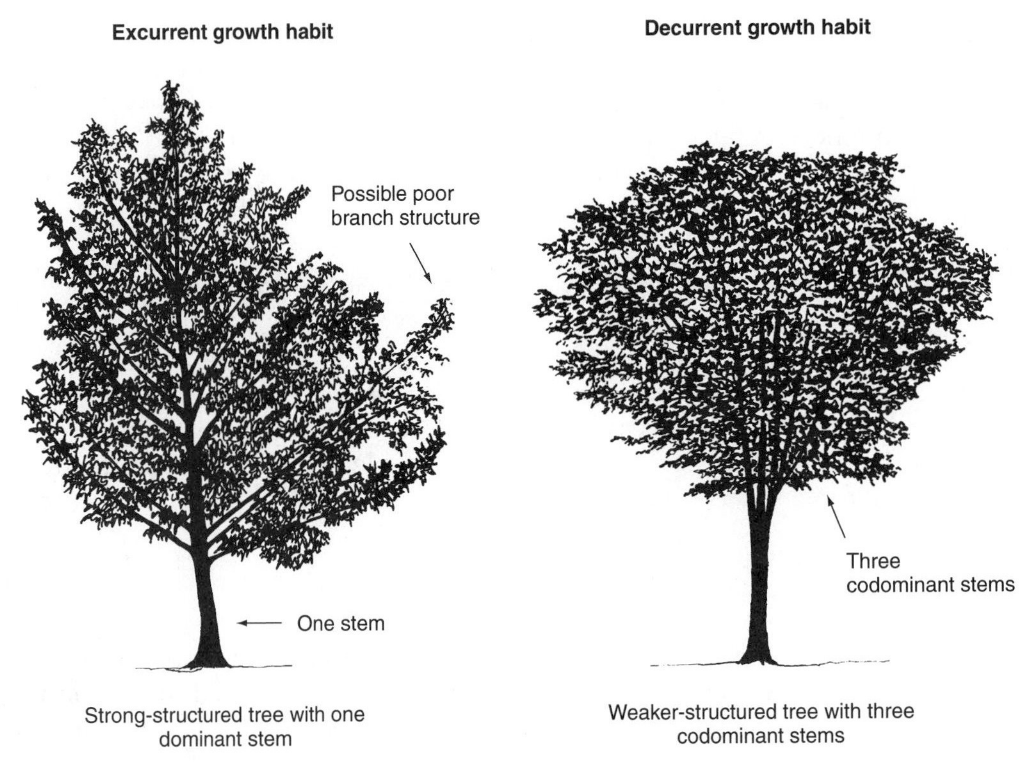

FIGURE 2-1. Most excurrent trees, including conifers, develop good structure with little pruning (left). However, most decurrent trees require regular training and pruning during the first twenty-five years in order to have them develop well. This prevents poor structure from developing such as that show at right.

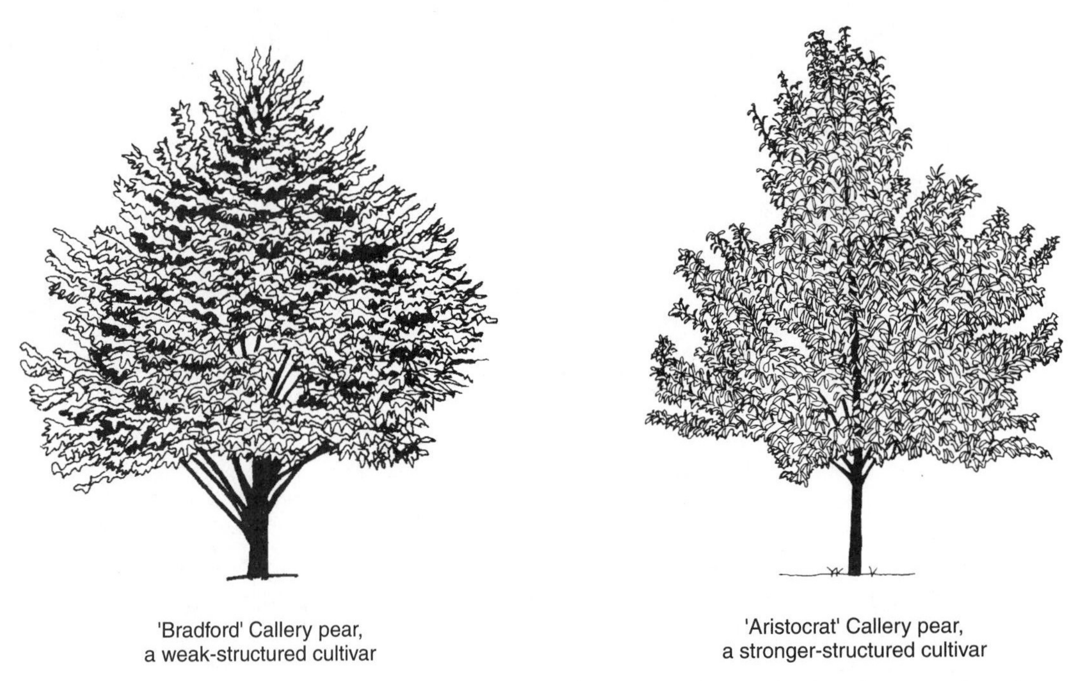

FIGURE 2-2. The 'Bradford' Gallery pear is a very popular tree cultivar for landscapes, but it can break apart in ice, snow, or wind storms (left). The 'Aristocrat' pear is a cultivar with a central leader and more horizontal branching habit (right). Along with other single-leadered trees, it is more resistant to storm damage than trees with several leaders.

point on the trunk, which makes it very susceptible to damage. Other cultivars of the same species grow naturally with a stronger structure, even with little pruning. For example, the 'Aristocrat' pear is a cultivar with a more-or-less dominant trunk and a more horizontal branching habit. Compared to the 'Bradford' Callery pear, it is more resistant to storm damage (but it is more susceptible to fire blight disease).

TREE FORM AND HABIT

Consider your design objectives, pruning budget, equipment, and crew capabilities before selecting a form of tree best suited for your streets. Although all trees need some pruning in the early years, the amount, type, timing, and technician skill level required often depend on **tree habit** and form (Table 2-2). The pruning requirements of three major tree forms—upright, pyramidal, and rounded—are compared here.

Upright

Many trees along streets, in parking lot islands, and in other high-traffic areas require periodic lower branch removal after planting to allow pedestrians and vehicles to pass safely. This maintenance requirement can be reduced by planting trees with an upright

TABLE 2-2. Advantages and disadvantages of planting trees with different canopy habits or forms.

Tree Habit	Advantages	Disadvantages
Upright	• Little pruning required to remove drooping branches • Trunk usually stays small or medium-sized	• Trees cast only a small amount of shade • Major branches on some trees can develop weak included bark and split from tree • Pruning occasionally needed to develop good branch structure
Pyramidal	• Strong trunk and branch structure often develops with little pruning • Lower branch removal can be performed by relatively unskilled labor • Lower branches are small and create only a small wound when removed • Trees cast a moderate amount of shade	• Lower branches often droop and need regular removal but they are small and easy to remove
Rounded	• On young trees, fewer branches droop compared to pyramidal trees • Trees cast abundant shade	• Highly skilled pruning crews needed • Regular pruning needed to develop strong structure • Branches that eventually droop, requiring removal, often are large and leave a large wound

or narrow canopy (note, though, that some trees may not be suited for planting along streets due to other characteristics). Branches on these trees grow mostly upright and do not droop into the path of pedestrians and vehicles (Figure 2-3 a–d). Compared to rounded or spreading trees, upright trees provide limited shade, due to the narrow canopy. The abundant shade cast by trees with a rounded canopy usually comes with a bigger price tag. That price tag is a higher pruning budget required to create a

(a)

(b)

(c)

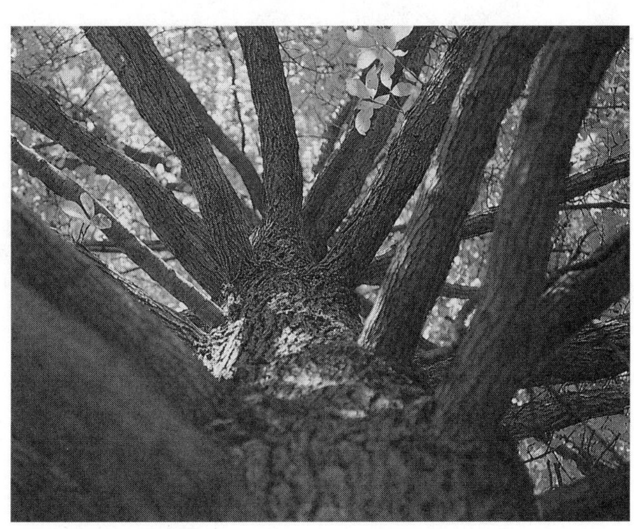

(d)

FIGURE 2-3. (a) Upright trees cast shade primarily in the early morning and late afternoon but require less pruning. (b) Spreading trees cast significant shade all day long. Large, low branches eventually droop and require shortening or removal. (c) Pyramidal-shaped trees require removal of drooping branches even when they are young but branches remain small in diameter (d), creating only a small wound when removed.

well-structured, rounded tree. The advantages and disadvantages of these two sides of the story must be balanced in your decision-making process.

Do not become complacent after planting trees with an upright growth habit. Some develop aggressive upright branches, with weak, **included bark** in the union (Figure 2-4 a–c), if corrective pruning is not done when the tree is young. This pruning should be directed at preventing development of **codominant stems** on the lower trunk. Without this, branches could split from the tree later. However, this characteristic may be typical of the tree and it may not be practical to try to correct it with pruning in all cases.

Despite the risk of developing included bark, many upright trees stay together and will not break apart easily, even without pruning. If the life expectancy of the tree at the site is less than twenty years, because of site constraints, you may not need to be concerned about trees with included bark. Many trees along streets planted in sidewalk cutouts and other small soil spaces, such as parking lot islands, live less than twenty years.

Rounded

Because fewer branches droop on young oval, spreading, or round-shaped trees (decurrent growth habit) than on pyramidal trees, these tree types might not be pruned in the first ten years after planting because there may be few complaints from citizens about drooping branches. However, without some pruning, tree structure may deteriorate due to **weak branch attachments** (Figure 2-5). During this ten-year period, trees can develop several to many aggressive, upright trunks, some with weak, included bark in the crotches. This could lead to breakage in storms as trees grow. In addition, without regular pruning, fast-growing branches that initially grew upright on the lower trunk can begin to droop, and may have to be removed. This often leaves a **large wound** in the trunk because low, fast-growing branches can develop large diameters (Figure 2-6 a and b). This large wound can weaken the trunk and initiate cracks and

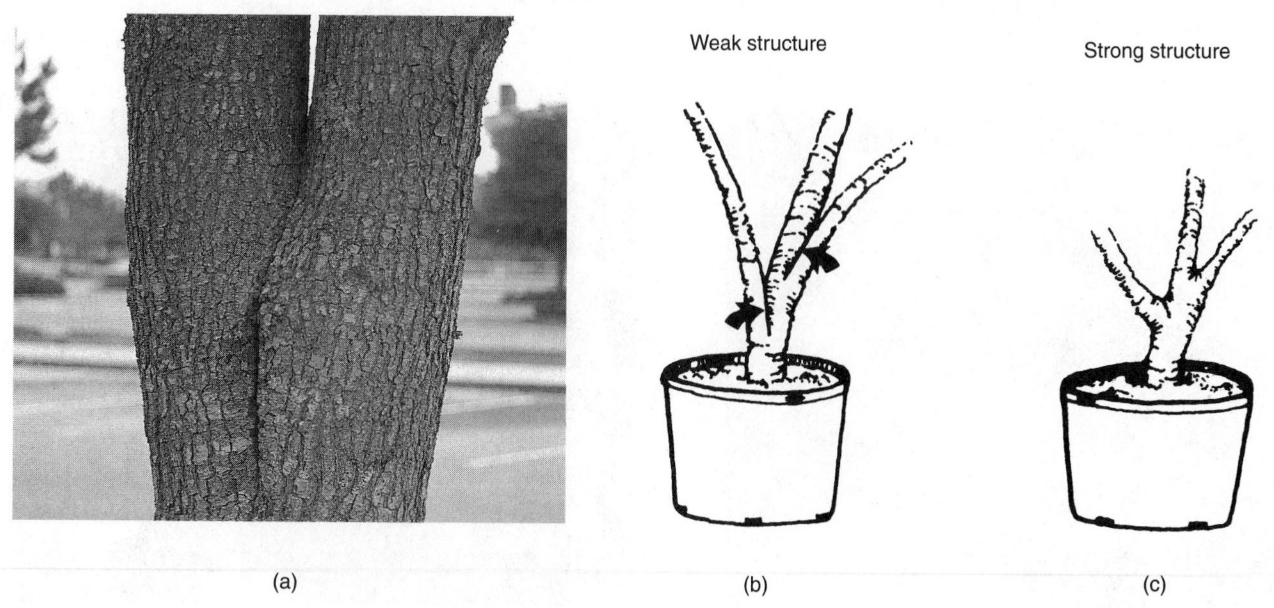

FIGURE 2-4. (a & b) Weak unions have bark included between the codominant stems. (c) Strong unions are wider, without included bark.

FIGURE 2-5. (a) Round-shaped (decurrent) trees may form poor structures and can break apart if not pruned regularly when young. (b) This often creates a large wound in the trunk. Trees with this extensive damage usually require removal.

(a)

(b)

(a)

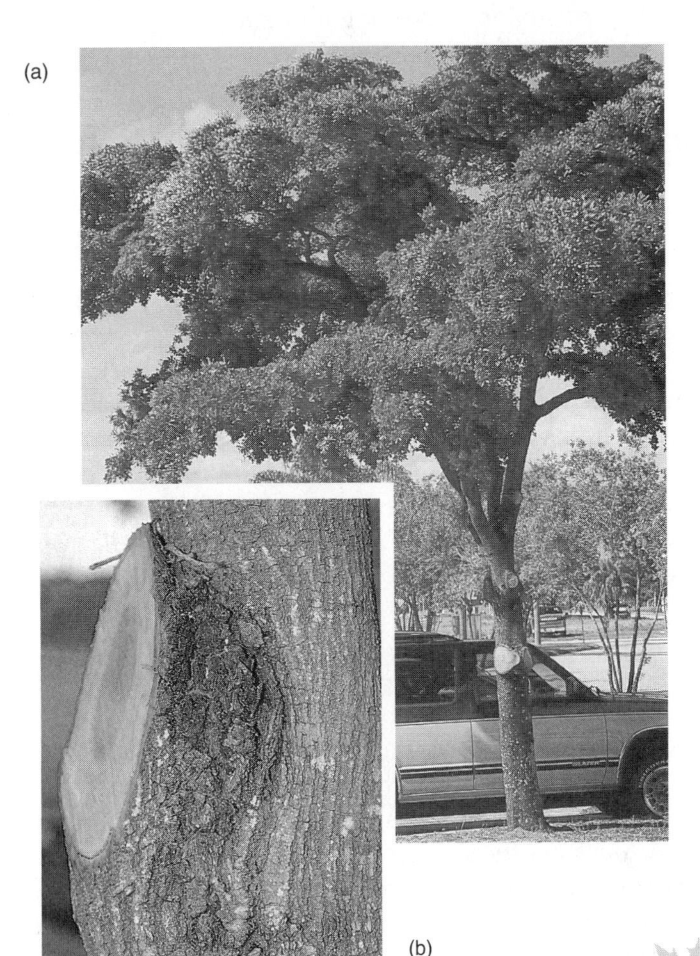

(b)

FIGURE 2-6. (a) Round-shaped trees often develop large low branches without regular pruning. As the tree grows, these droop into the way of traffic and must be removed to allow for vehicle passage. (b) The large wound created by branch removal can initiate decay and cracks in the trunk.

decay. Therefore, preventive pruning to develop good structure when trees are young is recommended on all trees, especially those with a rounded canopy shape. Pruning at two, five, and ten years after planting should be considered a minimum requirement. Pruning at one, three, five, and ten years would be even better.

Pyramidal

Trees with a pyramidal form (excurrent growth habit), such as deodar cedar, southern magnolia, and pin oak, frequently have a dominant central trunk far up into the canopy. They eventually cast abundant shade. Trees with a dominant trunk are considered to be strong and durable in urban landscapes. Little pruning is required to create this strong tree structure. However, lower branches on pyramidal trees often droop toward the ground. These have to be removed regularly to allow pedestrian and vehicular passage beneath the canopy.

Several prunings may be required, but most of the work can be performed by crews on the ground. Little tree climbing is required. Crews that remove only drooping, lower branches need fewer skills and less specialized training than those pruning higher in the tree. The diameter of drooping branches on pyramidal-shaped trees is usually relatively small, resulting in only a small wound when removed. If double leaders or multiple trunks form in the canopy, they can be easily corrected by removal when lower branches are pruned.

Tip: *Planting quality nursery trees is the first step to preventing formation of codominant stems.*

AT THE NURSERY

Once an appropriate species or cultivar of tree has been chosen for the site, proper tree selection from the nursery can also eliminate early pruning requirements in the landscape. Unfortunately, there are many trees planted that are of poor quality with poor structure and have defects (Figure 2-7) that could cause the tree to split apart in a storm as it grows older (Figure 2-5). Defects such as clustered branches and included bark often begin in the nursery, and they should be prevented or corrected there. Major branches that split from trees can cause serious damage to the tree or injure people and property. Some of the pruning needed to correct structural defects, such as a double trunk, in the landscape can be avoided by purchasing **good-quality nursery trees** with strong structure. Good-quality trees with an excurrent growth habit (Figure 2-1) probably will need less pruning to create good structure than decurrent trees for several years after planting.

IN THE LANDSCAPE

Maintenance practices can influence tree form and pruning needs in the landscape. For example, irrigating a newly installed tree improperly can result in dieback of trunks and branches (Figure 2-8). If the tree survives, several sprouts are forced to grow from latent buds and develop into **multiple trunks** and these often become equally dominant. These are referred to as **codominant stems** or trunks. The result is a weak structure. The tree can be pruned to rebuild a **good structure** (covered later in this book), but this takes time and money. The money saved in not providing irrigation will be spent pruning to

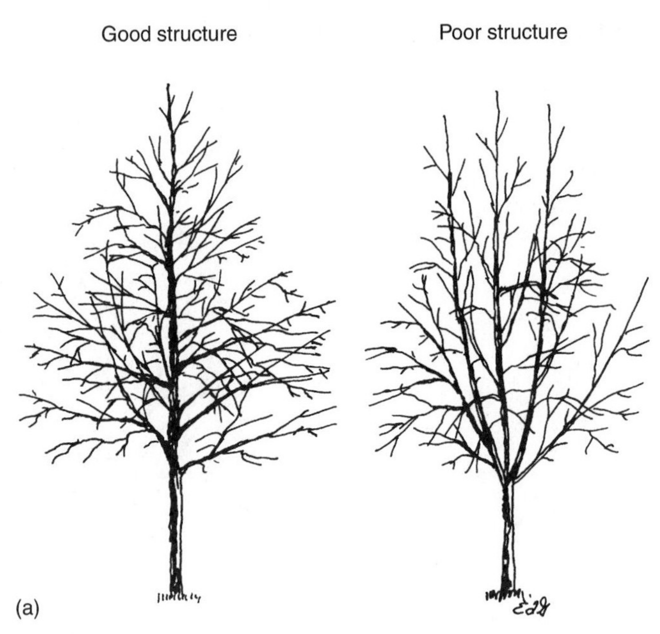

FIGURE 2-7(a). High-quality medium and large-maturing shade trees have one dominant leader (left), whereas poor-quality trees have several codominant stems.

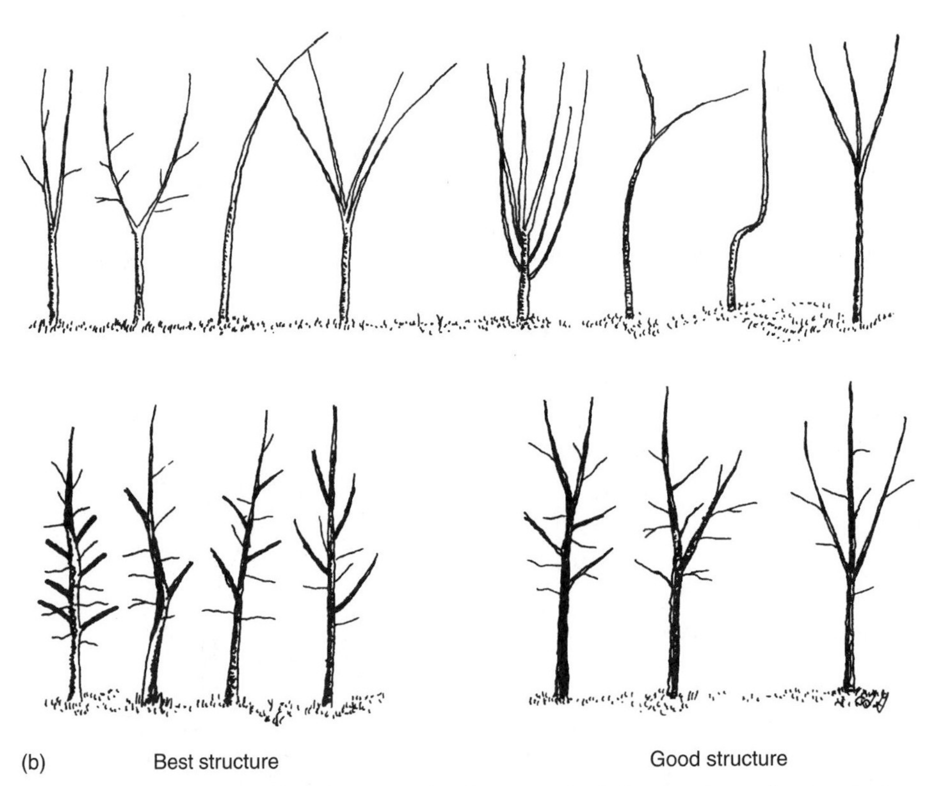

FIGURE 2-7(b). Poor quality shade trees have forked, bent, or deformed trunks (top). Good quality shade trees have one trunk into the top half of the tree (bottom right). The best quality shade trees have one trunk to the top of the canopy (bottom left). Note: For illustrative purposes, only the larger branches are shown. The small branches are not shown.

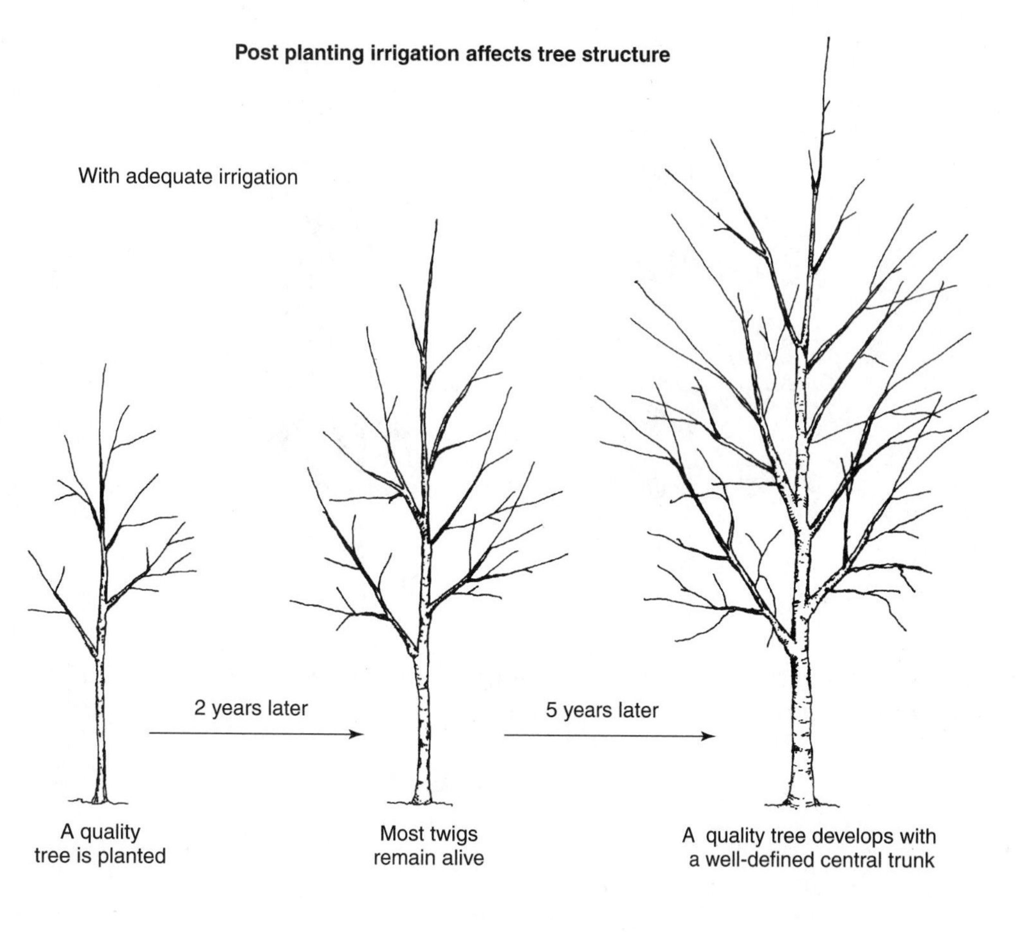

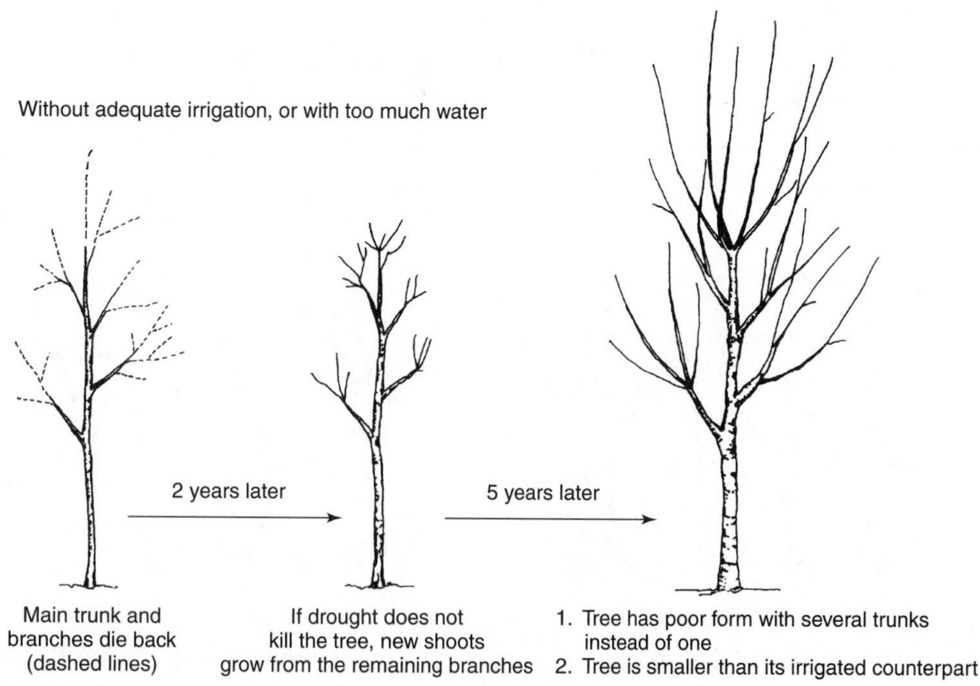

FIGURE 2-8. Good-quality nursery stock receiving adequate irrigation during establishment develops into a well-structure tree (top). A poorly formed tree with multiple trunks often develops without adequate irrigation (bottom). (Struve, 1994).

a good structure. Tree vigor can be lost if weeds and turf are allowed to grow close to the trunk (Green and Watson, 1989). Leaders develop poorly on less vigorous trees as several branches or stems compete to become the leader (Struve, 1994). The result is often weak form with several codominant trunks.

AT THE CONSTRUCTION SITE

Good decisions on tree selection by developers can have a dramatic impact on tree health and pruning needs later (Figure 2-9). Trees to be saved during construction should be capable of contributing safely to the value and aesthetics of the site when the development is completed. Trees with poor form, major faults and defects (see Tables 1-1, 1-2 and 1-3), and those with poor chances for survival due to root damage should not be left standing at the completion of a construction project. They are likely to become liabilities and require expensive remedies later. These remedies could include extensive irrigation, mulching, or pruning in an effort to correct a defect, or removal. Unsafe trees could also cause significant personal risk or property damage if parts of them **fail** and fall to the ground. Save the trees with **good form** and remove those with **poor form** (Figure 2-10) to minimize the need for pruning later. **Live crown ratio** should be at least 0.6, meaning that there should be live branches in the top 60 percent of the tree.

CHECK YOUR KNOWLEDGE

1) Excurrent trees such as many pines (*Pinus*) require less structural pruning because they:
   a. develop a central leader with little or no pruning.
   b. have small diameter branches.
   c. have main branches clustered together in whorls.
   d. have a cone shape.
2) An example of a tree with a weak structure is:
   a. Aristocrat Callery pear (*Pyrus calleryana*).
   b. black gum (*Nyssa sylvatica*).
   c. baldcypress (*Taxodium distichum*).
   d. Bradford Callery pear (*Pyrus calleryana*).
3) To avoid the problems associated with aggressive, upright branches on trees such as Bradford pear:
   a. prune aggressive branches to an outside facing bud or lateral branch.
   b. remove double leaders and branches with included bark.
   c. remove upright branches leaving those that are more horizontal.
   d. choose a tree that has a more desirable form.
4) What is the first step in preventing formation of codominant stems?
   a. keep branches smaller than about half the trunk diameter
   b. thin the outer portion of the canopy to allow light to reach the interior foliage on the main stem
   c. plant quality nursery stock with good form
   d. keep the tree vigorous with appropriate fertilization and other cultural practices

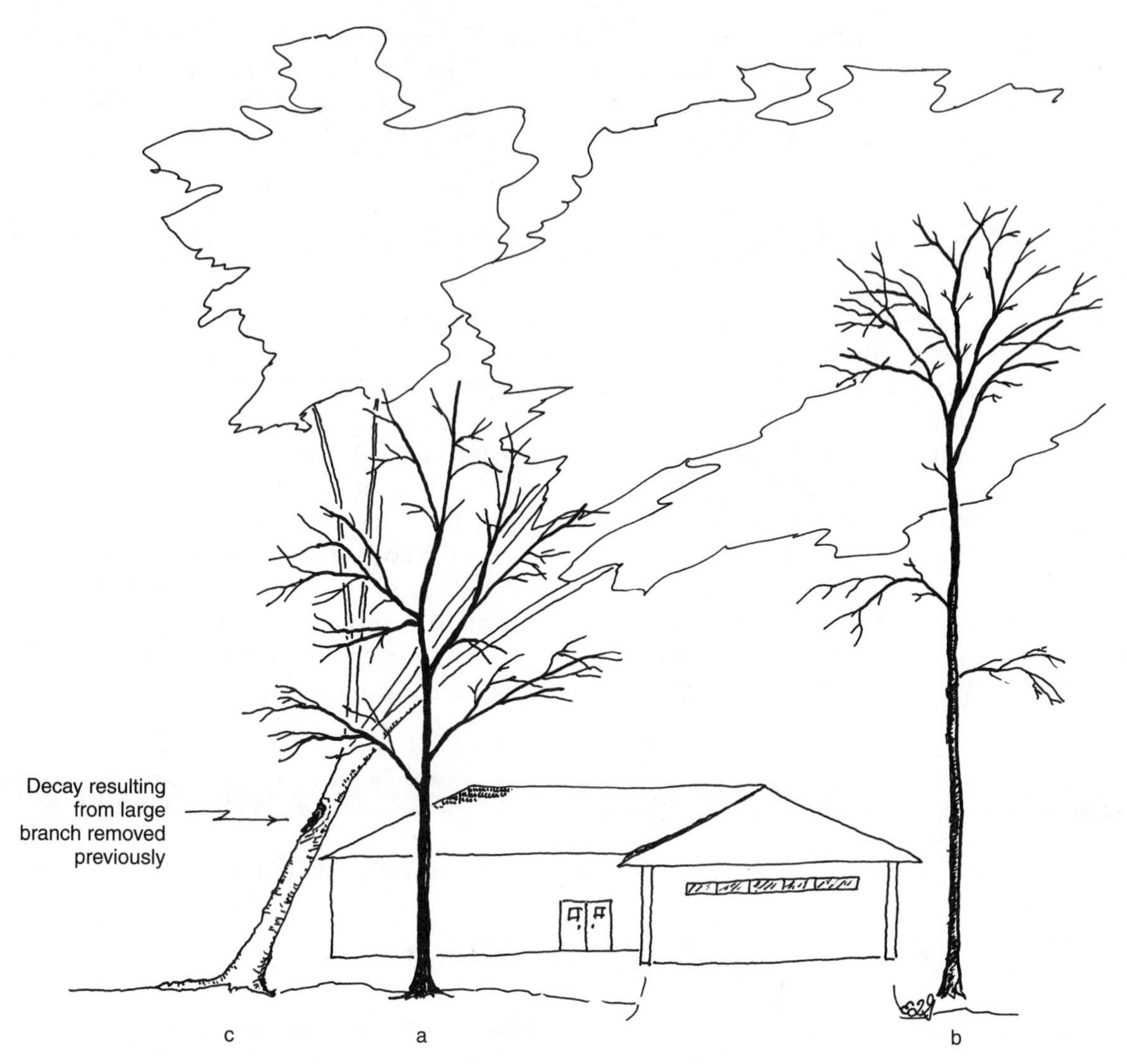

FIGURE 2-9. *Tree 'a' is most suited for saving because: 1) large trees are tough to save—smaller ones, like this one, do best. 2) small trees recover from root damage best. 3) there is one trunk and well spaced branches. 4) there is no included bark or other defects in the branch union. 5) the trunk is straight up. *Tree 'b' is OK but not a great tree because: 1) its very tall with most branches only at the top 2) the trunk is thin-tree may bend over or blow over if nearby trees are removed. *Tree 'c' is poorly suited to save because 1) large trees are tough to keep alive after roots are damage during construction 2) tree is leaning badly 3) trunk forks in the bottom half of the tree 4) tree hangs over the house 5) there is included bark in the bark unions 6) the trunk is decayed at the point where tree was pruned years ago 7) most foliage is toward the top of tree.

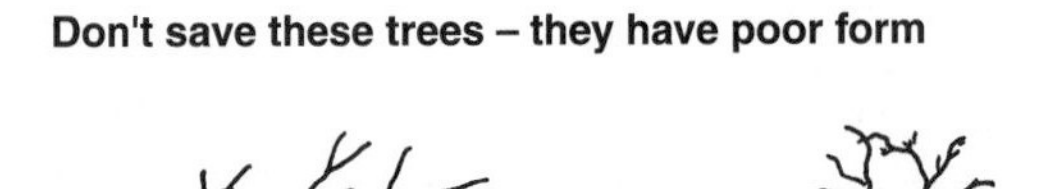

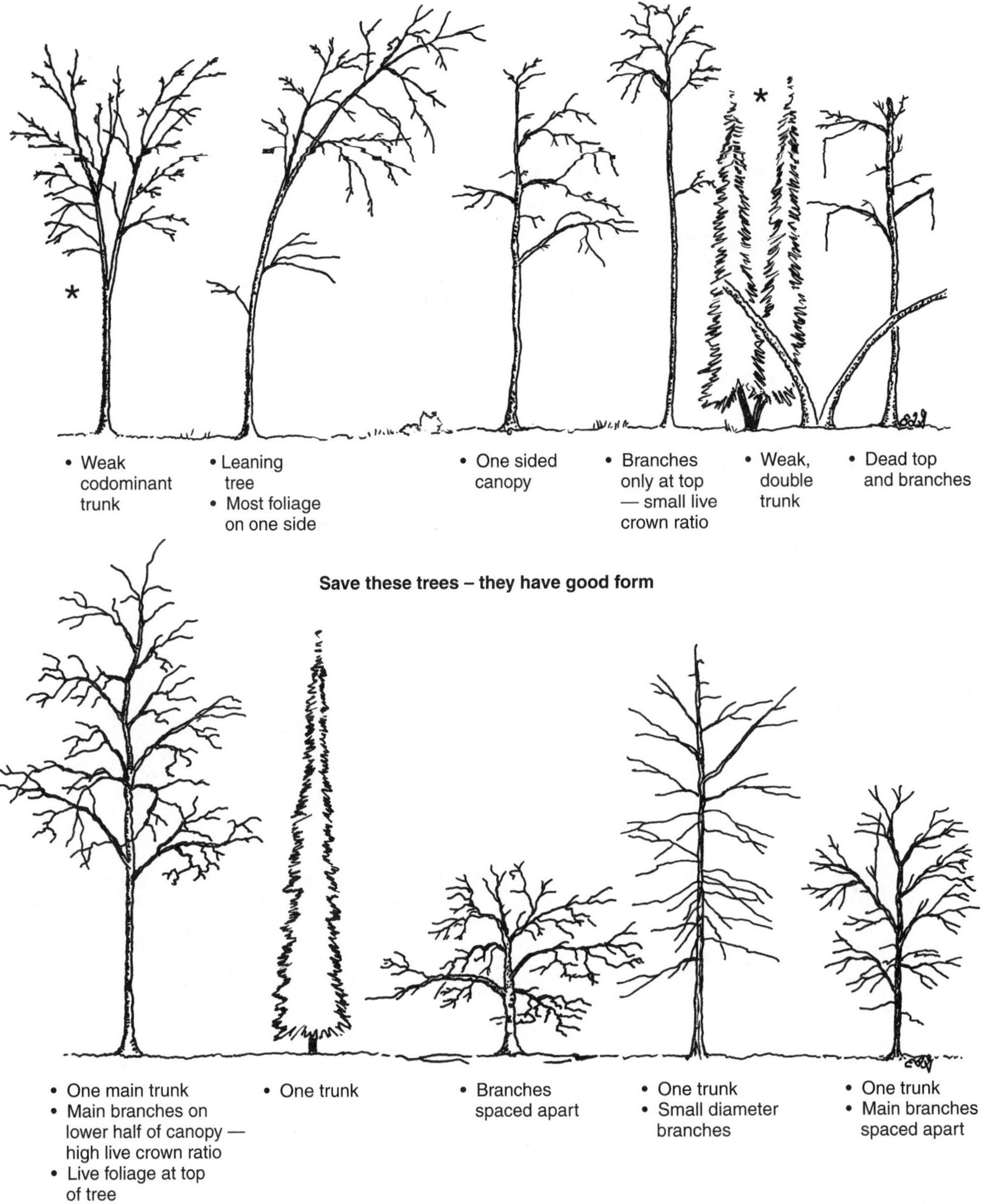

FIGURE 2-10. Don't save trees with poor form (top). Keep those that have better form and fewer defects (bottom).

5) Trees with a round-shaped canopy:
   a. often develop large low branches.
   b. usually have small-diameter branches.
   c. are stronger than pyramidal-shaped trees.
   d. remain fairly strong until maturity.

6) Which tree would you choose to save on a construction site?
   a. a tree with one trunk up to the top of the canopy and live branches in the upper and lower half of the canopy
   b. a tree that has several large stems and trunks with included bark in the lower half of the canopy
   c. a very large tree close to the building
   d. a small- to medium-sized tree with most of the foliage toward the top of the tree

Answers: a, d, d, c, a, a

CHALLENGE QUESTIONS

1) What can be done prior to construction to reduce pruning and tree liability issues after construction of a new building is completed?
2) Describe how landscape management practices following transplanting can impact tree pruning later.
3) Contrast pruning needs of excurrent with that of decurrent trees.
4) What are five landscape design strategies that can reduce tree pruning?
5) Compare the advantages and disadvantages of planting upright trees with that of planting round-canopied trees.

SUGGESTED EXERCISES

1) Go to a nearby site where trees have been cleared for buildings and construction is underway or recently completed. Identify which trees should not have been saved and why.
2) Go to a nearby nursery and select good- and poor-quality trees based on trunk and branch form and root systems. Have participants determine which are the good-quality trees and which ones represent poor quality. Then discuss each tree's quality with the entire group.
3) Travel to a nearby shopping center or commercial landscape. Critique the design of the spaces created for trees. Determine how the spaces and the entire site could have been better designed to suit trees and meet the objectives of the property owner. What pruning issues are present at the site?
4) Go to an undeveloped piece of land with large trees that is about to be developed. Help the developer choose the best trees to save from a trunk and branch structure standpoint.

CHAPTER 3

TREE STRUCTURE

OBJECTIVES

1) Contrast forest growth with open-grown tree structure and form.
2) Determine the role branch size plays in decay prevention and strength of attachment to the trunk.
3) Evaluate branch unions and forks.
4) Present the factors contributing to branch union strength.
5) Show how trees resist the spread of decay from branches to trunk.
6) Determine why all branches on most nursery trees should eventually be removed in the landscape as the tree grows older.
7) Compare the strength of single-trunked trees with that of multi-trunked trees.

KEY WORDS

Aggressive limbs
Bark inclusions
Branch angle
Branch bark ridge
Branch collar
Branch protection zone
Dominant trunk
Heartwood
Stem bark ridge

FOREST-GROWN TREE FORM VERSUS OPEN-GROWN FORM

Most shade tree species used in urban and suburban landscapes evolved close to each other over millions of years in a forest. Although cultivars dominate landscapes in some regions, these too were derived from forest-grown species. In this setting, the trunk typically grows upright, more-or-less straight reaching for sunlight. Sunlight reaches the tree mostly from above. Large-maturing trees spaced closer than 20 feet apart in landscapes often grow like this as well. As in the forest, lateral branches on the lower portion of the trunk are shaded, die, and are shed from the tree. The tree forms a **dominant trunk** with small diameter side branches (Figure 3-1). Most of the major branches are located in the top half of the tree. Trees with this structure evolved mechanisms for preventing decay and for developing strong structure.

Many open-grown landscape trees spaced more than about 20 feet apart develop a wider spreading canopy than they would in the forest. This occurs because the tree has access to sunlight from all sides. Few trees, other than those in prairie or desert climates, evolved in an open landscape setting where lateral branches grow in an aggressive spreading fashion (Figure 3-2). This is a fairly unfamiliar form that most trees are not well adapted to sustain. Included bark, a serious defect, is not uncommon in the union of the large branches on these trees (Figure 3-3). Therefore, many trees in an open landscape setting, which is typical of most planted trees, need training by pruning. This helps them develop strong architecture, that is a dominant trunk, which is compatible with urban and suburban landscapes. Trees with a strong structure can live longer than those with poor form.

Low, **aggressive limbs** on many landscape trees get in the way of people and vehicles in urban landscapes. Small lateral branches growing downward from these low limbs are typically removed to allow for clearance beneath the canopy. This is only a temporary fix. Removing small lateral branches from along the limbs forces them to grow too long causing the limbs to eventually sag under their own weight. A large pruning wound results when they need to be removed from the trunk later. This could cause health problems for the tree and decay and cracks in the trunk. Most limbs in the lower 15 to 20 feet of shade trees will be removed eventually. Preventive arboriculture keeps them small by pruning lateral branches from them *and* shortening them to minimize the wound when they are eventually removed.

On some tree types, removing small lateral branches from the base of the limbs causes main limbs to grow more upright, at least for a number of years. This encourages main limbs to grow faster because they have access to plenty of light at the top of the canopy. As these upright limbs develop, the main trunk or leader slows down its growth and may be lost altogether (Figure 3-2, center and right-center). The tree essentially forms several codominant stems instead of the one trunk common in forest-grown trees. The lack of one dominant trunk is considered a weaker form on many shade trees.

Notice in Figure 3-2 that low main limbs c, d, and e are still on this tree in middle age. Small lateral branches at their base were removed as the tree grew (see dotted lines). This forced the limbs to grow longer because their tips had access to plenty of sunlight. As the tree grew older, these fast-growing low limbs gradually sagged and were removed (see right drawing in Figure 3-2). This left large wounds that could cause root problems, initiate cracks in the trunk, discoloration, and decay. A better management practice is to train trees to a form that resembles a compromise between the forest-grown form and the open, unpruned landscape form. Later chapters describe how to do this.

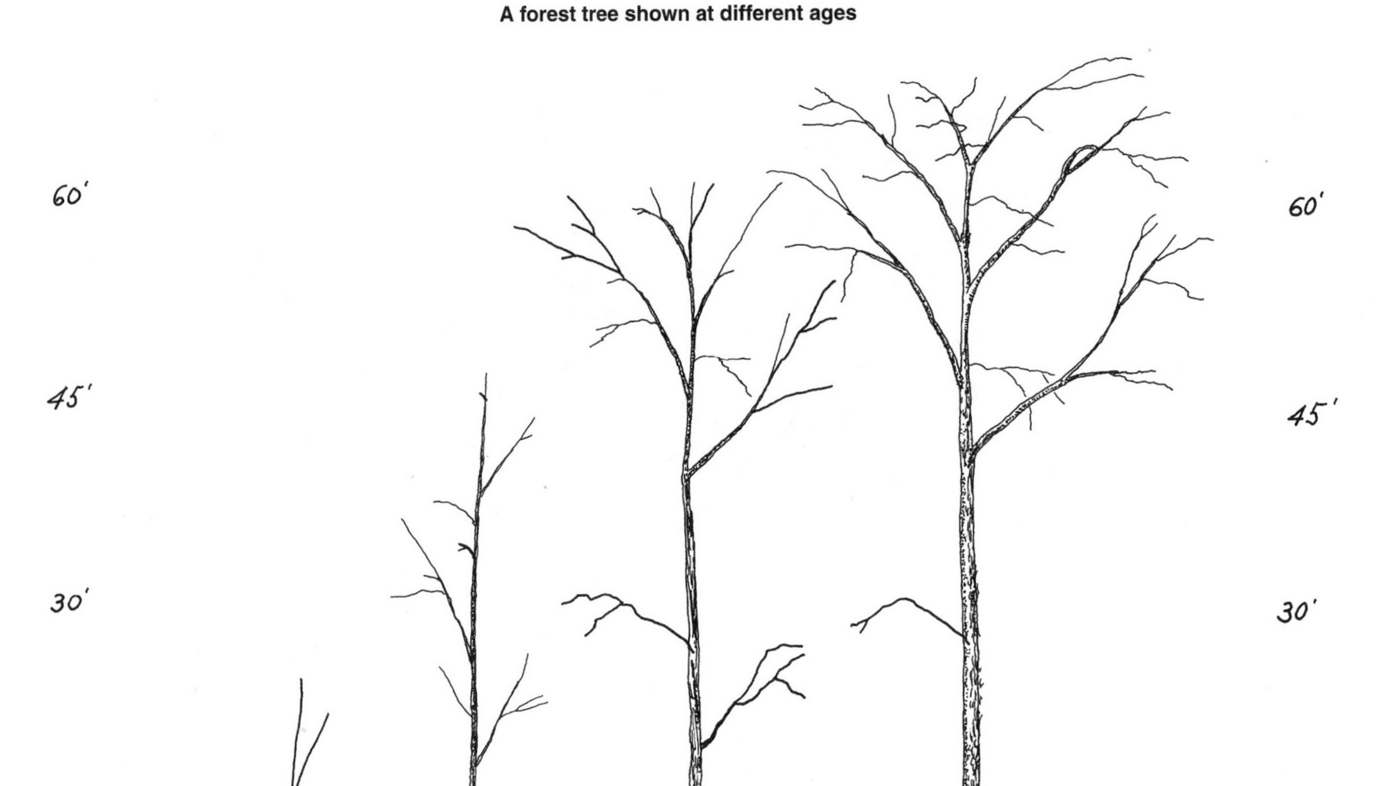

FIGURE 3-1. This figure shows the same forest-grown tree at five different stages from seedling to maturity. The trunk grows upright toward the sunlight because the canopy is shaded from the sides by surrounding trees (not shown). Most branches are in the top half of the tree. The dominant trunk form is considered a strong structure. Not all trees get this tall, but most grow with this type of habit in the forest.

Tip: *Prevent branches from growing larger than half the trunk diameter so they remain well attached to the tree.*

BRANCH ATTACHMENT

A **branch collar** sometimes forms at the base of a branch where it joins the trunk (Figure 3-3). The collar often appears as a distinct swelling on the bottom, sides, and top of the branch base, especially when the branch is much smaller than the trunk. Small branches on many tree species have collars. The collar contains trunk tissue as well as branch tissue combined in a more-or-less overlapping fashion. Damage to the collar from improper pruning can initiate decay or cracks in the trunk below the injury. To

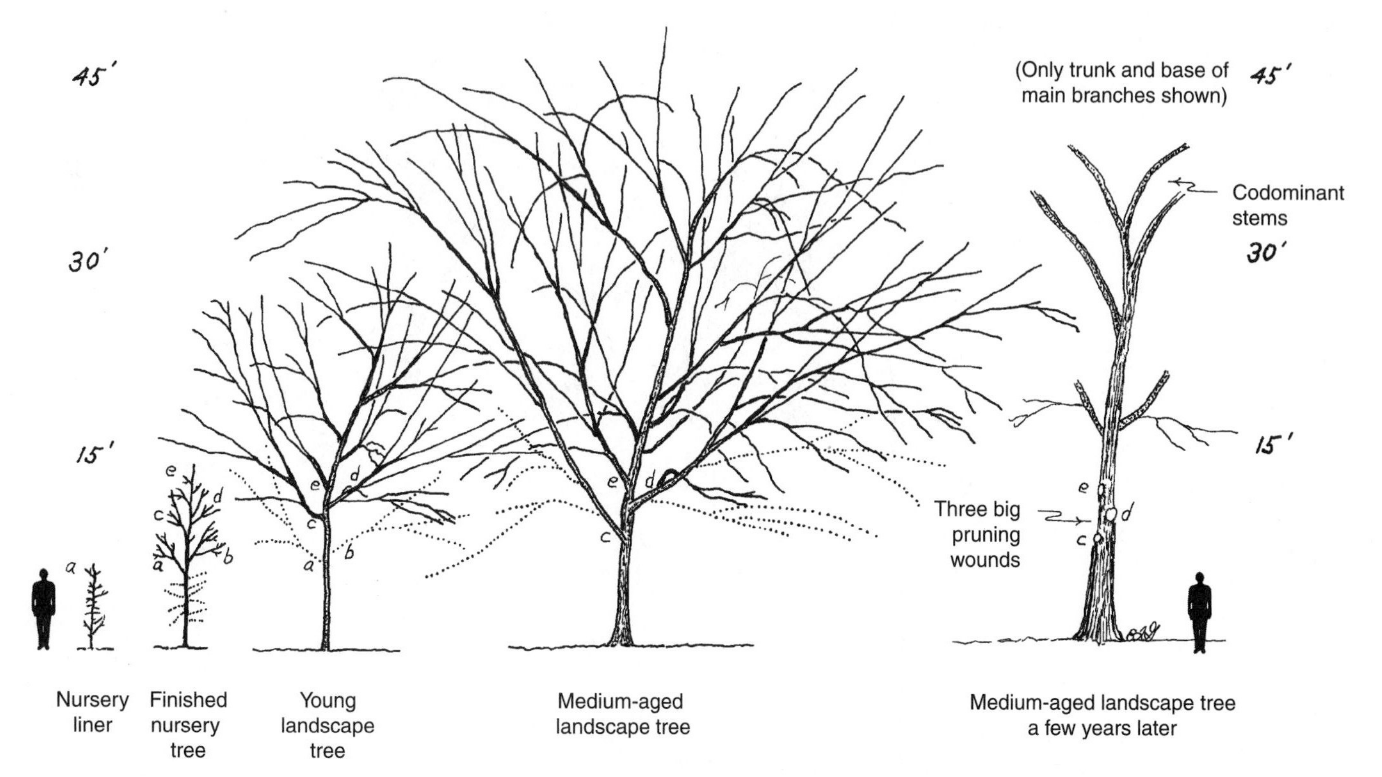

FIGURE 3-2. This figure shows one open-grown landscape tree at five different ages. Five branches are labeled from a to e. Note that all branches and stems that were on the tree in the nursery were eventually removed from the tree in the landscape. This is very common in most landscapes. This co-dominant stem, spreading form is considered weaker than the dominant trunk form. On most nursery trees, vitually all the branches are temporary in the sense that they will be removed from the tree later. Therefore, these should not be allowed to grow to a large size because a large wound is created when they are removed. This causes troubles later. Removing only low, interior branches as shown by the dotted lines forces the tree to grow into this poor form.

prevent this, the outside edge of the branch collar should be located before using a tool to remove the branch.

Collars vary in shape and size, depending largely on the size of the branch and on the species. There is variation among trees of a species, and even among branches on the same tree. On the larger branches of some trees, collars are more difficult to locate usually because they are not present. After studying Figure 3-3, and by looking at a variety of branch collars on trees in the woods, nursery, and landscape, collars will become more visible to you. Trees such as sycamore, honeylocust, holly, magnolia, red maple, dogwood, black olive, ficus, and crape-myrtle have distinct collars that are easy to recognize. Others, like oaks and elms, may have less visible collars, which can be difficult to locate because there is no distinct swelling.

Branches growing at a rate slower than the trunk will stay much smaller than the trunk. As a result trunk wood grows around the base of the branch forming a highly visible collar. This helps secure the branch to the tree. The edge of the collar may be one to several inches away from the trunk on slow-growing branches. Trunk wood cannot grow

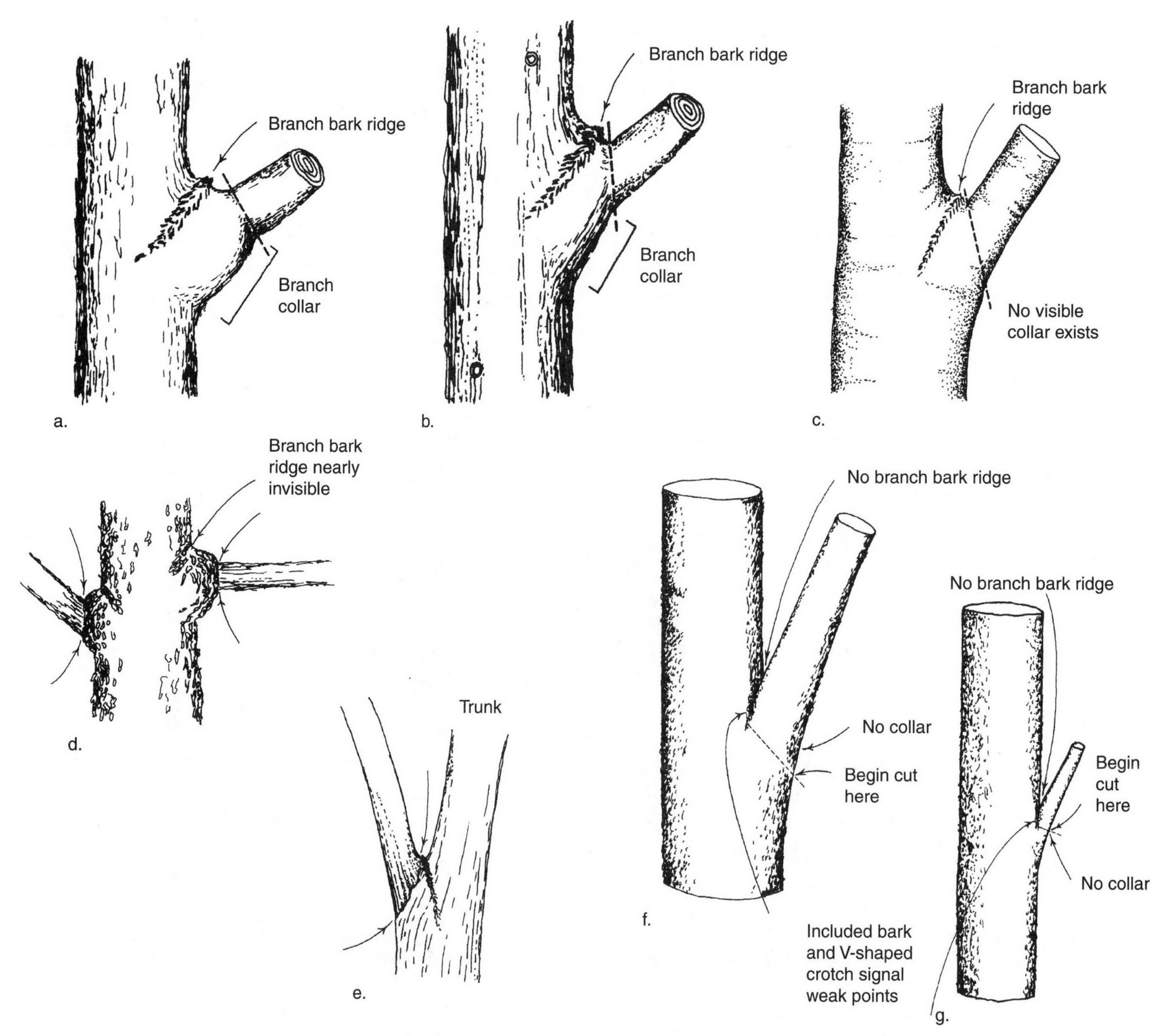

FIGURE 3-3. Branch collars and branch bark ridges take on a variety of shapes and sizes on different species and ages of trees. When removing a branch from the trunk, cut along the dashed line on the branch side of the swollen collar tissue (a and b). Many branches do not form a distinct collar so there is a smooth transition from trunk onto the branch (c). On older trees, trunks form large collars around the base of small branches (d). On some species, this usually indicates a slow-growing or dying branch. Make the pruning cut between the two arrows (d and e). Branches with bark pinched between the trunk and the branch (included bark) have no collar and no branch bark ridge and therefore are pruned differently (f and g). Notice that the crotch with included bark is shaped like the letter V, indicating a weak crotch. Make the final pruning cut along the dashed line. The cut should end in the crotch at the point where the branch tissue actually meets the trunk tissue, anywhere from several inches to several feet down into the crotch. The cut may have to be finished with a chisel to prevent injuring the trunk tissue. The U-shaped crotches shown at top indicate a strong attachment. Dashed lines indicated where to make final pruning cuts.

around the base of a branch that is growing at a rate similar to the trunk growth rate. This can cause the branch to be less well secured to the tree because there is no overlapping tissue forming a collar. You will learn in later chapters how structural pruning focuses on keeping branches less than approximately ½ the trunk diameter and preventing the formation of included bark. This can be accomplished by regularly removing secondary branches from a parent branch you want to keep small.

Branches that grow at a rapid rate and become larger than about ½ the trunk diameter often do not have visible collars. The collars do not wrap on top of or under the branches (Figure 3-3, right). In fact, they become codominant stems rather than true branches (though they can originate from a lateral bud). Codominant stems grow at about the same rate, and together they can take over the role of the trunk (Figure 3-4, bottom). They sometimes are oriented upright, especially when they are young and shaded due to crowding in the nursery. When this occurs, included bark often appears in the crotch, indicating that the stems or branches are poorly attached to the tree. Branches that grow at a narrow angle in relation to the trunk often lack visible collars and can develop included bark as well, even if they grow very slowly and are much smaller than the trunk. Some lindens, trumpet tree (*Tabebuia*), and mahogany are especially prone to this defect (Figure 3-3, bottom right).

Fast-growing branches on young trees frequently grow upright, forming a narrow angle in relation to the trunk. Bark can become pinched or included in the crotch. This phenomenom has led to the misconception that the **branch angle** is related to the strength of attachment, which is untrue. For example, a large-diameter branch growing at a wide angle can be as poorly attached to the trunk as a similarly sized branch with a narrow crotch angle. It is the presence of included bark that indicates a poor attachment. Although included bark is most common on upright branches growing at a narrow angle in relation to the trunk, on some trees it develops in the union of branches growing at wider angles.

BRANCH AND STEM BARK RIDGE

The **branch bark ridge** is composed of rough, typically darkened, raised bark, which is formed in the angle at the union of a branch and the trunk and extends down both sides of the trunk (Figure 3-3). Together with the collar, the portion of the ridge pushed up in the crotch provides guidelines to proper pruning. A number of trees, such as crape-myrtle, some eucalyptus, some elms, and some oaks, do not have a distinguishable branch bark ridge and some tropical trees also lack a visible ridge. However, birches, honeylocust, lindens, many oaks, maples, figs, and many other trees have a prominent ridge at most crotches. The **stem bark ridge** is the raised bark in the crotch between two more-or-less equally sized (referred to as equally dominant or codominant) stems (Figure 3-4, bottom).

A branch union with included bark does not have the traditional bark ridge. One of several forms can occur in a union with included bark. If included bark was present from the time the branch originated from the parent stem, no raised tissue may be present along the sides and on top of the union. Branch and parent stem bark will appear to dive into the union. The entire length of the union appears as a closed crack. If included bark developed some time after the branch formed, a ridge might be present along the

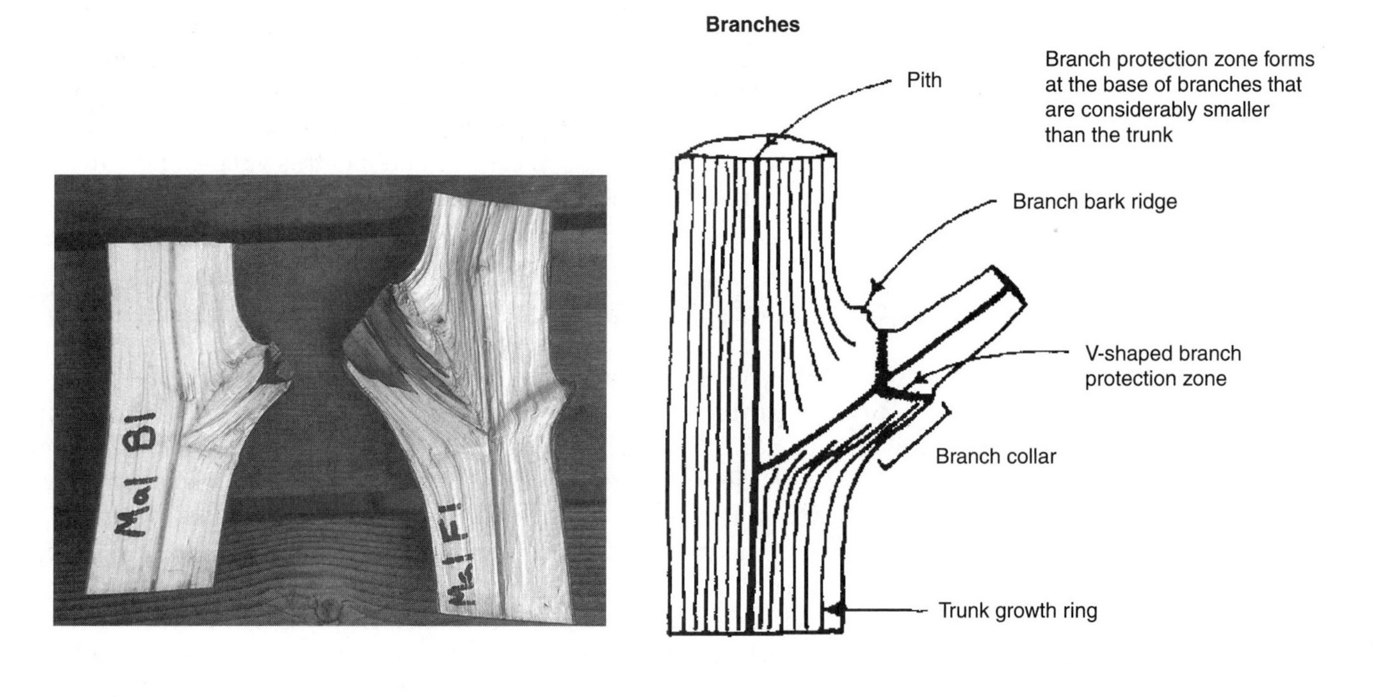

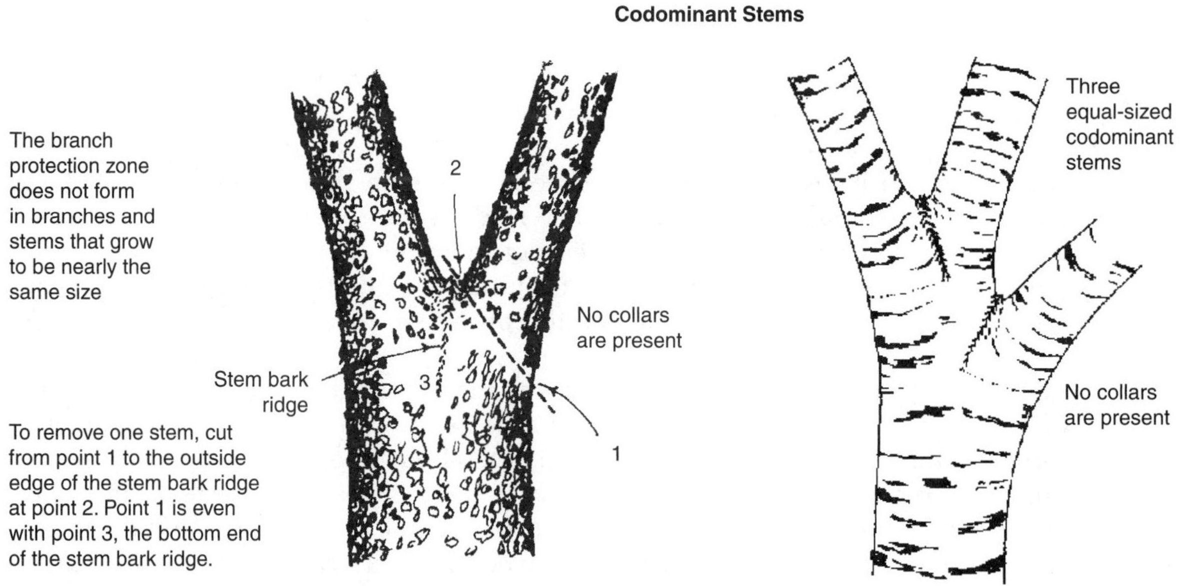

FIGURE 3-4. Section through the pith of a 7 year old tree (top right). If branches are allowed to grow larger than about half the trunk diameter (measure directly above the branch), they become codominant stems (bottom). No protection zone forms in codominant stems. Photo: Section through maple branch union 6 months after removing a branch and a stem. Branch protection zone formed on the branch that was small relative to trunk size (left). The cone-shaped discolored region is clearly visible to the outside of the protection zone. Protection zone did not form on the codominant stem that was removed 6 months ago. Discoloration is clearly visible well back into the trunk (right).

lower sides of the union. However, bark toward the top of the union will dive into the union without a raised ridge. If included bark was developing earlier but is not forming now, the portion of the union containing the inclusion will have a raised ridge, but there will be a very subtle valley in the peak of the ridge. This is very difficult to illustrate, but with some practice on real trees you can see and feel this valley. There may be a raised bark ridge in the top of the union if included bark is not currently forming.

BRANCH PROTECTION ZONE

Chemicals, including phenols, resins, and terpenes, are generated in a narrow zone of cells called the **branch protection zone** within the branch collar. They inhibit the spread of organisms and decay from a branch into the trunk (Figure 3-4, top). This zone is not visible from the outside of the tree. The tree expends energy to build the protection zone, but in so doing it retards the spread of decay. If the collar is injured or removed during pruning, the protection zone can fail to develop, allowing organisms of decay to enter the trunk. If the collar and protection zone are not injured, the tree will benefit because decay will often be limited to the dead branch core and will not move into trunk tissue. The branch core is more-or-less cone shaped with the pruning cut surface representing the base of the cone. It extends into the trunk toward the point where the branch first emerged from the trunk.

Several anatomical features in the branch base aid the branch protection zone in restricting movement of discoloration and decay into the trunk. These include shorter and narrower xylem vessels, more deposits in vessels, and an abrupt turning of the vessels where the branch meets the trunk. These combine to slow down the movement of water and elements through the branch union. These features do not occur at the base of branches that are large compared to the trunk, nor in codominant stem bases. Branches larger than about half the trunk diameter measured directly above the branch union typically lack the ability to form an effective protection zone (Eisner and Gilman, 2002).

The branch protection zone does not form in the darkened tissue in the center portion of branches. This tissue has been referred to as **heartwood** and in many trees begins to form in branches by the time they are five to about fifteen years old. The negative implications for the tree should be clear when removing large diameter branches with heartwood, so consider all other options first. Because discoloration and decay could spread easily into the trunk through heartwood when a large branch is removed, give careful consideration before proceeding. There may be better options such as shortening the branch back to a large lateral branch.

Tip: *Prevent stems and branches from growing more than half the diameter of the trunk so the branch protection zone can form properly.*

When a tree forks or a branch or codominant stem grows to nearly the same size as the trunk, there will be a smooth transition from the branch to the trunk (Figure 3-4, bottom). The protection zone will not form at the base of these branches or stems. Consequently, decay organisms can spread into the trunk when one of these large branches (or codominant stems) is removed from the tree by pruning. For this reason, it is very important to recognize codominant stems and forks in the lower part of the tree and remove or shorten one of the two stems when they are small.

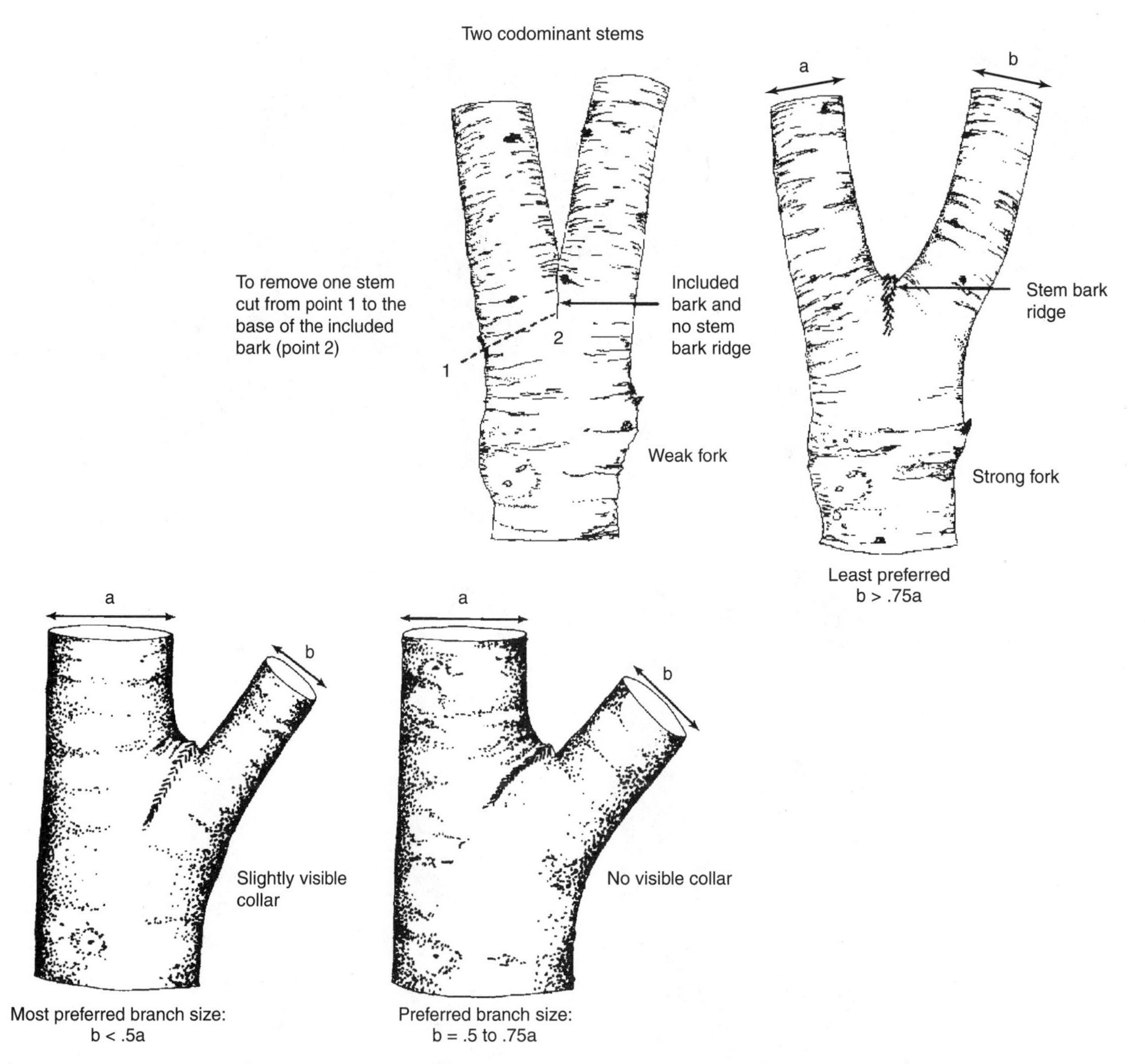

FIGURE 3-5. A fork without included bark (top right) is stronger than one with included bark (top left). However, forked trees are weaker than trees with branches much smaller than the trunk. Branches less than ½ the trunk diameter (bottom left) are preferred over larger limbs (bottom right).

To help prevent the formation of codominant stems and encourage formation of protection zones, allow branches to grow no more than ½ the diameter of the trunk. This is accomplished through regular pruning (called subordination; see Chapter 8) to slow the growth rate of fast-growing branches and stems (Figure 3-5, bottom). This will keep the pruned branch smaller than the main trunk and help ensure that the protection zone and collar at the base of the branch form properly. It can also help prevent formation of included bark, a condition to avoid. These pruning techniques are described in later chapters.

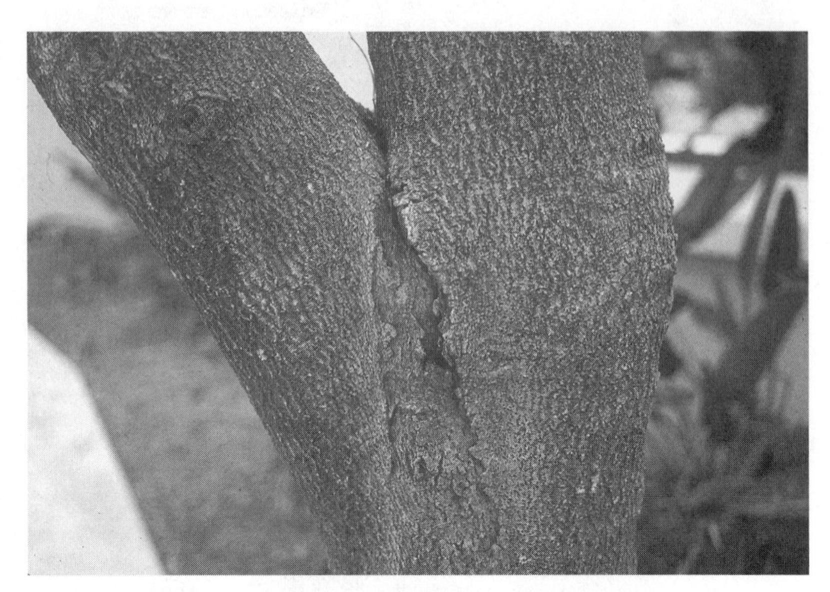

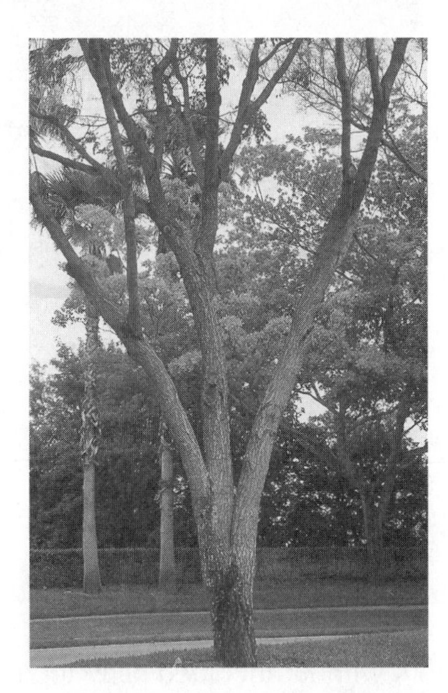

FIGURE 3-6. Because bark is included in the union, these stems are not well secured to each other. The tree on the lower right was overlifted. No pruning was done to remediate the 3 codominant stems.

STRENGTH OF BRANCH AND STEM ATTACHMENT

Branches or stems with no included bark in the union are less likely to split from a tree than those with included bark (Figure 3-5). The presence of included bark could mean that a collar has not formed around the branch (Figure 3-6). Trunk decay below the branch could result from injury caused by the branch and trunk or codominant stems rubbing against one another in the wind. Bleeding may be associated with included bark. Columns of decay inside the trunk can extend for many feet below the crotch.

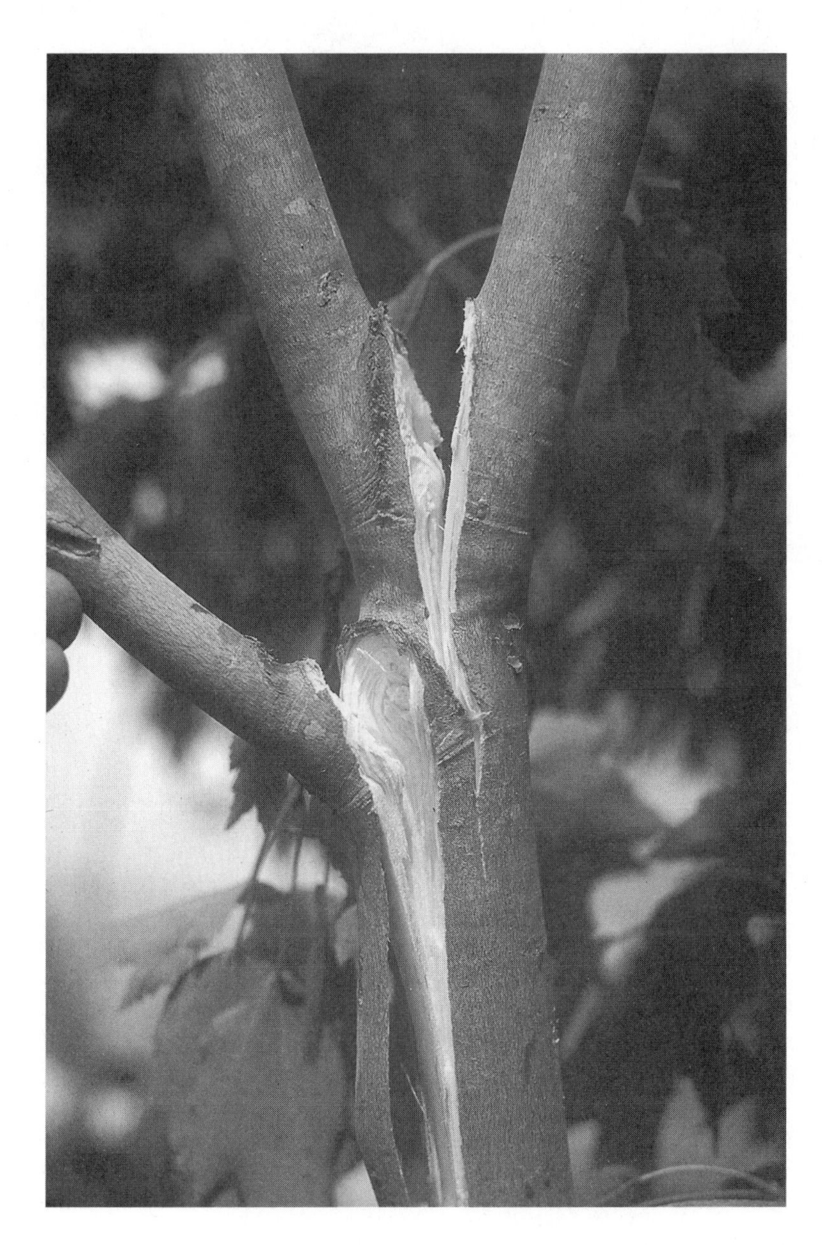

FIGURE 3-7. A small diameter branch (lower union) attached to a much larger trunk is better secured to a tree than a codominant (upper union). Twice as much load (56 kg) was required to split the small branch from the trunk than to separate the upper codominant stems (27 kg).

Codominant stems can be fairly well secured to a tree provided there is no included bark in the crotch (Figure 3-5). One of the main objectives of pruning trees is to prevent included bark from forming, or to reduce growth rate on branches with included bark in the union. Codominant stems and large branches with included bark may be the single most dangerous condition (other than decayed roots) posed by trees in urban landscapes.

A branch that is small compared to the trunk diameter is usually well secured to the tree (Figure 3-7). The same branch growing on a trunk that is only slightly larger than the branch has a weaker attachment. When the branch is the same size as the trunk, we call them codominant stems and these are not as well connected, especially on young trees. This is another reason to prevent branches from growing to more than about half the trunk diameter. Branch breakage was found to be three times more common when foliage was on the tree than in the dormant season (Chris Luley, personal communication).

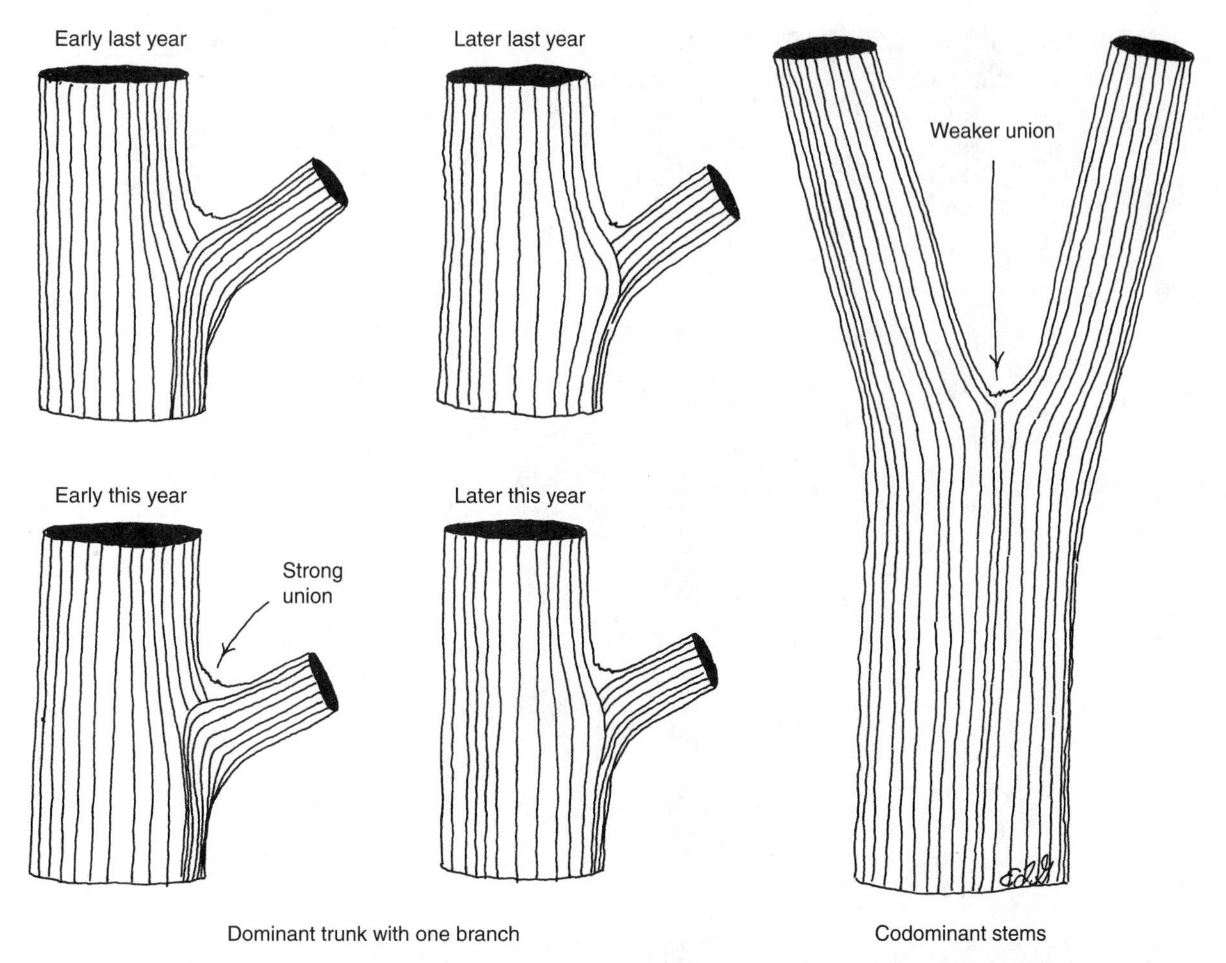

FIGURE 3-8. Branch and stem unions with bark removed. Note that early in the growing season, branch wood grows partially over last year's trunk tissue at the base of the branch (left). Later in the season, trunk wood grows over this branch wood (center). This makes for a strong union. There is no overlapping wood in the union of codominant stems (right).

Except for epicormic sprouts that form from superficial latent buds in the bark, a branch that is small compared to the trunk is usually deeply embedded in the trunk. That is what gives the branch union its strength. This deep embedding occurs when wood from the branch in the spring grows partially over last year's trunk tissue at the base of the branch (Figure 3-8 left). Then the trunk wood on some trees overlaps the branch base later in the growing season. The annual overlapping of wood results in a strong union.

When bark is included in the branch union, or when there are codominant stems, wood does not overlap in the union (Figure 3-8 right). If there is a **bark inclusion** (included bark) in the union, the cambiums of the codominant stems (or branch and trunk) are growing against each other in the inclusion. This results in the codominant stems (or branch and trunk) pushing themselves away from each other. Trees can self-destruct in this manner by splitting along the bark inclusion that is essentially a crack in the tree. This pushing action, combined with the lack of overlapping wood, causes weak unions (Figure 2-5).

Weak unions are especially prevalent on open-grown landscape trees (Figure 3-2). That is why they require training and pruning—to help prevent this condition. Weak unions with bark inclusions can also occur on trees that were in a forest when they were young, but are now exposed to more sunlight because surrounding trees were removed.

CHECK YOUR KNOWLEDGE

1) The structure that helps prevent spread of discoloration and decay from a branch into the trunk is called a:
   a) branch bark ridge.
   b) growth ring.
   c) xylem ray.
   d) branch protection zone.

2) A smooth transition, without a swelling, from the base of a branch onto the trunk usually indicates the collar is:
   a) present.
   b) not present.
   c) not injured.
   d) lacking trunk tissue.

3) Decay is most likely to enter the trunk if you remove a:
   a) small branch.
   b) large branch.
   c) dead branch.
   d) large codominant stem.

4) A crotch with included bark will usually NOT have:
   a) a "U" shape.
   b) decay.
   c) a branch core.
   d) a "V" shape.

5) What do you call the natural boundary that forms to help prevent decay from entering the trunk following a properly executed branch removal?
   a) branch collar
   b) wall 4
   c) branch protection zone
   d) ray cells

6) Trees in open landscapes require pruning whereas trees in the forest do not because:
   a) forest trees have tapered trunks.
   b) light reaches open-grown trees from all sides.
   c) open-grown trees are shorter than those in the forest.
   d) landscape trees grow faster.

Answers: d, b, d, a, c, b

CHALLENGE QUESTIONS

1) What are two reasons for preventing branches from growing more than about half the trunk diameter?
2) What is one method of keeping branches smaller than half the trunk diameter?
3) Differentiate between typical tree form in the landscape with typical form on a forest-grown tree. Why are these so different on many decurrent trees?

SUGGESTED EXERCISES

1) Remove a small branch from a tree. Be sure there are plenty of lateral branches attached to the small branch. Attempt to pull several lateral branches about one-quarter-inch diameter from a larger branch. Now pull a quarter-inch branch from a branch about the same size (you are essentially pulling codominant stems apart). Can you feel how much easier it is to pull the codominant stems apart? Why is that? What does that say about pruning strategies?
2) Find a tree about 5 inches trunk diameter that you can mutilate. Remove a small branch that is less than half the trunk diameter from the trunk making an appropriate removal cut. Remove a codominant stem making an appropriate reduction cut. One year later remove about a six-inch-long section from the trunk centered on each crotch. Cut through the crotch to expose the transverse section (i.e., split the sample through the pith of the stem and trunk with a chisel). Clean the surface with a sharp blade or sand it smooth. Compare what you see. Why does the discolored wood on the small branch sample extend only a short way back inside the tree? What stopped its spread?

CHAPTER 4

TREE BIOLOGY

OBJECTIVES

1) Show how root, branch, and trunk biology should influence pruning strategies.
2) Determine where energy reserves are stored in trees.
3) Describe the tree maintenance practices that reduce energy storage capacity.
4) Describe compartmentalization of decay in trees.
5) Differentiate between good compartmentalizers of decay and poor compartmentalizers.
6) Show how certain pruning cuts can injure trees permanently.
7) Learn how to maximize energy reserves in trees.
8) Develop pruning strategies based on biology to maximize health.

KEY WORDS

Apoplast
Best management practices
Callus
Cambium
Closure crack
Compartmentalization
Decay
Discoloration
Good compartmentalizer
Growth ring
Phloem
Photosynthesis
Poor compartmentalizer
Radial crack
Rays
Reaction zone
Ring crack
Root defects
Starch
Stomata
Symplast
Topping
Transpiration
Tyloses
Walls 1, 2, 3, and 4
Woundwood
Xylem

INTRODUCTION

A basic understanding of tree and shrub biology can make it much easier to grasp the idea of how pruning can impact plant health. The relationships among the important parts of a plant are presented in this chapter. These basics are not taught in many biology or landscape courses because there are few places to find this information. For greater detail on woody plant biology, see Shigo (1991) or Lonsdale (1999).

ROOTS

Roots are usually the forgotten part of the tree. They typically comprise from one-third to one-fifth of the total dry weight of the tree. Their tips extend to about three times the edge of the canopy (Figure 4-1). Roots deflected by structures such as rocks, foundations, streets, and sidewalks can reduce the uniformity of the root system and decrease tree stability. Nursery containers can also deflect and deform root systems (Figure 4-2). Deformed root systems can result in unstable trees that can fall over. Roots damaged or severed during construction or utility installation can also result in tree instability due to loss of upright support, decay, and internal defects such as cracks.

Pruning a tree with **root defects** might help compensate for a deformed or injured root system. For example, a tree left in a nursery container too long might not develop roots on one entire side of the tree after it is planted in the landscape. This occurs because few lateral roots form on the outside portion of a curved root segment (Figure 4-2). Trees with one-sided root systems can fall over. Thinning the canopy of a tree with a deformed root system to reduce the force of the wind against it could allow the tree to remain standing longer. However, pruning the canopy of a

FIGURE 4-1. Tree roots spread way beyond the branches to about three times the canopy spread. More than half of the root system is located beyond the edge of the canopy. Few roots grow deep into the soil on most sites.

tree with this root defect should never be considered a long-term solution to root deformities.

Tip: *Damaged or deformed roots can have a large impact on tree stability.*

Roots perform functions other than stability. Some growth regulators are produced in the roots. Roots take up water and elements. They also form associations with organisms, such as mycorrhizae, that can help the tree in many ways. Roots fend off diseases and insect attack, and they store some of the energy produced in the leaves.

WOOD AND CANOPY

Water entering roots moves to the canopy through the xylem (Figure 4-3). **Xylem** makes up the woody part of the tree that begins on the inside of the cambium. It extends through to the pith or the trunk center. Xylem is made up of living and nonliving cells in roots, trunk, branches, and foliage. It can be thought of as a pipeline from roots to leaves. In dicot plants, a new layer of xylem is produced by the secondary growth system, the **cambium,** each year (or several times each year in some tropical trees) forming a new **growth ring** in trunks and branches. Xylem in foliage is produced by primary growing points in buds.

As water evaporates from the foliage (this process is called **transpiration**) through openings in leaves called **stomata,** it pulls adjoining water molecules with it. This pulling action helps draw water up the trunk and into the leaves. In addition, some trees may be capable of exerting a pumping action to push water up the tree. Most water moves up through the **apoplast** or the network of open, dead conducting elements in xylem. Elements (also called nutrients) such as magnesium and potassium move from the soil up to the foliage in the xylem cell sap.

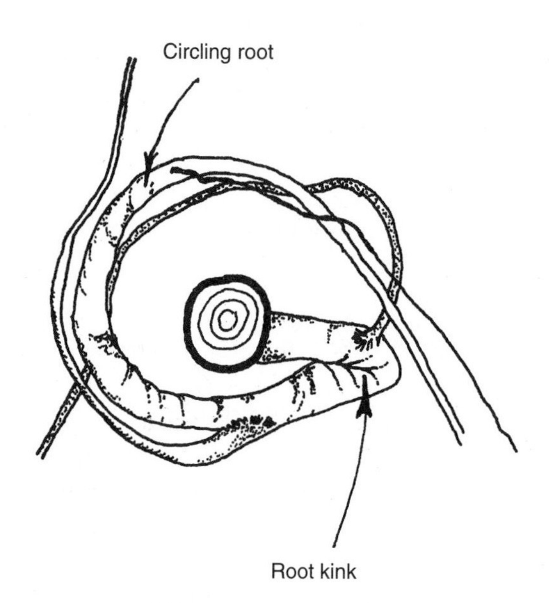

FIGURE 4-2. Roots circling near the top of the container or kinked roots can slow growth or even kill the tree as it grows. Root defects such as these can result in a one-sided root system in the landscape leading to tree instability.

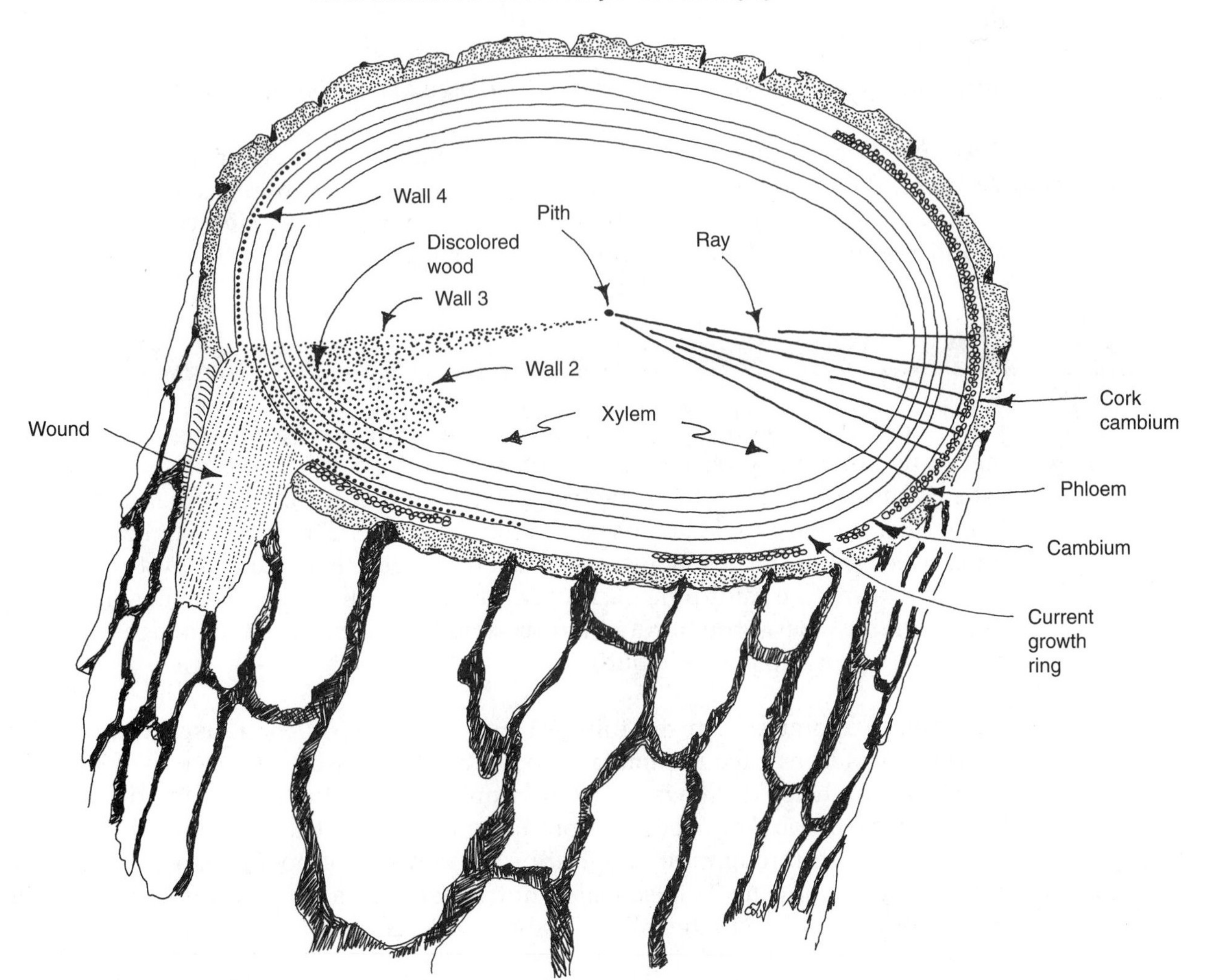

FIGURE 4-3a. Woody tissues are arranged in the following order from outside to inside the tree: bark, cork cambium (produces bark), phloem, cambium (produces xylem and phloem), and current growth ring of the xylem. When a tree is injured, it can respond by plugging the xylem ends (plugging makes up wall 1—not shown). Injured trees also react to form wall 4. Wall 4 may look like dots (look closely along cambium near the wound) or a solid dark line along the cambium in recent wounds. Wood often discolors behind the wound to the inside of wall 4. This discolored wood may reach the pith quickly if wall 2 is weak. Rays make up wall 3 (only some of the rays are shown in the drawing).

Photosynthesis in foliage, twigs, and other green plant parts produces sugars (and other components) that are used by the tree to carry out its many functions. Sugars are moved about the plant in a layer of cells called the **phloem.** Phloem is made up of living cells located just outside the cambium. The cambium produces the phloem in trunks, branches, and roots. The tree usually expends energy moving sugars, growth regulators, proteins, and essential elements up and down the phloem to other locations in the plant.

Once sugar arrives at a location, it is used to carry out normal processes or it is stored. It is stored as **starch** in the network of living cell contents in the xylem called the **symplast.** Starch is a chain of sugars linked together. Starch is considered the money, or the energy, in the tree bank. The bank is the living xylem or wood in branches, stems,

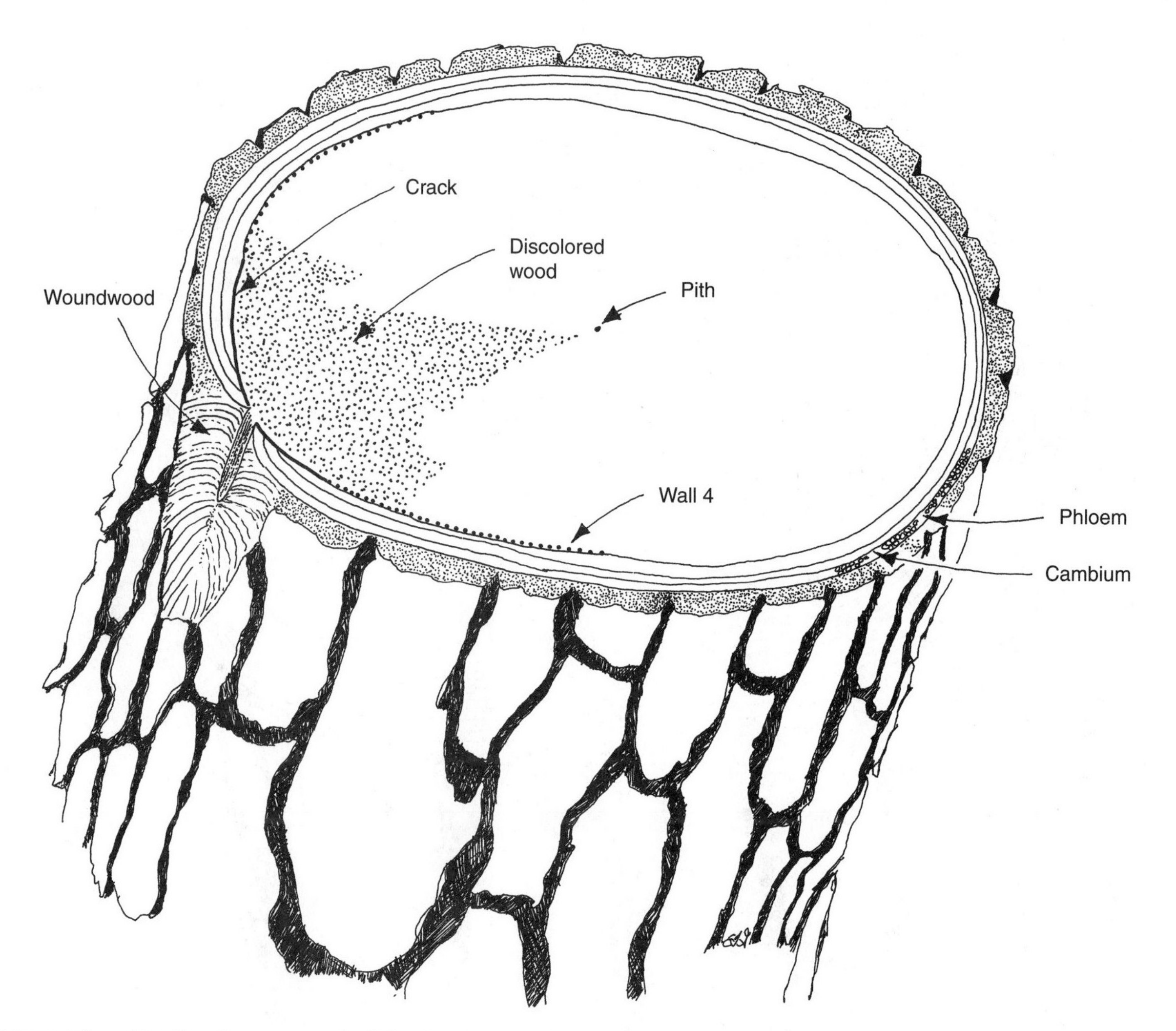

FIGURE 4-3b. Wall 4 has extended further around the trunk two years after the injury. The amount of discolored wood has also increased. Two new growth rings formed and woundwood has almost closed the wound. A crack is developing along the portion of wall 4 closest to the wound. (Rays are not shown).

trunk, and roots. If there is less stored starch, there is less stored energy in the bank. Trees need stored starch to carry on normal functions, especially to break dormancy.

Rays are long groups or plates of living cells that extend from the phloem into the xylem toward the center of the trunk. On their way to the xylem, they cross the cambium. They can extend several inches or more up and down the trunks, branches, and roots. Sugars move from the phloem into the energy bank (xylem) through the rays. On healthy trees, rays are rich in starch.

Improper pruning such as **topping** (Figure 4-4), cutting roots, flush cutting, removing large branches, and injuring the trunk and main branches can make the energy bank considerably smaller. If the bank is made smaller, less energy can be stored. This makes the tree more susceptible to attack by agents such as insects and disease. Improper pruning can also make it tough for trees to recover from floods, droughts, or other stresses.

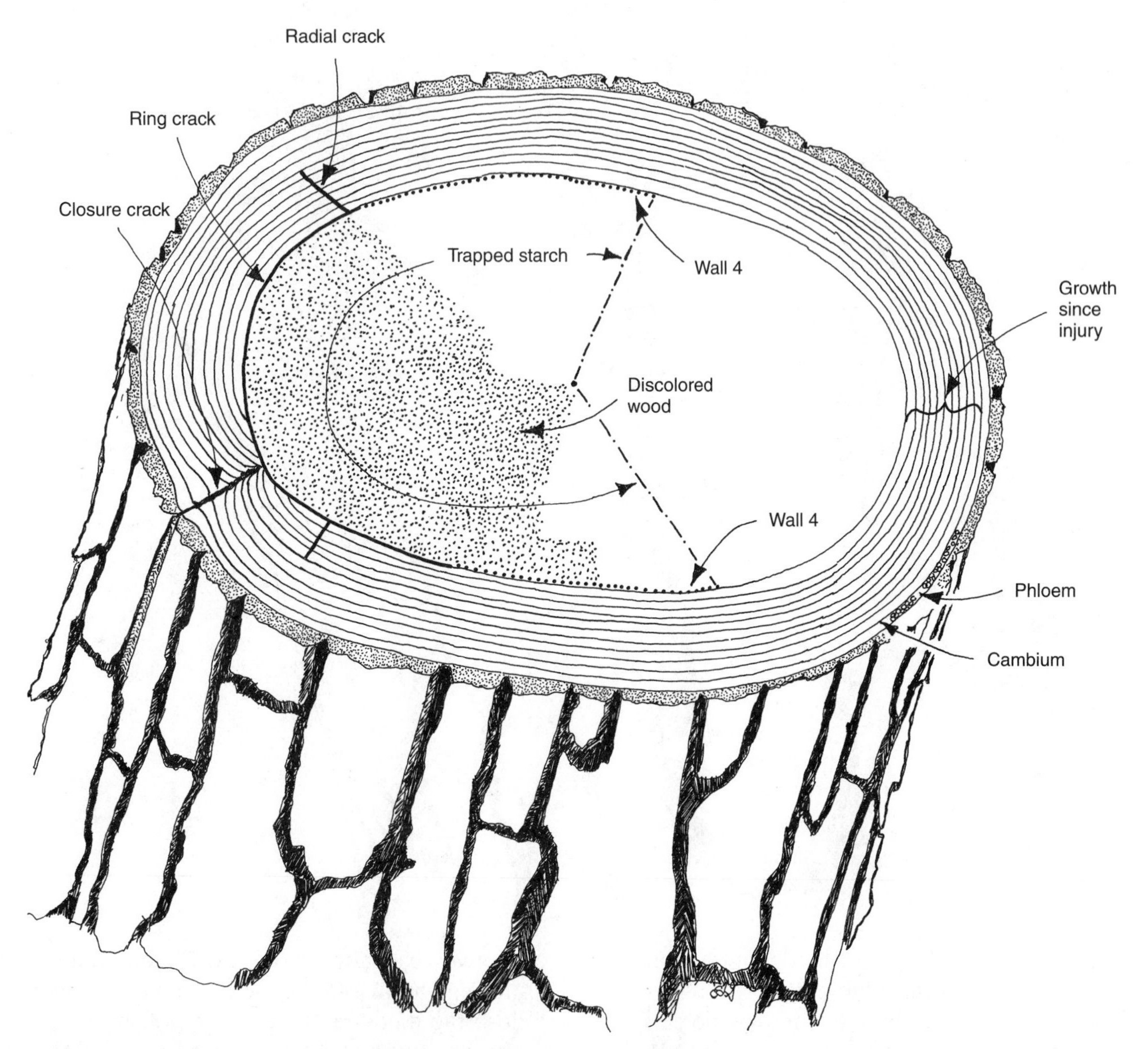

FIGURE 4-3c. Wall 4 extends further around the trunk compared to two years after injury. It is now more than half way around the trunk ten years after injury. However, note that it stayed in the same position in the trunk—it does not move out with the new cambium. It may continue to develop along the growth ring that was injured ten years ago. The ring crack may continue to grow along wall 4. Two radial cracks are forming from the ring crack. The wound is closed but a closure crack extends from the point of the original wound to the bark. Ten growth rings are visible outside wall 4. This corresponds to one each year after injury. The starch stored between wall 4 and the pith and between the dashed lines cannot move easily out to the phloem through the normal pathways. If the tree is to use this starch it must reach the phloem in other ways. The discolored area is much larger that it was eight years earlier. (Rays not shown).

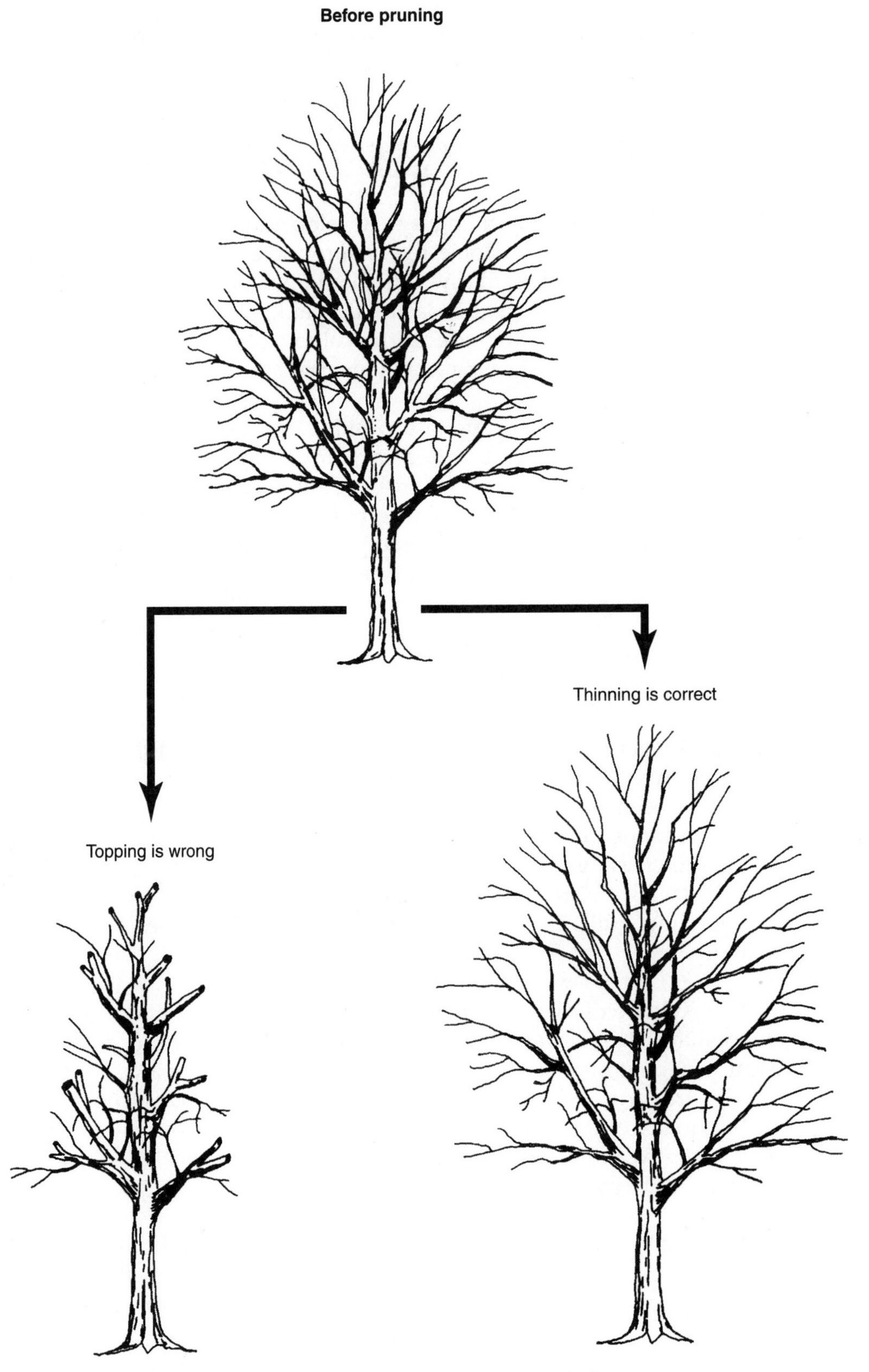

FIGURE 4-4. Never top a tree. Instead, prune it to retain its natural shape.

 **Tip:** *Improper pruning techniques and mismanagement can reduce the available storage space for energy reserves.*

COMPARTMENTALIZATION OF DECAY IN TREES

One of the objectives of pruning is to develop and maintain strong architectural structure. In order to meet this objective, appropriate live stems and branches should be shortened or removed regularly. This should be done in a manner that does not measurably reduce energy storage capacity (the bank) of the tree. The bank becomes smaller because improper pruning reduces the volume of healthy wood that can store starch. This can lead to a tree starving itself if energy needs outpace reserves.

Injuries result from inappropriate pruning cuts, storms, ice, snow, animals, wind, excess weight, temperature extremes, trunk wounds, disease, chemicals, and other stresses. A tree reacts to injury by creating boundaries around it. The boundary-setting process that resists loss of normal wood function and resists the spread of discoloration and decay has been referred to as **compartmentalization. Discoloration** is the orderly response of the tree to microorganisms resulting in darkened wood but no strength loss. **Decay** is the orderly breakdown of tissue resulting in strength loss. The rate of the discoloration and decay process depends on the severity of the wound, position in the tree, size of the wound, time of year, species, tree age, and the types of infecting microorganisms.

Compartmentalization may be described using the CODIT (Compartmentalization Of Decay In Trees) model (Shigo, 1991). Four different boundaries (also called walls) in trees have been presented in CODIT: wall 1, wall 2, wall 3, and wall 4. Each forms in a different manner and protects the tree in unique ways (Figure 4-3). The walls are numbered in increasing order of their ability to retard movement of decay organisms. For example, wall 3 is stronger than wall 1. Wall 1 may or may not be present at the time of wounding. Walls 2 and 3 are present in the tree at all times. Wall 4 forms in response to injury.

Wall 1: Xylem vessels immediately above and below an injury plug with crystal-like **tyloses** and chemicals when a tree is injured. This plugging response forms wall 1. Some plugging may occur normally without injury. Plugging forms a weak boundary in some trees such as hackberry (*Celtis*) and poplars (*Populus*), but is stronger in others such as many of the oaks (*Quercus*) (Table 4-1). Because wall 1 is weak, decay in some trees can advance rapidly up and down the trunk from the injury resulting in long columns of decay and hollow branches and trunks.

Wall 2: The growth rings make up wall 2. The transition from one growth ring to the next retards advancing decay organisms. Decay organisms often have a tougher time moving across growth rings (wall 2) than they do up and down the stem (wall 1). The functioning of wall 2 can be demonstrated when you view a cross section of a trunk or branch that was injured previously (Figure 4-3a). A darkened region often appears to stop its advancement inward toward the pith at the boundary of one growth ring with the next. This is wall 2 working.

TABLE 4-1. Trees vary in ability to compartmentalize decay.

Poor compartmentalizers	
Aesculus	*Myrica*
Betula	*Peltophorum*
Brachychiton	*Persea*
Celtis	*Pinus virginiana*
Delonix	*Populus*
Eucalyptus (some)	*Prunus*
Erythrina	*Quercus laurifolia, nigra, leavis, shumardii, palustris*
Fagus	*Salix*
Ficus benjamina	*Schinus*
Good Compartmentalizers	
Acer x *freemanii, rubrum, saccharum*	*Robinia pseudoacacia*
Albezia saman	*Taxus baccata*
Bucida buserus	*Sweitenia*
Bursera simaruba	*Tabebuia*
Castanea sativa	*Ulmus americana*
Juglans	
Pinus rigida	
Quercus geminata, macrocarpa, petraea	
Quercus robur, rubra, virginiana	

Wall 3: The rays make up wall 3. They have plenty of decay fighting capability because they are rich in starch. The rays (wall 3) retard decay from spreading around the trunk. Discoloration and decay have a tougher time moving across wall 3 than walls 1 and 2. The strength of wall 3 can be demonstrated by viewing a cross section of a trunk or branch injured several years ago. Notice that there is a clear demarcation between darkened tissue and normal light colored wood (Figure 4-3a). If walls 2 and 3 fail and decay organisms break through, the affected trunk or branch can become hollow. Wall 4 forms the outside edge of the hollow.

Wall 4: This is the strongest boundary that retards spread of discoloration and decay in trees. This **reaction zone** forms from the cambium along the edge of the outermost growth ring present at the time the tree was injured. It begins at the point where the tree was injured and it may extend all or part way around the tree (Figure 4-3). Wall 4 stays in the same position in the tree but may extend further around or up and down the trunk with time. It does not move out with the new cambium. There may be numerous wall 4s in a tree, depending on its wounding history. They often appear as crescent-shaped dark lines when viewed on a wood cross section. Wall 4 forms the edge of a hollow.

Wall 4 develops in response to many different types of injuries (Table 4-2). It can take several years for wall 4 to reach the other side of the trunk—or it may never reach that far. Wall 4 extends above and below the injury essentially in the shape of a pipe. It may develop a few inches or many feet above and below the injury.

TABLE 4-2. Causes of wall 4 formation in trees.

Canopy reduction	Mechanical injury to roots, trunk, or branches
Dieback	Removing large diameter branches
Drought	Removing codominant stems
Flood	Removing collars
Flush cuts	Storm damage
Improper pruning cuts	Topping or rounding over a tree

Wall 4 prevents discoloration and decay organisms from moving into the wood produced after the injury occurred. This means it is extremely difficult for discoloration or decay to move from inside wall 4 to the outside of wall 4. Although this task appears simple, it is *vital* to the longevity of trees. Imagine if decay organisms could spread into wood formed after injury—trees could not live to become old majestic masters.

The obvious advantage of wall 4 is that it retards decay, but there are two very important disadvantages. The first disadvantage is that sugars cannot move across wall 4; that is, sugars have a more difficult time moving in or out of the portion of the bank (xylem) surrounded by wall 4. As a result, some stored starch can get trapped in the rays and xylem located inside of wall 4 (Figure 4-3c). However, the starch is available to decay organisms. The tree may have wasted the effort required to produce the sugar and store it as starch. It earned the money (made the sugars), deposited the money in the bank (stored it as starch), then could not withdraw some of the money (starch was trapped inside). You can imagine how much stress this causes the tree by imagining your stress after you earned money, deposited it in a bank, and could not get some of it back. Creation of wall 4 also makes the energy bank smaller so less starch can be stored in the future. This occurs because wall 4 essentially shuts off new deposits of sugars into the walled off portion of the xylem.

The second disadvantage is that a crack can form along wall 4. This separation or delamination is called a **ring crack** and it may follow wall 4 all or part way around the trunk (Figure 4-3c). One or more secondary cracks, called **radial cracks,** can form from the ring crack along a ray. Another serious crack is the **closure crack** that occurs as the **callus** and **woundwood** attempt to grow over and close the wound. This crack often extends from the point of injury out to the current location of the bark. Sometimes this crack never closes. Even if it closes, the crack remains along with its associated weakness (Lonsdale, 1999). Cracks in trees cause weakness that can make them susceptible to breaking. In fact, cracks are probably of more concern than the decay that results from injury (Shigo, 1991).

Trees vary in their ability to form walls 1, 2, and 3 (Table 4-1). These walls are weak in trees that are **poor compartmentalizers** of decay. Hollow trunks result from weak walls. These three walls are stronger in trees that are **good compartmentalizers.** Following injury to a poor compartmentalizer, wall 4 may reach the opposite side of the trunk quickly, within a few years. Because walls 2 and 3 are stronger in a good compartmentalizer, wall 4 may only need to be produced part way around the trunk.

BEST MANAGEMENT PRACTICES BASED ON BIOLOGY

It is vital to care for trees in a manner that minimizes injury. Injured trees produce reaction zones, called wall 4s, to protect themselves, but this can cause serious problems for trees. The tree care management practices discussed next take into account the tree's internal response to treatment. As with medicine, just about every practice or treatment is a balance between advantages and disadvantages. The **best management practice** is the best available treatment, considering the benefits and drawbacks, based on current knowledge in the field.

Trees should not be topped or rounded over because this exposes wood to infection and initiates wall 4 and cracks in the wood. Wall 4 quickly forms completely around the cut stem following a topping cut. It can extend several inches or 20 feet or more below the cut depending on species of tree and cause root decay (Figure 4-5). This can trap stored starch inside the tree making it less available to the sprouts that result. Some of the energy needed for sprout growth may come from other portions of the tree resulting in even lowered reserves. One recent survey found that trees topped in the landscape break and fail more often than other trees (Karlovich et al. 2000). Some of this was probably due to cracks formed along wall 4s causing fractures and weakness in the wood.

On mature trees, perform canopy reduction (covered in Chapter 13) only after every other option has been considered because reduction cuts can initiate wall 4 and cracks. On young and medium-aged trees, reduction cuts *can and should* be used to reduce the length of codominant stems. This improves the structure and the potential life span of the tree. Reduction cuts on these younger trees are considered a best management practice because removing a codominant stem or large branch entirely (if it does not have a protection zone) could cause more problems than reducing its length.

If clearance is required under the canopy, raise the canopy slowly over a period of years, not all at once. Removing too many low branches at one time can initiate formation of wall 4s, decay, and cracks in the trunk. Gradual removal will allow the tree to adequately adjust to less live foliage. Do not remove too much live foliage all at once, especially on mature trees, because this can cause root problems, sprouts, wall 4 formation, stress, decline, or even death.

Energy reserves can be kept high in trees by removing the least amount of live branches necessary to accomplish the desired goal. Removing too many live branches can cause stress in trees by reducing sugar production and reducing storage capacity by initiating wall 4s. Many trees sprout in response to overpruning because they are attempting to replace the stored energy removed with pruning. However, live branch pruning is an essential ingredient to forming good structure, so it is necessary procedure in an urban tree care program.

Tip: *Keep energy reserves high by removing the minimum amount of live branches to accomplish your objectives.*

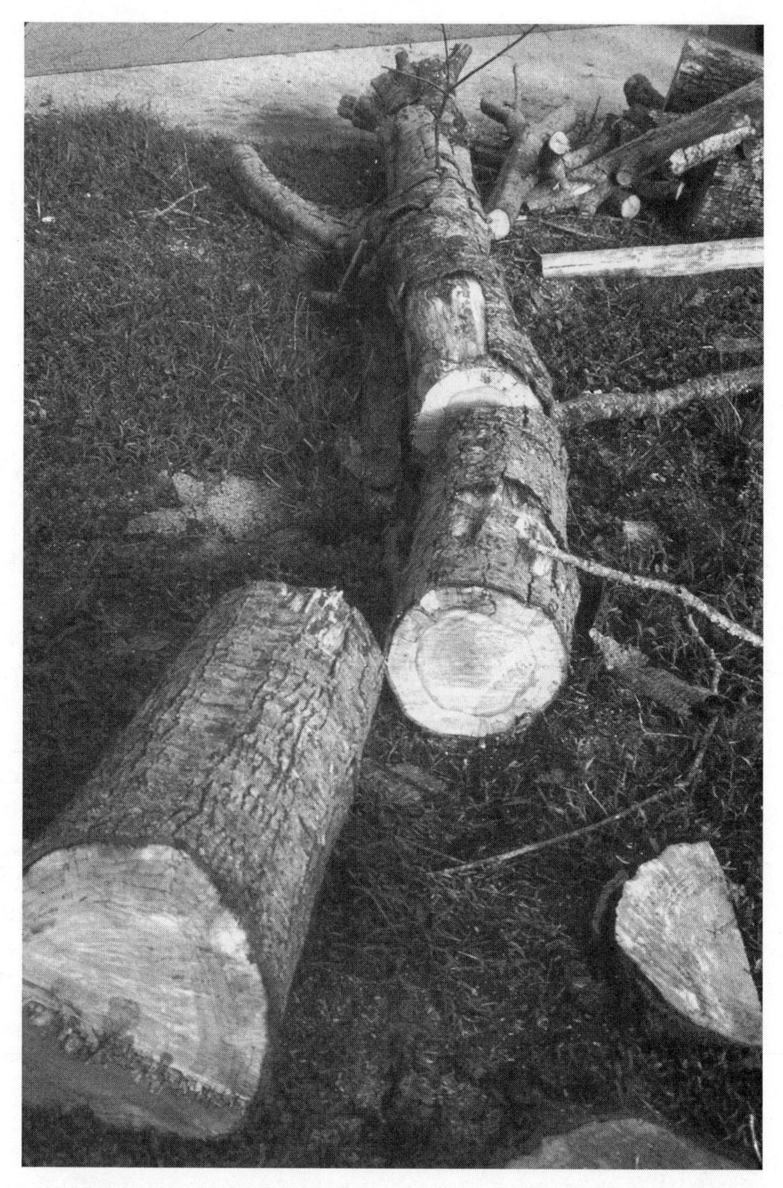

FIGURE 4-5. The tree was topped by cutting through the main trunk seven years ago (see top of photo). It was cut up into sections to view internal conditions. Wall 4 can be seen as a dark circle at each cut section from the top of the tree (top of photo) to the root system. This makes it clear that topping seven years earlier caused wall 4 to form as a cylinder through the entire length of the tree. The starch stored inside wall 4 can become trapped and unavailable to the tree.

High energy reserves in nursery trees mean rapid growth in the nursery. To keep energy levels high, live branches on the lower portion of the trunk are left on the tree but are shortened during the first few years of production. This keeps them small but allows them to produce sugars to feed the trunk and roots. The result is rapid increase in root growth and trunk diameter. However, if live low branches become too vigorous, a large wound results when they are removed prior to sale. This could cause cracks in the trunk that will have little effect in the nursery but could be the source of trunk cracks after the tree is planted in the landscape.

CHECK YOUR KNOWLEDGE

1) A tree capable of forming strong boundaries that resist decay can be referred to as a:
   a. poor compartmentalizer.
   b. good compartmentalizer.
   c. healthy tree.
   d. strong-structured tree.

2) What CODIT wall is responsible for preventing decay from spreading into new wood produced after injury?
   a. wall 1
   b. wall 2
   c. wall 3
   d. wall 4

3) What wall slows the spread of decay in toward the pith?
   a. wall 1
   b. wall 2
   c. wall 3
   d. wall 4

4) Which is the strongest wall?
   a. wall 1
   b. wall 2
   c. wall 3
   d. wall 4

5) What is the appropriate order of tissues in dicots moving from the bark to the pith?
   a. cambium, phloem, xylem, cork cambium
   b. xylem, cambium, phloem, cork cambium
   c. cork cambium, phloem, cambium, xylem
   d. phloem, cork cambium, cambium, xylem

6) A best management practice is:
   a. the best treatment you can think of now.
   b. a best technique used by business to manage people.
   c. the practice or treatment you have used for the past ten years.
   d. the best available treatment based on current knowledge.

7) What forms along wall 4 that can cause trees to fall apart or break?
   a. a ring crack
   b. a radial crack
   c. a closure crack
   d. included bark

Answers: b, d, b, d, c, d, a

CHALLENGE QUESTIONS

1) Describe how starch can get trapped in the xylem and is make unavailable to the tree.
2) Describe to a colleague how hollow trunks form in trees.
3) From the tree's perspective, compare the advantages and disadvantages of wall 4.
4) Canopy reduction is performed to reduce tree size. What is a serious downside of this technique to consider when evaluating treatment alternatives?

SUGGESTED EXERCISES

1) Find a fresh woodpile with clean end cuts or make some clean cuts with a saw. Locate samples illustrating operation of walls 2, 3, and 4.
2) In a woodpile, find a wall 4 in cross section on a branch or trunk. Determine how many years ago the injury took place that caused formation of that wall 4. Can you find any other wall 4s on the same sample? Sometimes, there are many on the same sample.
3) In cross section, see if you can find a radial crack originating from a ring crack. Cut the sample into several cross sections to see how far the crack extends up or down the sample.

CHAPTER 5

PRUNING CUTS

OBJECTIVES

1) Describe the three basic types of pruning cuts on live branches and how trees react to each.
2) Show how to remove dead branches and those with included bark.
3) Differentiate between a flush cut and a collar cut.
4) Contrast small branch removal with large branch removal.
5) Evaluate pruning cut accuracy from wound closure pattern.
6) Prune branches without stripping trunk bark.
7) Determine when pruning wound dressings are useful.

KEY WORDS

Collar cut
Drop-crotch cut
Drop cut
Flush cuts
Frost cracks
Heading cut
Lateral cut
Node
Pinching
Pollarding cut
Reduction cut
Removal cut
Roundover
Sunscald
Terminal bud cluster pruning
Thinning cut
Topping
Woundwood

INTRODUCTION

There are three basic pruning cuts (Figure 5-1). (1) A **reduction cut** reduces the length of a branch or stem back to a live lateral branch large enough to assume the apical dominance—this is at least one-third the diameter of the cut stem. This cut has also been named a **lateral cut** or a **drop-crotch cut.** It was called a **thinning cut** in the national pruning standard ANSI A300, 1995 (the standard will be covered in Chapter 15). The current version of the pruning standard drops the thinning cut terminology because it is easily confused with the pruning type called thinning. (2) A **heading cut** reduces the length of a stem or branch back to a predetermined height regardless of position of lateral branches. A type of heading cut called a **pollarding cut** might be considered a fourth type of cut. It will be described in Chapter 10. Trees respond differently to these cuts (Table 5-1). (3) A **removal cut** removes a branch from the trunk or parent branch. This was also called a thinning cut in ANSI A300, 1995 and is called a **collar cut** by some arborists.

REDUCTION CUTS

A reduction cut reduces the length of a stem or branch by removing the terminal portion back to a branch or lateral of equal or smaller diameter. The cut is made to a living side (lateral) branch that is large enough to assume apical dominance. This size varies with species, but should be at least ⅓ (minimum) to ½ (preferred minimum) the diameter of the cut stem (Figure 5-1, top). To make the cut, bisect the angle between the branch bark ridge and an imaginary line perpendicular to the stem or branch to be removed (Figure 5-2). Do not cut parallel to the branch bark ridge as this removes too much wood creating a weak point.

It is hard to correctly prune a young tree without making reduction cuts. These cuts are used on larger trees as well. For this reason, it is a very important skill to master. Reduction cuts are essential to direct the growth of a young tree, and the only recommended method of reducing the size of the tree. Many of the cuts demonstrated on young and medium-aged trees in the following chapters are reduction cuts. Because they cut back to a natural barrier (the branch protection zone), removal cuts are preferred over reduction cuts on older trees. Reduction cuts cause injury to the cut stem or branch. This injury could extend down the cut stem on trees that compartmentalize decay poorly. This disadvantage is often offset by the improved structure that results from appropriate reduction techniques. Trees, especially young ones, that compartmentalize decay well are able to retard adverse effects from reduction cuts.

HEADING CUTS

Reduction and removal cuts are made back to an existing branch or stem also called a **node;** heading (also called **pinching,** shearing, tipping, rounding over, and topping) cuts a branch or stem between nodes (Figure 5-1, bottom). In addition, if you cut back a stem to a living side branch less than ⅓ the size of the cut stem or if you cut back to a bud, then the cut is considered a heading cut.

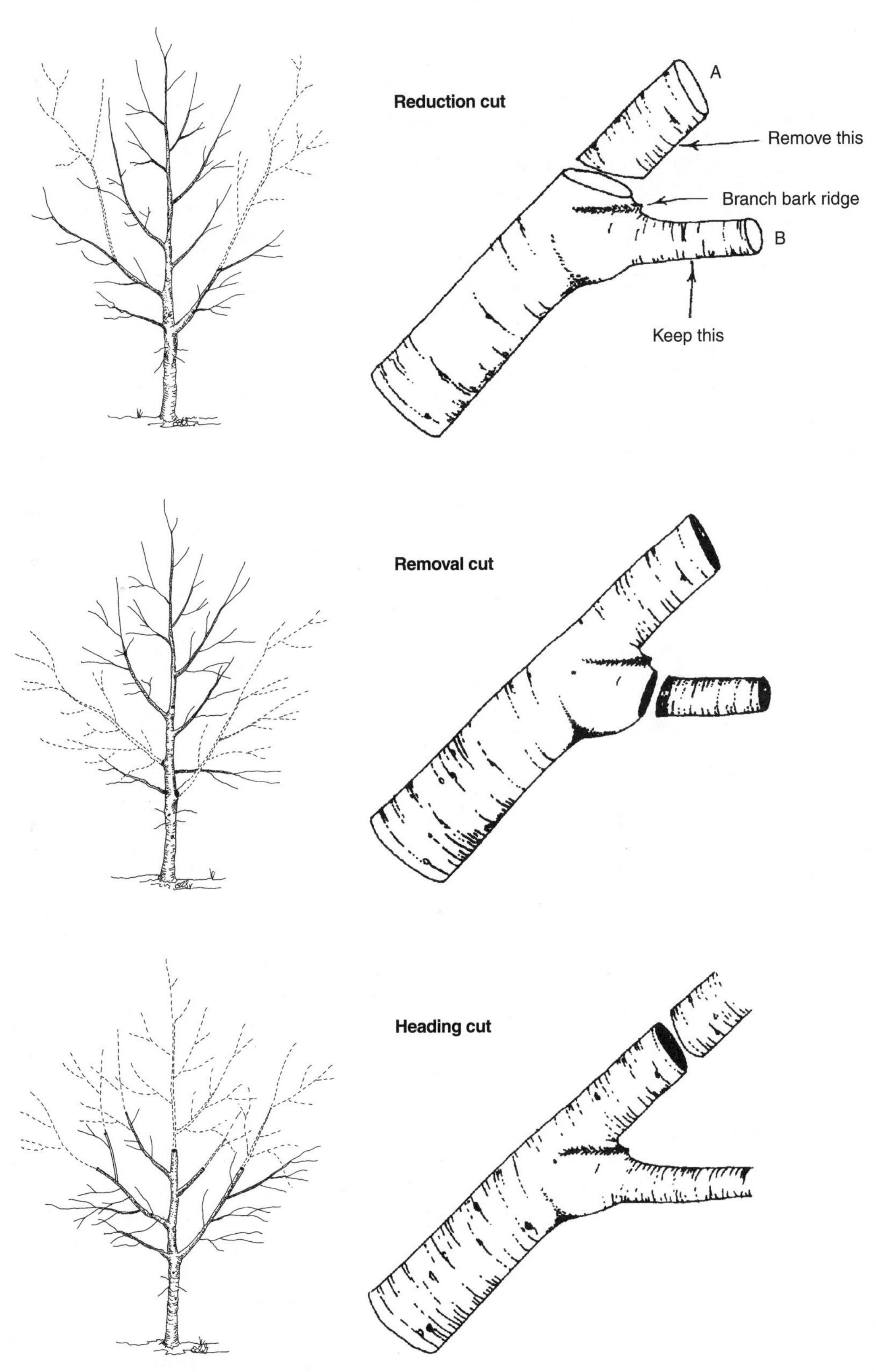

FIGURE 5-1. Reduction, removal, and heading cuts (see dashed or dotted lines on full tree drawings at left) each elicit a different response. A reduction cut (top) is made back to a branch (B) no smaller than about ⅓ (minimum) to ½ (preferred) the diameter of the cut stem (A). A removal cut is made back to the collar, whereas a heading cut is made at no particular point or back to a bud.

TABLE 5-1. Comparison of tree response to various pruning cuts.

	Tree Response								
Pruning Cut Type	**Invigorates Existing Interior Foliage**	**Induces Sprouting**	**Canopy Density**	**Canopy Height**	**Canopy Width**	**Susceptibility to Future Storm Damage**	**Susceptibility to Sun Injury**	**Susceptibility to Stem Decay and Cracks**	**Cuts Back to Natural Boundary**
Reduction	yes	usually	reduced	reduced	typically reduced	low to medium	medium to high	medium to high	no
Removal	yes	may	reduced	usually not affected	may be slightly reduced	low	low	low	yes
Heading	no	yes	reduced, then increases*	reduced	typically reduced	high	high	high	no

*Canopy density increases quickly after pruning.

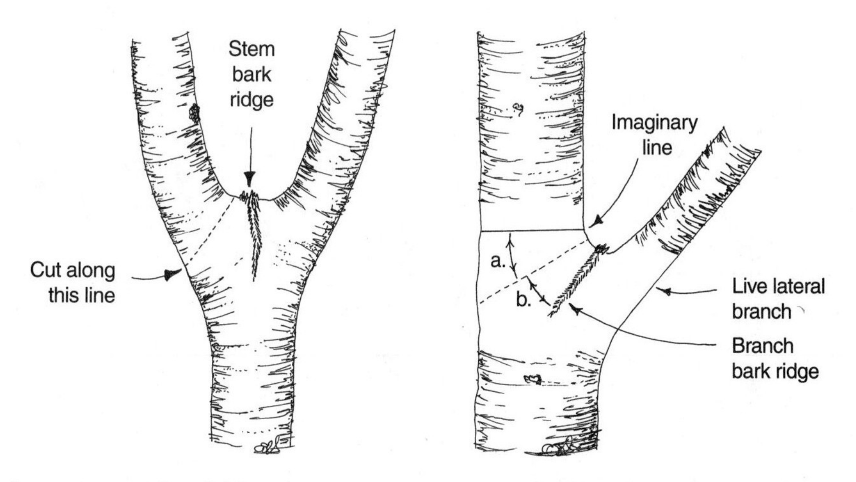

FIGURE 5-2. A reduction cut should bisect the angle formed by the branch bark ridge and an imaginary line perpendicular to the cut stem (right). Angle 'a' should equal angle 'b'. Making a cut parallel to the branch bark ridge leaves the union weak because it is too thin. Same goes for removing a codominant stem (left).

Young trees are headed in nurseries to slow growth on stems and branches competing with the main leader, create branches in desired locations (Figure 5-3), direct growth, increase branching on a leggy stem on young nursery trees, or create and maintain special effects. Heading is used by tree growers and gardeners on young saplings to create a compact or balanced canopy, or to thicken the appearance of the canopy. Heading is also regularly used by growers to keep in check the length of temporary branches along the lower trunk. The leader on very young saplings is headed on certain trees to create a set of branches at this point. Heading trees in the landscape or nursery is not recommended unless you are training a young tree and understand how to prune the tree afterward (see "Trees with an Unbranched Trunk" in Chapter 8) or if you are pollarding (see Chapter 10).

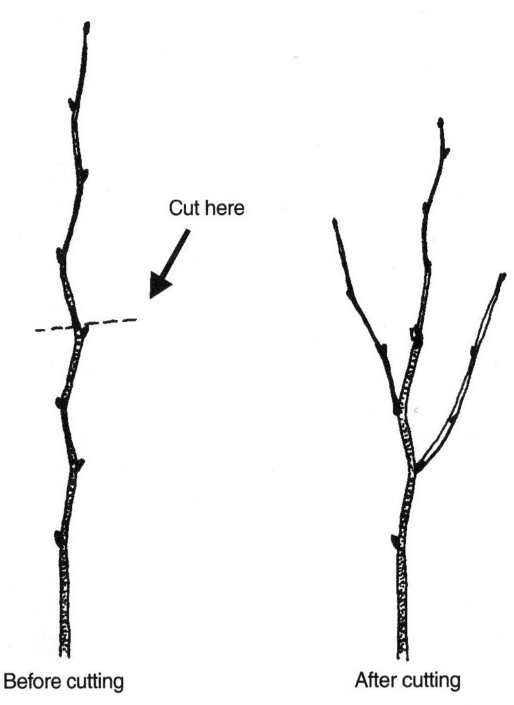

FIGURE 5-3. Heading, pinching or tipping a stem will force two or more buds located back from the cut to generate shoots. Such cuts are appropriate for cutting one- or two-year-old branches when training young trees.

Young, vigorous trees and shrubs easily compartmentalize following the heading of small-diameter branches no more than one or two years old. Heading older tissue is not recommended because it can initiate decay and cracks in the cut stem and ruins tree structure. Heading is usually performed just prior to growth in the spring or early in the growing season in order to stimulate lateral buds to develop into branches.

Removing several inches from the tips of emerging branches (pinching) on young trees prior to a growth flush creates a denser tree by increasing the number of branches. The thumb and index finger can be used on young, soft tissue; pruning shears are used on hardened wood. Pinching is usually appropriate only on small-growing trees, trees that are still young in the nursery, or those that are maintained at a certain size (hedges, topiaries, espaliers) with regular shearing or clipping. The power shearing **roundover** pruning that cuts through branches an inch or more in diameter on large-maturing trees sometimes seen in landscapes and some nurseries is the equivalent of topping and is not normally recommended. It slows growth and ruins good tree architecture. Sometimes the need for pinching or tipping in the nursery can be avoided by using a compact or dwarf cultivar of the tree you are growing or by providing more light to the canopy by spacing trees farther apart in the nursery.

New branches arise from buds located just behind the end of the headed branch. When one-year-old stems are headed, new branches grow more-or-less in the direction the bud is pointed, and the outermost bud (the one toward the tip of the cut branch) has the most influence on the direction of future growth. On some young trees, you can control the orientation of the new branch by carefully selecting the pruning cut's location. Cut so the outermost bud is pointed in the direction you want the branch to grow. However, on many trees growth on headed branches is vigorous and upright. Heading branches only on the canopy edge creates dense growth on the outside of the canopy, which shades the inner canopy foliage. Existing leaves and branches on the interior of a tree that is headed can become shaded out and die. This creates poor taper on the limbs, which means they are about the same diameter all along their lengths. Branches with poor taper are weak and can snap.

Pinching the new shoots on some trees, such as Japanese maple and fringe tree, stops active growth of those shoots. No new branches will be generated until the next growth flush, when buds are well developed. This could be later in the growing season or even next year, depending on the tree species and the vigor of the plant.

Unfortunately, the trunk or large-diameter branches are sometimes headed to indiscriminate lengths in order to bring a large tree to a desired size. This is called heading or **topping** (Figure 4-4). Heading or topping is not recommended. Rounding over the canopy is inappropriate for most trees unless it is performed at least annually. Heading large-diameter branches often causes decay and cracks in the cut branch stubs and trunk, depletes energy reserves, causes sunscald on the trunk and branches, weakens roots, destroys the tree architecture and structure, causes vigorous sprouting, attracts boring insects, and wastes energy, because the removed branches have to be disposed of. The vigorous sprouts that result are poorly attached to the tree and easily broken. A weak, potentially dangerous tree is thus created, especially if it will be allowed to grow large after it is topped.

If a tree needs regular pruning to keep it small, the wrong tree was planted in that site. Consider cutting it down and replanting with an appropriately sized tree, or else move the obstacle you were pruning the tree away from and let the tree grow.

Trees have only one type of defense against the wood-rotting effects of heading and reduction cuts: that is, plugging the conducting elements in the stem. This plugging is the wall 1 described in the previous chapter. Because plugging is a relatively weak defense, the wood often discolors and can decay behind these cuts. Cracks can form in the wood. On the other hand, removal cuts back to a natural boundary layer in the branch collar, namely, the branch protection zone. Because this is a relatively strong defense boundary, decay often will fail to spread into the trunk. Unfortunately, the reduction cut is sometimes used synonymously with a removal cut. Although reduction and removal cuts both thin the canopy and cuts are made back to nodes, the tree defense system responds to a reduction cut more like it responds to a heading cut. In addition, a tree pruned with reduction cuts looks very different from one pruned only with removal cuts. Be sure to make this distinction between these two cuts.

Sprouting is minimal if the pruning is only light to moderate. Trees pruned with removal and reduction cuts usually look better after pruning than those receiving heading cuts. Heading cuts are best saved for training young trees in the nursery, not for those in the landscape.

REMOVAL CUTS

Locating the Right Spot to Make the Cut

A removal cut removes a branch back to its parent stem or trunk (Figure 5-1, center). The part that remains has a larger diameter than the part that was removed. Examples include removing a limb from the trunk, cutting a lateral branch from a limb, or cutting a smaller branch from a larger one. Details of these procedures are covered in the next section.

There is no universal angle to the trunk at which to make a proper cut. The choice of the angle is determined by the location and shape of the collar if one is present and the branch bark ridge. Figure 5-4 shows that most cuts are made at an angle, going down and away from the trunk. Some will be made parallel with the trunk. Occasionally, if the collar is oriented in a certain way, cuts may be made so the bottom of the cut angles toward the trunk.

Before the invention of chain saws, many pruning cuts on large branches were done properly (perhaps inadvertently). This was because the correct cut outside the collar was the shortest and quickest cut. Cutting through the collar required a longer cut, which was not commonly done. Unfortunately, chain saws have made it easier to cut through the branch collar. Cutting through the collar, can initiate decay and cracks in the trunk because the branch protection zone is removed.

Final cuts to remove lateral branches from parent branches or branches from trunks must be made just beyond the swollen collar, if present, at the base of the branch. In addition, always make the cut to the outside of the branch bark ridge or stem bark ridge (Figure 5-4). Because the swollen collar typically extends beyond the branch bark ridge on top of the branch, to simply cut to the outside edge on the branch bark ridge may not be correct. If a swollen collar is present, be sure to cut just beyond the swelling, on top of the branch. One way to assure this is to cut where the top of the branch makes an abrupt turn toward the branch bark ridge. The top of the branch is usually fairly straight. Begin the cut where this straight line turns abruptly upward. The proper cut is never made on the trunk side of the ridge.

Never make a pruning cut flush with the trunk because this cuts into the collar and makes a larger wound (Figure 5-4). Cutting into the collar opens the trunk to decay, causes cracks, and increases the likelihood of canker disease infection. Flush cuts damage trees even though the wound may close faster than a correctly executed cut (Figure 5-5, left). Properly executed cuts close in a circular fashion (Figure 5-5, right) on small branches but may be oval on some larger branches.

On the other hand, you should not leave a branch stub by cutting too far beyond the collar (Figure 5-4, far right). A stub longer than about ⅛ inch is probably too long for a small-diameter branch, whereas one longer than ¼ inch in a large branch is unnecessary. Branch stubs are susceptible to wood-decaying organisms, especially while the cut is

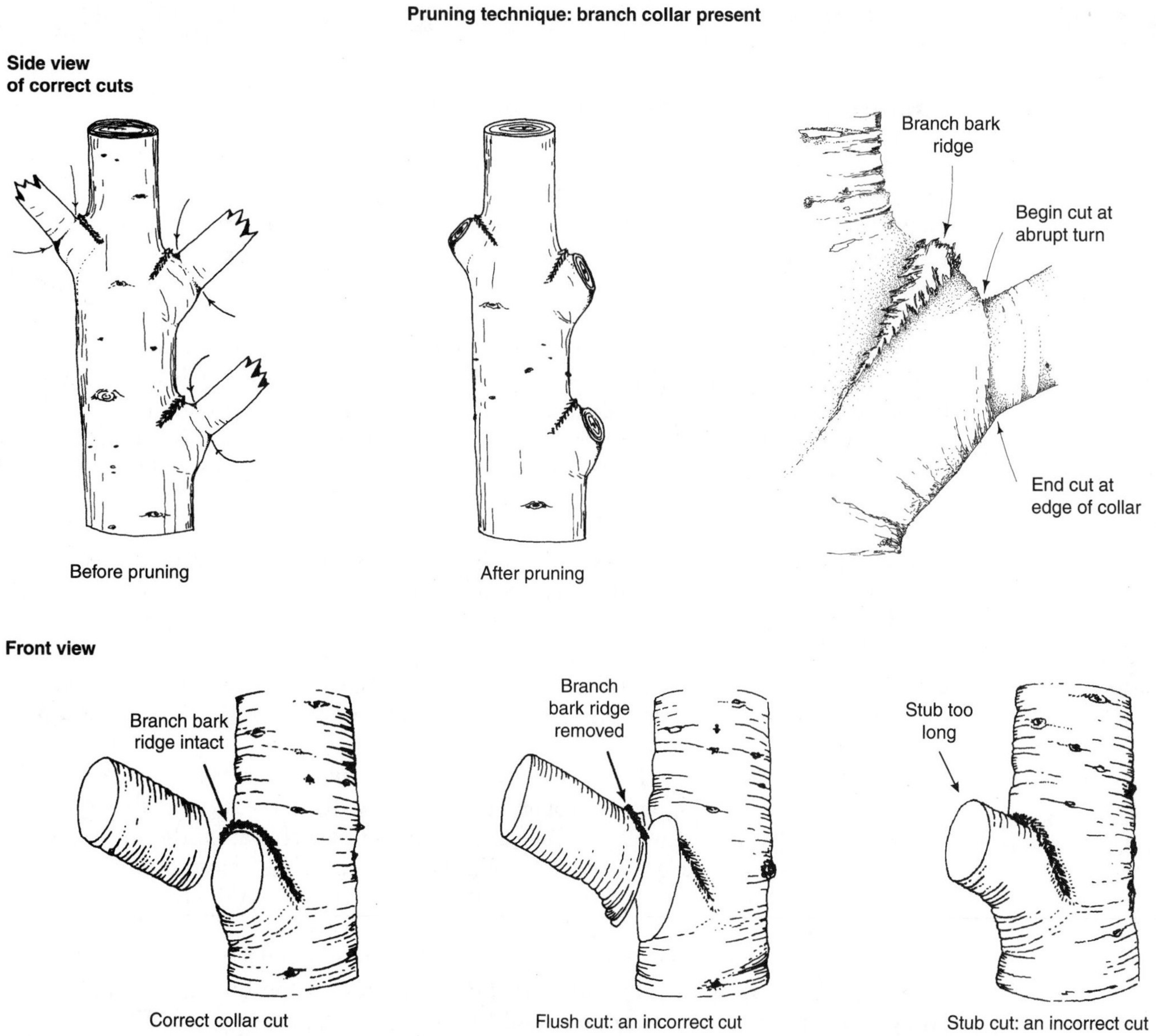

Pruning technique: branch collar present

Side view of correct cuts

Before pruning

After pruning

Branch bark ridge

Begin cut at abrupt turn

End cut at edge of collar

Front view

Branch bark ridge intact

Correct collar cut

Branch bark ridge removed

Flush cut: an incorrect cut

Stub too long

Stub cut: an incorrect cut

FIGURE 5-4. The branch collar forms a swelling at the base of the branch (top right). A correct pruning cut is made between the arrows at the edge of the collar. Notice that the arrows at the top of the branches are located at the point where the branch top makes an abrupt turn toward the branch bark ridge. A properly executed cut will leave the entire collar on the trunk (bottom left). Never make a flush cut (bottom center), and do not leave a branch stub (bottom right).

open to the air, before **woundwood** completely closes over it. A proper cut without a stub will close over more quickly. If left on trees, decay beginning in stubs can break through the branch protection zone and move into the trunk, causing trunk rot and creating a potentially weak tree. Do not leave stubs, living or dead, on trees. *The swollen collar left on the trunk after a proper pruning cut is not a stub.*

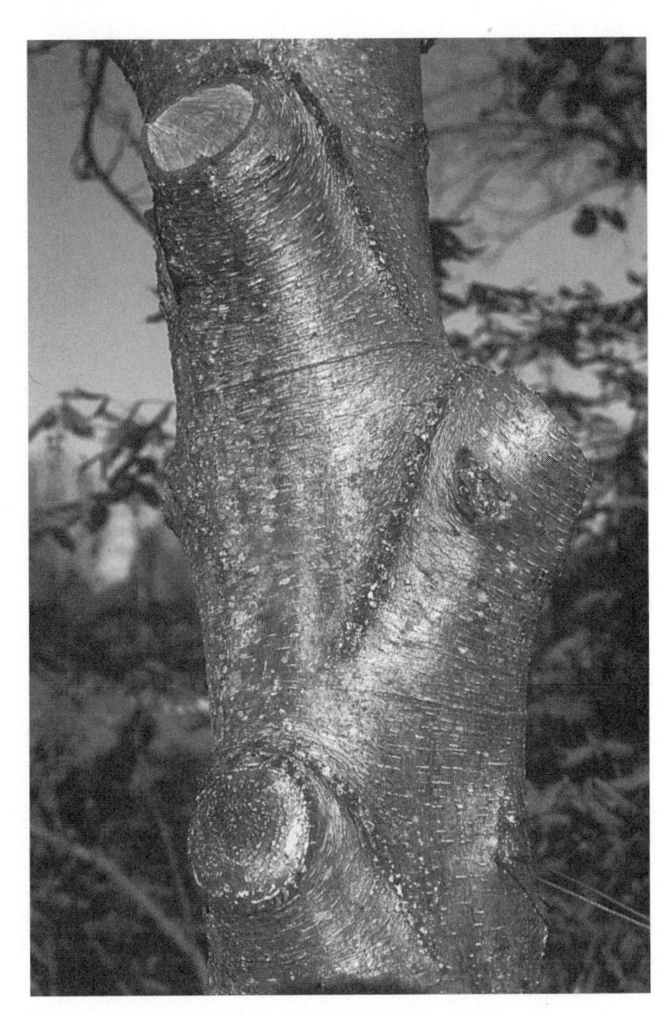

FIGURE 5-5. This cut (left) was made too close to the trunk. The edge of the branch bark ridge was removed on top of the cut, and the collar was cut into on the bottom. These three cuts (right) were made properly. Notice the intact branch bark ridge and the circular cross-section of the cut. The circular, closed pruning wound at the bottom indicates that the cut was made properly.

It is usually easy to see where the cut should be made at the top of a branch, but the collar on the underside of the branch may be hard to locate. Many branch unions have no visible collar. Figure 5-6 helps show the angle of the proper pruning cut when the boundaries of the branch collar are not easy to identify. This is often the case when the branch is larger than about a third to a half the trunk diameter. Some trees almost never form visible collars around healthy branches.

When bark is included in the union between the trunk and the branch or between two stems, the union is weakened and is shaped like the letter **V,** not a **U**-shape. A **U**-shape is common on many well-formed, strong unions. When this occurs, the pruning cut must be made with a saw from the bottom of the branch or stem up toward the top (Figure 5-7, bottom).

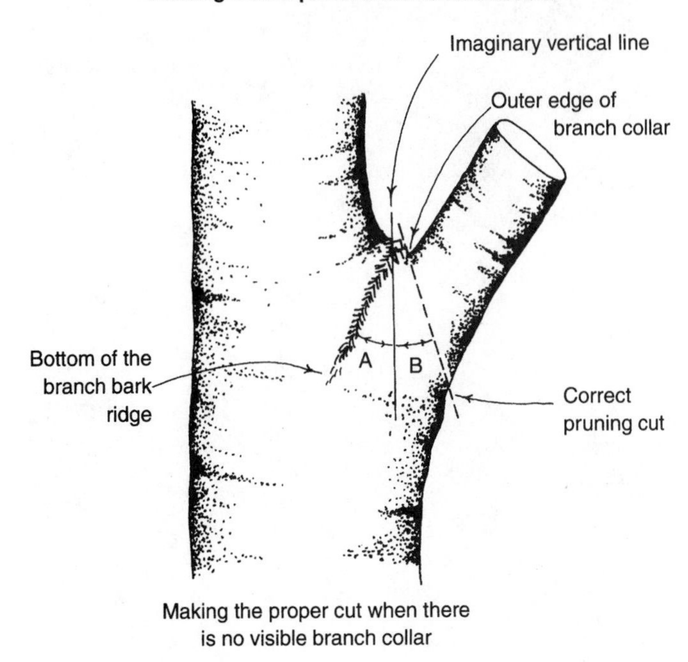

FIGURE 5-6. If the bottom of the branch collar is hard to see, estimate angle B by drawing an imaginary vertical line as shown (parallel with the trunk). Beginning on top of the branch, at the outer edge of the branch collar, make a pruning cut so that angle B is greater than or equal to angle A. The cut is likely to end near the bottom of the branch bark ridge. Use the above as a guide line. If callus forms uniformly around the cut made with angle B > A, then cut in that manner.

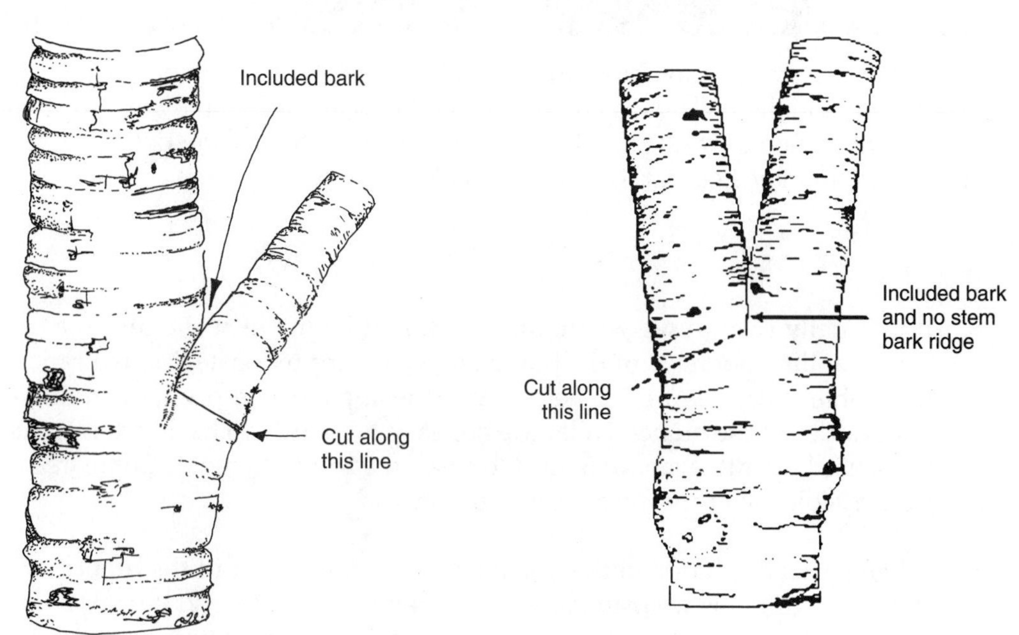

FIGURE 5-7. Remove a branch or stem with included bark by cutting as far down into the union without injuring the trunk that will remain.

Making the Cut

Small branches. Branches less than ½ inch in diameter can be cut with a hand pruner or lopper (see "Pruning Tools," Chapter 6). Rest the blade (not the anvil) at the edge of the collar and cut so that the blade cuts across or up through the branch. This reduces the likelihood of injury to the branch collar. This technique also prevents the branch union from splitting.

Branches between ½ and 1 inch in diameter can be cut with a lopper, but a saw will do much less damage to the collar and is recommended. The cut can be made from the top of the branch down if the branch is at a wide angle with the trunk (Figure 5-8, left). The pruning cut must be made from the bottom of the branch toward the top on branches that form a narrow angle with the trunk (Figure 5-8, right). If the saw is forced into the union of branches with tight angles, trunk bark may be injured above the branch and the cut will be made through the collar, damaging the tree.

Branches more than 1 inch in diameter. Use a saw, not a lopper, as loppers damage the collar when they are used to remove live branches. Branches that are too heavy to be held with your hand will often split from the trunk when the saw is about ⅔ through the branch. This strips the bark from the collar and the trunk below the branch as the branch falls, causing serious and irreparable damage to the tree. To prevent this, make three cuts with a saw when removing branches larger than about 1 inch in diameter (Figure 5-9). This has been called a **drop cut.**

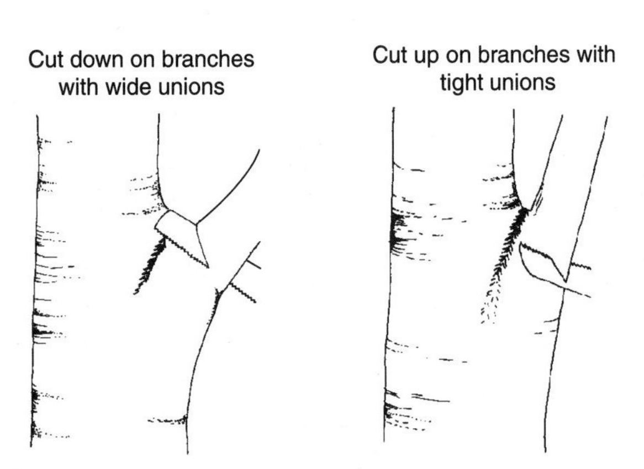

FIGURE 5-8. Some limbs can be cut from the top of the branch toward the branch bottom (left). If the crotch is tight and the saw will not fit from the top at the correct cutting angle, make the cut from the bottom up (right).

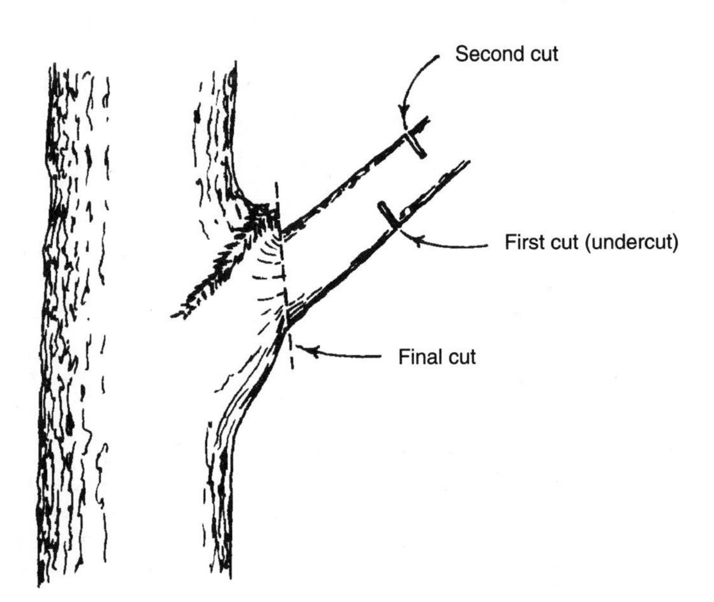

FIGURE 5-9. A drop cut. To remove a branch over 1 inch in diameter, make the first cut on the underside to the depth at which the saw nearly binds (about ⅓ of the way through). Make the second cut through the branch at a point 1 to 2 inches beyond the first cut. The final cut is made outside the branch bark ridge and the branch collar. To prevent the chain saw from pulling out of the climber's hands, make the second cut directly above the first cut.

The first cut is made on the lower side of the branch, 12 to 15 inches away from the trunk. For this technique to work properly, the cut must be made about ⅓ up through the branch or to the point just before the branch weight begins to bind the saw. The second cut is made downward from the top of the branch within 2 inches of the first cut. As the saw cuts about ½ to ¾ of the way through the branch, the limb will split between the two cuts without tearing bark on the trunk. The remaining stub can be easily supported while it is cut from the tree, or can be cut in a manner to prevent bark tearing under the cut. If a branch is large and could damage the trunk, other trees, landscape plants, or structures located under the tree, lower the branch with ropes.

Learning How to Make the Right Cut

To teach yourself how to remove branches at the correct spot, over the next several years you should watch how callus and woundwood grow around pruning cuts. Callus can usually be recognized by its lighter color as it emerges from below the bark surrounding the wound made by the saw. If the pruning cut is properly executed, callus and woundwood will develop as a nearly circular ring around the wound, resembling a donut (Figure 5-10, left; Figure 5-5, right). Callus or woundwood shaped like the letter **U** or an inverted **U** could be an indication that an improper cut was made into either the top (**U**) or the bottom (inverted **U**) of the branch collar. An inverted **U** pattern can form when large branches are removed, even with a correct cut, because trunk tissue may not form a complete collar around the bottom of the branch. Woundwood missing from the top *and* bottom of the pruning cut or elongated oval-shaped woundwood tissue can indicate a tree's response to a flush cut (Figure 5-10, center).

Some trees form the **U**-shaped pattern of woundwood even following a proper cut (Figure 5-10, right). The reason for this is not clear. One explanation could be that the trunk tissue at the top of a live branch fails to grow to the branch side of the branch

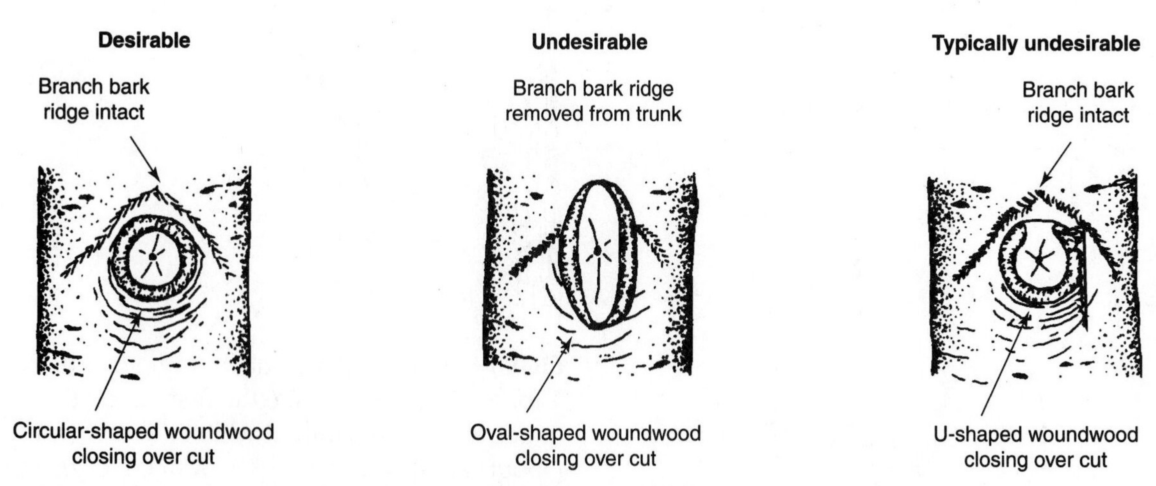

FIGURE 5-10. Woundwood will form in a nearly circular fashion following a correctly executed final pruning cut (left). Cutting into the collar creates an oval woundwood pattern (center). Flush cuts often begin closing in this manner several months after pruning. Occasionally, woundwood forms slowly at the top of a properly executed cut, creating a **U**-shaped pattern (right). A **U**-shaped pattern may also form if the branch bark ridge is removed.

bark ridge. Therefore, woundwood originating from trunk tissue has to grow across the branch bark ridge and out to the location of the pruning cut, beyond the ridge, to reach the edge of the cut. Woundwood around the rest of the cut (the sides and bottom of the **U**) is located closer to the cut surface, so it becomes visible sooner than tissue at the top of the branch. More research is needed in this area.

Removing Dead, Dying, and Diseased Branches

Removing dead branches is almost always a good idea because removal allows the trunk wood to close over the cut branch. *Do not remove the collar growing around the dead branch stub* because it contains live trunk wood (Figure 5-11). Removing the collar will open the trunk to decay organisms by cutting through a natural barrier and may cause other problems such as cracks inside the tree. The swollen collar at the base of the dead branch stub can grow quite large if the branch remains on the tree long after death. Some people may object to leaving this large protrusion from the trunk; however, the tree will benefit from leaving the collar intact. The tree owner will also benefit, because proper branch removal helps to maintain a safe, healthy tree.

Lonsdale (1999) reports that tools are not generally regarded as agents of disease transmission. Some practitioners agree (Tom Smiley, Bartlett Tree Expert Co. personal comm.). After removing branches from trees or leaves from palms infected with wilt diseases such as *Fusarium* wilt, some arborists disinfect pruning tools with a solution of one part bleach or alcohol to nine parts water to help prevent disease transmission to other plants. Bleach can cause severe corrosion of tools. Some people use Lysol™

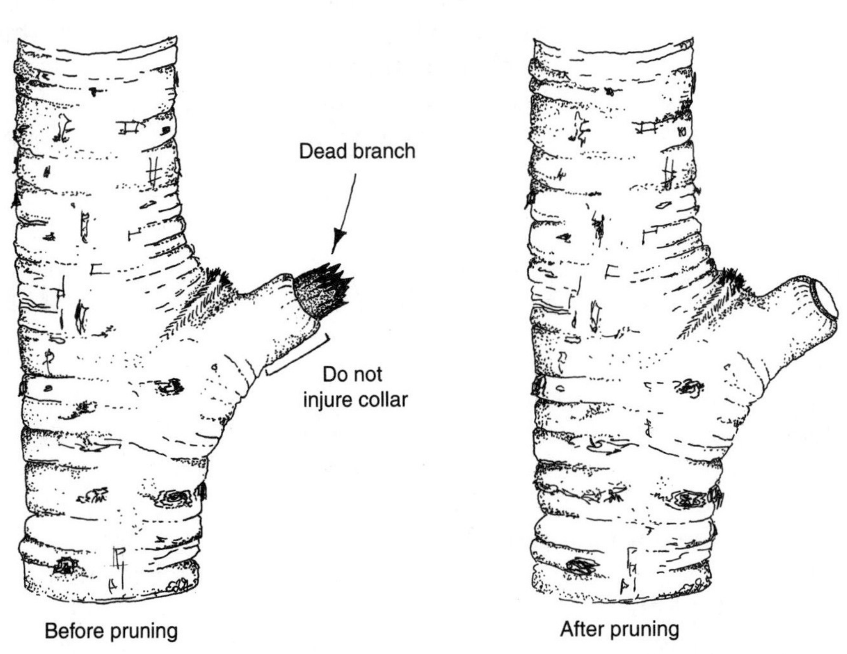

FIGURE 5-11. Do not cut into the swollen collar growing around the dead branch, even if it is large. Removing the collar from around the dead branch will injure the trunk since this is composed of trunk wood. Cutting into trunk wood initiates compartmentalization, decay, and possibly cracks in the trunk.

liquid or spray, or antifreeze (an alcohol with a rust suppressant) instead. After removing debris and dirt, dip the pruning tools in the solution for at least 30 seconds. Otherwise, wilt diseases could spread easily to other trees on infected pruning tools. Chain saws cannot really be sterilized.

Flush Cuts

For many years the standard pruning practice was to remove branches with a final cut flush with the trunk or main branch (Figures 5-5, 5-10). However, this practice has been proven extremely harmful to the long-term health of trees. The tree expends energy attempting to fight microorganisms that begin invading the wound site. This depletes starch levels in the tree, reducing its ability to overcome future stresses. Implications of making a flush cut can become evident as a tree ages. A look behind a flush cut made several years earlier will reveal cracks, wood discoloration, and in many instances, decay. This is not surprising, because trees have evolved through time without experiencing this type of injury and so they have not developed good defense mechanisms against flush cuts.

Flush cuts open the trunk tissue to decay by removing the wood in which the branch protection zone forms (Figure 3-4, top). There is no scientific evidence that this decay process can be stopped or slowed by *any means* other than the natural processes inherent in the tree. Only properly executed pruning cuts minimize decay and cracks.

Insects and microorganisms often attack the trunk above or below a flush cut. Sudden changes in temperature can cause the wood near an old flush cut to split and peel. So-called **frost cracks** often originate from the tree's response to old flush cuts and other old wounds to the trunk and root system. Injured bark near a flush cut can be mistaken for **sunscald** or sunburn. Sunscald is caused by a warming of the trunk in winter followed by a quick freezing. It also happens in summer when sun shines directly on thin bark. Trunk cankers, dead spots on trunks, and declining root systems can also be associated with flush cuts. Excessive sprouting often can develop from around the flush cut, particularly if the cut was made while the shoots were elongating or the leaves were growing.

TERMINAL BUD CLUSTER PRUNING

Terminal bud cluster pruning is a new technique used on young nursery trees. Although not traditionally referred to as pruning, it can be used on oaks, ashes, and other trees that form a cluster of buds at the tip of the twig to prevent formation of undesirable multiple leaders and clustered branches. To use this technique, all buds except for the largest one, referred to as the terminal bud, are removed from the end section of the dormant twig. This causes lateral buds, referred to as intercluster buds, along the dormant twig to emerge as shoots to become lateral branches (Figure 5-12). Branches typically emerge at a wider angle than they would without this treatment. This encourages development of a dominant central leader and distributes branches along the stem. Terminal bud cluster pruning should promote rapid growth because no living tissue is removed.

8) If tree A and B are identical, and tree A is pruned primarily with reduction cuts and tree B is pruned primarily with removal cuts, both trees will:

   a. be equally susceptible to stem decay.
   b. be equally susceptible to sun injury.
   c. have the same height.
   d. have invigorated interior foliage.

9) Which of the following is FALSE about reduction cuts and removal cuts?

   a. Both can thin the canopy.
   b. Both cut back to a natural boundary.
   c. Both can reduce susceptibility to future storm damage.
   d. Both invigorate existing interior foliage.

10) Terminal bud cluster pruning is a new technique used primarily on nursery trees to:

   a. reduce height.
   b. develop and maintain a dominant leader.
   c. reduce branch number.
   d. minimize formation of included bark.

Answers: d, a, c, c, a, c, a, d, b, b

CHALLENGE QUESTIONS

1) List as many consequences of topping as you can.
2) Compare a tree pruned primarily with reduction cuts with the same species pruned primarily with removal cuts.
3) Describe why you would use fewer reduction cuts on mature trees than on young and medium-aged trees.

SUGGESTED EXERCISES

1) Take a walk in a nearby landscape and find as many branch unions with a branch bark ridge as you can. Now find some branch unions with included bark. Compare them to learn the difference.
2) Take a walk in a natural woodland. Do you find included bark in branch unions as frequently as in a landscape setting where most trees are planted? Why?
3) Study unions with included bark from the outside of the tree. Try to guess how far down into the union the included bark occurs. Then cut through the union in cross section in several locations to determine the extent of the included bark. Relate the outward appearance of the branch union with what you found inside.
4) Make a flush cut and a correct removal cut on a tree. Compare the two in detail and discuss what happens behind the cuts during the next ten to fifteen years. Return to the cuts several years latter and dissect the trunk at the point of the cuts. Discuss the differences between the tree's reaction to the two types of cuts.

CHAPTER 6

PRUNING TOOLS

OBJECTIVES

1) Describe tool selection and usage.
2) Choose a tool you are capable of handling.
3) Choose the appropriate tool for the job.
4) Determine when climbing spikes are appropriate.
5) Show how to use tools safely.

KEY WORDS

Aerial lift
Anvil pruner
Bypass pruner
Carpenter's saw
Chain saw
Extension pruners
Hand pruners
Hedging shear
Loppers
Pole saw
Pruning carts
Shears
Throw line

Safety equipment including ear protection, chaps, safety glasses, gloves, and hard hats, is recommended to protect chain saw operators from injury. Safety glasses help keep sawdust and chips out of the eyes, gloves prevent minor cuts and abrasions, and hard hats minimize head injuries. Ear protection helps prevent hearing loss while using power equipment such as chain saws. Chaps help protect legs from injury.

PRUNING SHEARS

One-handed pruning shears (hand pruners) are designed to cut small live branches up to about ½ inch in diameter (Figure 6-1). When they are used to cut through larger branches, the surrounding bark is often damaged. The tool could be damaged as well. When pruning is conducted frequently on young nursery trees, only a hand pruner is needed because only small branches need to be removed. Regular pruning prevents unwanted branches from growing larger than this.

Hand pruners come in a variety of shapes and sizes, but most belong to two types. One (the **bypass** or scissors type) passes a curved blade past a hooked **anvil** (Figure 6-1, left). The cutting blade is sharpened on the side facing away from the anvil, never on the anvil side (Figure 6-2). Sharpening on the anvil side of the blade causes the branch to get stuck in the pruner. If needed, the anvil is sharpened on the surface facing away from the blade. The other (**anvil pruner**) type of hand pruner has a straight blade sharpened on both sides that cuts against a flat, straight, metal anvil. The curved bypass

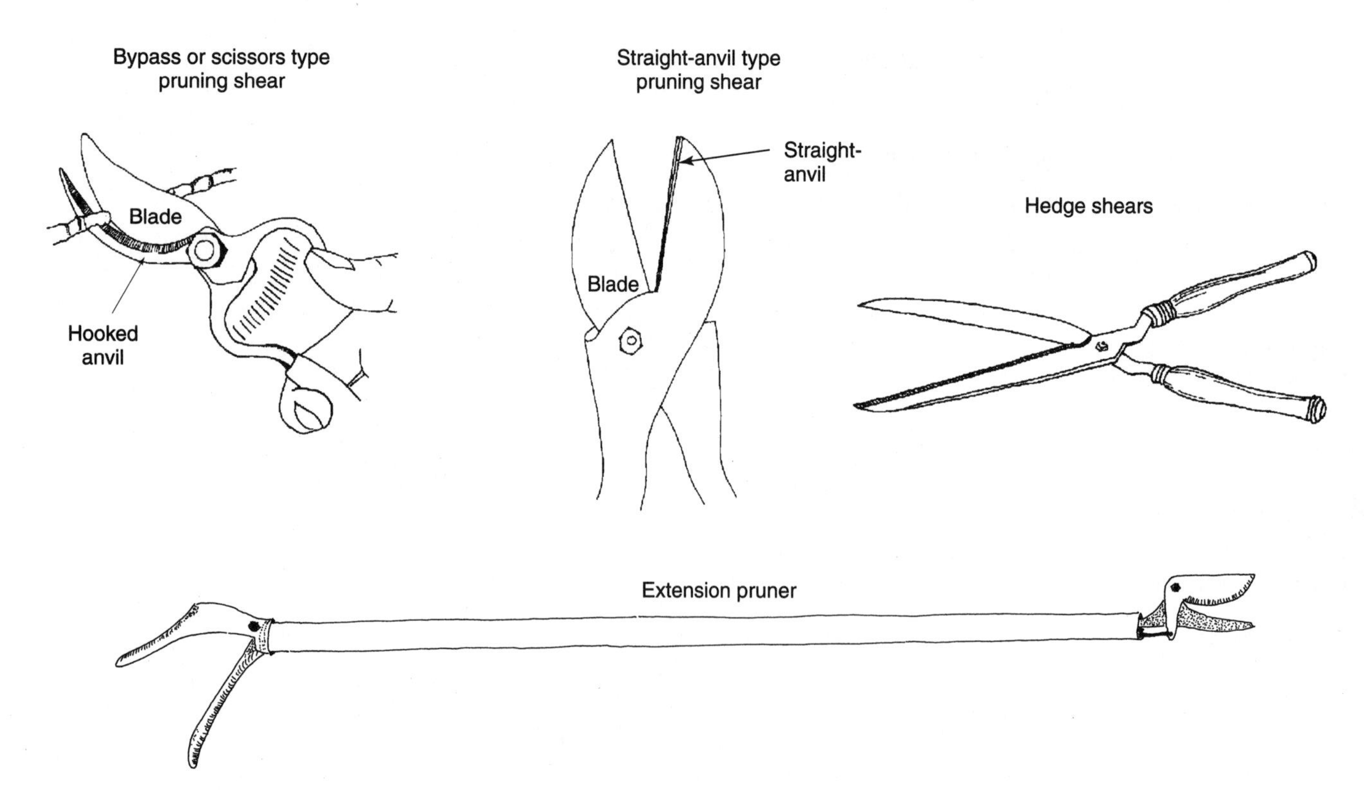

FIGURE 6-1. Pruning tools come in a variety of styles. The bypass type (left) is best for pruning live branches less than ½ inch in diameter. Hedge shears are suited for trees, and shrubs maintained as hedges; they are also used to shorten lower temporary branches in the tree nursery. Extension pruners are useful for reaching the upper canopy of nursery trees and for pruning shrubs in the landscape.

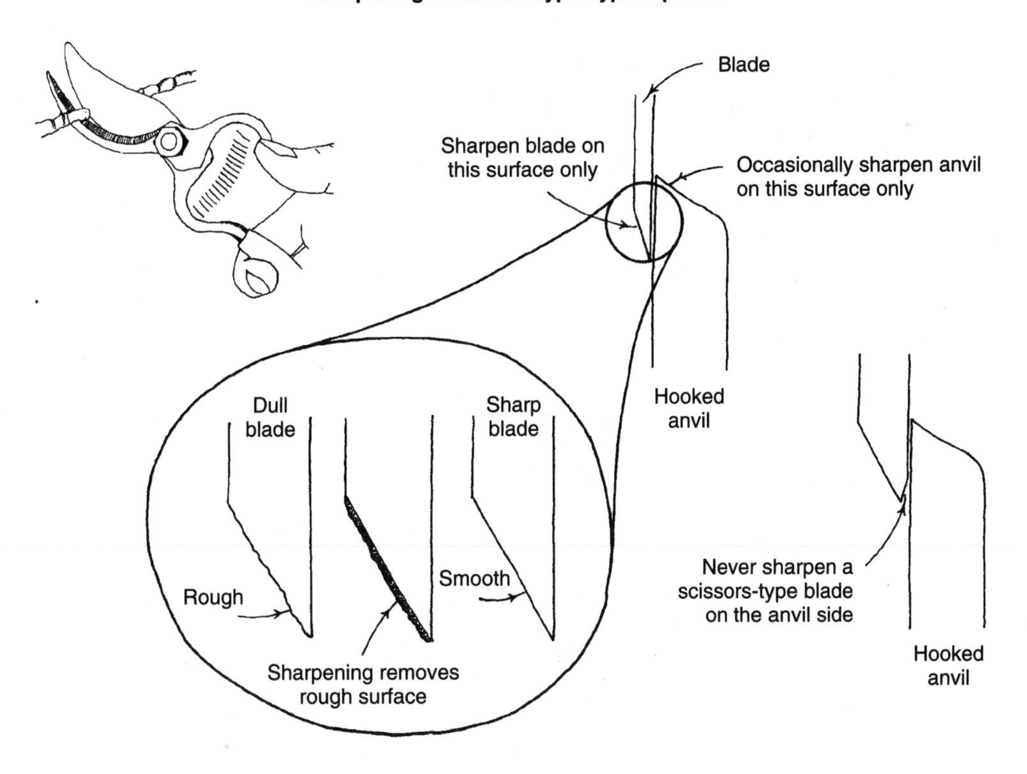

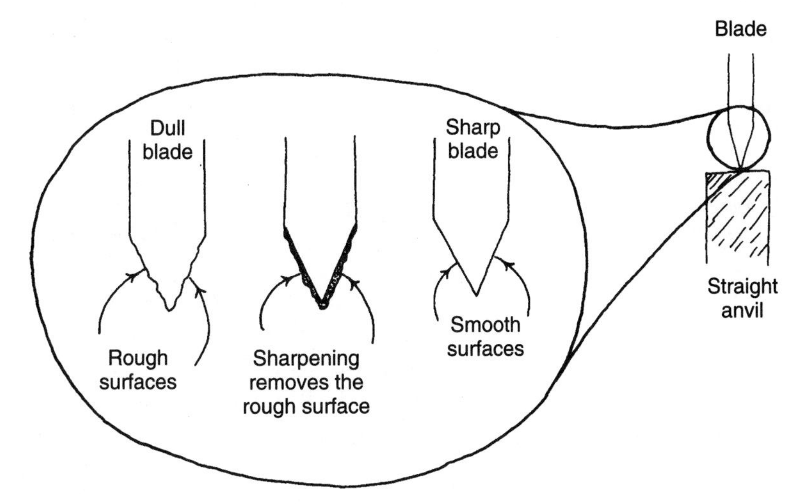

FIGURE 6-2. The two pruner types should be sharpened in different ways. The bypass (scissors) type must be sharpened only on the side of the blade facing away from the anvil. If it is sharpened on the anvil side, branches will bind and become stuck between the blade and anvil, causing the pruners to work incorrectly. On the other hand, the straight-anvil type blade should be sharpened evenly on both sides of the cutting blade. A fine sharpening stone works best to sharpen pruners.

pruner makes a more precise cut and may damage less tissue if properly used. Injury to the branch collar and crotch can be minimized if the blade cuts up or diagonally across the branch. The A300 Pruning Standard adopted by the tree care industry discourages the use of straight-anvil type pruners on living trees (American National Standards Institute 2001). A third type of pruner works like a scissors with two sharpened blades. Some people prefer these because they are lightweight and inexpensive; they are only suited for small twigs.

A recent, welcome addition to the tool arsenal is the bypass pruner with integrated extension pole (Figure 6-1). This **extension pruner** is especially useful for tree growers pruning trees taller than 6 feet. It is easy to operate and efficient, and most professional models stay sharp for at least one day's work. The standard, light-weight model is 6 feet long, but other models have telescoping extension tubes that allow you to prune higher in the tree. The telescoping models tend to be heavy resulting in operator fatigue when used for extended periods.

Extension pruners are also handy for shrub maintenance in landscapes. They can be used to reach plants without stepping into and compacting the garden soil or crushing flowers. It is easier to see what branches to prune because you are able to stand several feet away from the plant while pruning it.

Tool maintenance is essential for making a good cut. A smooth, accurate cut is impossible to make with a dull blade. A dull blade will injure the branch collar because more force is required to pass the blade through the branch. The muscles in your hand can be easily fatigued and injured because of the extra force required to cut with a dull blade.

Sharpening frequency depends on usage and the hardness of the steel in the pruner blade. Some blades are made from soft steel that is easy to sharpen, but they require frequent sharpening because they dull quickly. Others are made from very hard steel, which require more effort to sharpen but they remain sharp for a long time.

LOPPERS

Loppers are also available as bypass and straight-anvil types. A **lopper** can be used to prune live branches up to about an inch in diameter. Although loppers will cut through larger branches, it is nearly impossible to make the proper cut without damaging the branch collar on a tree or leaving a stub. Because of the force required to make this large a cut with a lopper, the bark on the anvil side of the cut often separates from the wood, injuring trunk tissue in the collar. Loppers are best suited for cutting tree branches once they are on the ground. In most cases for live branches a fine pruning saw is preferred over loppers. Loppers are very useful for reducing the size of shrubs.

HEDGING SHEARS

A **hedging shear** is a useful tool for pruning trees in a production wholesale nursery. Instead of using a hand pruner that cuts only one branch at a time, a hedging shear can increase efficiency by cutting through several branches. It should only be used to shorten lower temporary branches on shade trees. It is an essential tool for creating uniform conical canopies on evergreens such as hollies. It can be used to clip one-year-old shoots on the outside edge of the canopy to increase the density of the canopy edge and shape the tree in the nursery.

A hedging shear is used occasionally on landscape trees to create and maintain a formal hedge, screen, topiary, or other architectural form. An annual shearing is required in most instances for this to work effectively on trees. Unless trees will be regularly (annually) sheared into a formal shape, shearing is not recommended since this is the equivalent to topping. It is the standard tool for maintaining hedges in many gardens worldwide. Hand operated hedge shears have mostly given way to electric and hydraulic devices.

HAND SAWS

Pruning saws are specially designed to cut through green wood ½ inch or larger in diameter. Different blade designs are available (Figure 6-3). A curved blade cuts through wood as you pull it toward your body. It cuts little or not at all when you push the blade away from you. A straight saw usually cuts as the blade is pushed away from you. It cuts less wood as you pull it toward your body. One of these designs may be better suited than the other for your physical capabilities. Saws that cut on the pull stroke are considered safer for working in trees. They lessen the likelihood of pushing yourself out of the tree. Never use a **carpenter's saw** to prune live branches because the saw will bind in the branch and will not cut. A carpenter's saw is designed to cut through dried lumber only, and it cuts on the push stroke.

There are two components to saw maintenance: tooth sharpness and tooth set. Sharp teeth allow the user to make a quick, smooth cut. If you can run your fingers up and down the saw teeth without getting cut, it is time to sharpen the teeth. A sharp saw will leave shallow abrasions in your skin as your fingers are pulled gently across the teeth.

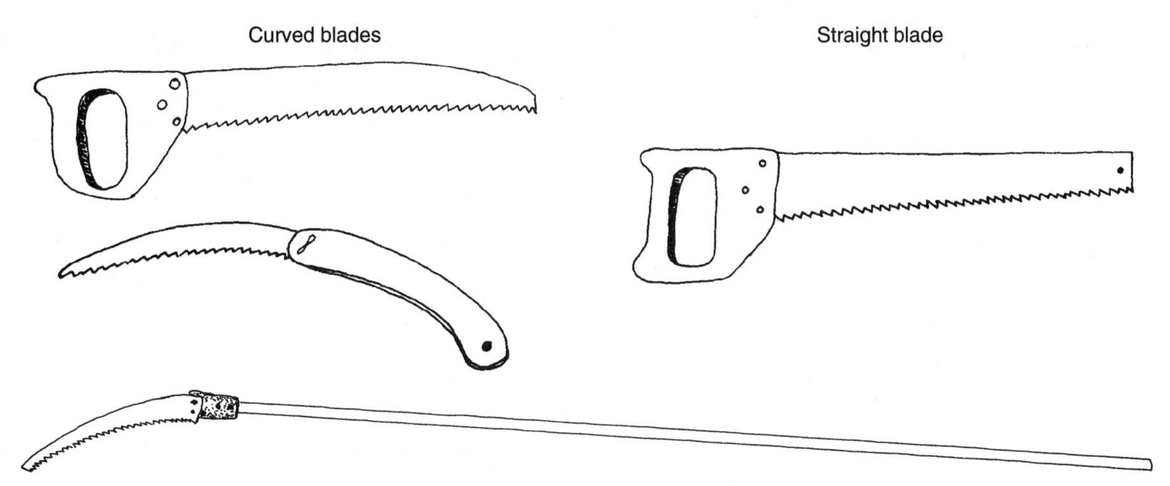

FIGURE 6-3. Saws with curved blades (left) cut on the pull stroke. The teeth are usually angled back toward the handle end of the saw. Many saws with a straight blade (right) cut on the push stroke. The handle on the straight-blade saw is designed to keep your hand in place as the saw is pushed through the wood. The newer, triple-sharp, Japanese-type blades cut entirely on the pull stroke, even when the blades are straight.

The teeth are alternately pushed (set) to one side and then the other along the length of the blade. Properly set teeth allow the saw to be pushed and pulled easily and without binding through the saw curf. The saw curf is the groove created in the branch as the saw cuts through it. If the saw binds and becomes stuck in the groove, the teeth may need to be set. Proper set and sharpness of the teeth are both crucial to safe and precise pruning. Saws are best sharpened by a professional service.

The latest blade design incorporates tri-edged teeth instead of the conventional needle-point teeth. These are marketed as razor, turbo, or Japanese-style saws and they are razor sharp. The saw is similar to a conventionally shaped tooth except that the end of each tooth has an added bevel. Cuts made with a tri-edged blade are cleaner and less ragged than cuts made with a conventional blade. They are especially useful for cutting small-diameter branches. However, in a large cut made with a tri-edged saw, the saw can bind because the teeth are not set to the side as on a conventional blade. They are difficult to sharpen and must be sharpened with a specially designed file. Many people simply replace dull blades with a new one.

CHAIN SAWS

U.S. Occupational, Safety, and Health Administration (OSHA) regulations require professional tree care workers to attend safety classes for **chain saw** operation. Homeowners and others who are not trained arborists have no business removing tree limbs that require a chain saw. My wife, a registered nurse, has treated too many homeowners who have fallen out of trees or were injured using a chain saw in a tree. Some people become paralyzed as a result of these accidents. Only trained arborists should use chain saws in trees. For a list of professional arborists in your area, contact the National Arborists Association (Amherst, NH). For a list of certified arborists in your area, contact the International Society of Arboriculture (Champaign, IL), or visit their World Wide Web site at http://www.natlarb.com.

POLE SAWS AND LOPPERS

A saw mounted on the end of a pole is referred to as a **pole saw** (Figure 6-3). The saw cuts as the blade is pulled toward the operator. A lopper can also be mounted on a pole and is operated from the ground using a rope. It can be used to remove small-diameter branches within about 15 feet from the ground. Do not operate a pole saw near a power line unless you are specially trained for this task. Each year, people are injured or killed by pruning with a pole saw near a charged electric line. The trick to cutting with a pole saw is to be sure the saw is sharp and use smooth strokes. This prevents the branch from pulling toward the operator.

Pole saws are very useful for removing branches in the canopy that are difficult to reach by climbing. Arborists often find them useful while they are in the tree. They are also used from the ground to remove low branches from the canopy and to make reduction cuts on small trees. But pole saws are one of the most abused and misunderstood tools in the landscape industry. They are often used by landscapers and other ground crews who have a poor understanding of trees. The most common abuse occurs when *only* low branches are removed. Following this treatment, many people consider this tree pruned to completion because the lower drooping branches are gone. But the important pruning is often neglected—the structural pruning in the canopy that creates strong architecture and reduces hazards.

CLIMBING AIDS

Unfortunately, spurs and spikes on the boots used to climb tree trunks and palms cause injury. When cambium is injured, the tree reacts, causing a net reduction in its energy reserves and possibly causing decay. In most circumstances spurs and spikes have no place in live tree care. They can be used to climb a tree that will be cut down or to rescue an injured worker in a tree. They are also suitable if the bark is thick enough to prevent damage to the cambium or if branches to be climbed are more than a **throw line** (a rope with a weight on the end) distance apart.

There are several methods to climb a tree without causing injury to it. A throw line can be thrown over a low branch. The arborist pulls up into the tree using a secured rope. Many arborists gain access to a tree that has no branches on the lower portion of the trunk using a cherry picker or hydraulic lift. Others use a specially designed ladder that is secured to the trunk. When these options are unavailable or when there are no branches within 60 feet of the ground, some arborists use a charged device originally designed for throwing a line from one ship to another. The device, costing several hundred dollars, is pointed into the tree and a .22-caliber charge shoots a weighted string accurately 80 feet or more from the ground.

AERIAL LIFTS

Aerial lifts are used to reach the canopy of small and large trees. They are equally useful for pruning a 15-foot-tall crape-myrtle as they are on a mature red maple. Lifts are manufactured to be mounted on a variety of trucks. Some can reach only to the top of a 20-foot-tall tree; others are large enough to allow easy access to the canopy of a 50-foot-tall tree. They allow the arborist to work on big trees in the portion of the canopy they should be working in—the outer portion. They provide limited access to the interior of the canopy.

PRUNING CARTS

Pruning carts have been designed to allow easy access to the tops of nursery trees. They are usually pulled behind a tractor. Most can hold two or more people so their feet are up to about 8 feet off the ground. The best ones allow adjustment so the people pruning the trees can place themselves at any of several heights off the ground. This is a wonderful labor saving device because it makes it much easier to appropriately prune trees larger than about 2 inches caliper.

CHECK YOUR KNOWLEDGE

1) Why is the bypass type pruner more appropriate than the straight-anvil type? The bypass type:
   a. crushes less tissue.
   b. is easier to sharpen.
   c. is lighter in weight.
   d. is less expensive.
2) Hedging shears can be used on:
   a. nursery trees.
   b. ornamental trees.
   c. shrubs.
   d. all of the above.
3) How does a carpenter's saw differ from a pruning saw?
   a. They are similar—there is no difference.
   b. A carpenter's saw is larger.
   c. Pruning saws are sharper.
   d. The teeth are oriented differently.
4) Loppers are usually not appropriate for pruning live branches on trees because they:
   a. are awkward to use.
   b. are only available as anvil type devices.
   c. damage the collar when removing a branch.
   d. are difficult to sharpen.

Answers: a, d, d, c

SUGGESTED EXERCISES

1) Remove one ¾-inch diameter branch from a tree using a lopper, remove another using a *new* tri-edged pruning saw, and another using a pruning saw with traditional teeth. Discuss the differences in the work required to remove these three branches. Compare the injury to the collar from each of the removal techniques.
2) Remove several branches using an anvil type pruner and several using a bypass type. Discuss the differences in the condition of the collar following removal.

CHAPTER 7

WHEN TO PRUNE

OBJECTIVES

1) Decide how often to prune and when to begin.
2) Design a pruning cycle to fit your program.
3) Minimize bleeding from pruning wounds.
4) Decide when to prune specific trees.
5) Prune to reduce pest problems.
6) Learn how to influence flowering.
7) Determine if pruning is needed at transplanting.

KEY WORDS

Auxins
Heavy pruning
Light pruning
Pruning cycle
Utility tree-care

START PRUNING EARLY

It is never too soon to prune. Start pruning early in the life of the tree in order to develop a strong framework of branches. Many trees with a decurrent growth habit, such as oaks, maples, and elms, benefit from pruning beginning when they are about one year old. Before you pick up a saw, decide on the function of the tree—for planting along streets; in parking lots; for shade, windbreak, or screen; or as a small multi-trunked specimen—and prune accordingly. Try to visualize the form of the tree ten or twenty years from now. Ask yourself where major permanent limbs will be located on the trunk and try to identify these as soon as they are formed on the tree. Slow growth on other branches by subordinating them with reduction cuts. In many instances, the first permanent branch is located 10 to 20 feet from the ground.

IN THE NURSERY

To promote fastest total tree growth and keep sprouting to a minimum, prune frequently and at regular intervals. Remove as little live foliage as possible at each pruning while meeting your objectives. Two- to three-year-old trees pruned for the first time may require removal of branches ½ inch in diameter or larger. This could remove a large portion of the crown and temporarily slow growth rate. If more than ¼ of the foliage needs to be removed on young nursery trees, you waited too long to prune.

Tip: *Prune regularly to promote the fastest growth.*

The number of prunings per year required to produce a good-quality nursery crop in the shortest possible time has not been determined for all species. It depends on the quality of tree you want to produce, tree species, climate, and perhaps other factors. In the warmer climates one pruning the first year followed by two in years two and three will develop a great quality tree (Table 7-1). If a good structure was developed the first three years, only one pruning may be needed in years four and five in the nursery.

This program lessens the amount of growth that occurs in branches that will be removed or shortened before harvest, thus preventing large pruning wounds. It also results in more growth in the most desirable branches, that is those in the permanent nursery canopy. With a regular pruning program, unwanted branches and those in the wrong position can be shortened or removed before they become large. One annual pruning may be adequate in colder climates. Up to three may be needed in the warmest regions of the world where growth rates are most rapid.

TABLE 7-1. Number of annual prunings recommended for nursery production of quality shade trees.

USDA Hardiness Zone	Number of Prunings
2–6	1–2
6–8	2
8–11	2–3

AT TRANSPLANTING

Older texts recommended that branches be removed at transplanting to compensate for root loss. This has been conclusively proven unnecessary for successful transplanting dicot trees of any size. Trees from 1 inch to 5 feet in trunk diameter are moved without pruning to compensate for root loss. Root initiation and root growth may be slowed by removing healthy branches following root pruning or transplanting. **Auxins** produced in the buds send a signal to the roots to initiate growth. Since buds are removed when you prune, root initiation can be reduced. In addition, some sugars and other phytosynthetic by-products produced in leaves are translocated to developing roots, so a transplanted tree with branches removed produces less root growth. Some palms are regularly pruned at transplanting (see chapter 14).

Many horticulturists recommend removing dead, diseased, crossed, broken, or insect-infested branches at transplanting. While this is a good idea, it should not be necessary if you purchased quality plants and they were transported and installed properly. Eliminate the need for major pruning after transplanting by purchasing well-structured trees that are free from insect and disease problems.

If there are no nurseries that carry a particular species with suitable trunk and branch structure, be prepared to develop good structure after planting. This can begin when the tree is planted. If you feel that extensive pruning is needed to correct some trunk or branch defects on a recently planted tree, and you will not be able to prune the tree for several years, prune it at transplanting to correct the problem (Figure 7-1). If you wait, it may never be pruned correctly. Although the tree may be set back slightly as a result of this pruning, its long-term health will be improved due to improved structure. If your tree management plan allows you to prune one or two years after planting, do not prune to correct structural defects at planting.

PRUNING CYCLES IN THE LANDSCAPE

Many trees in urban and suburban landscapes are not pruned until they create a problem. Most tree care programs are reactive, not preventive. Under a reactive program, tree care managers and others try to fix problems as they arise. Tree health (and sometimes the health and well-being of people and property) can suffer as a result.

A preventive pruning program places trees under the care of a professional. This professional develops a plan that should include a pruning program designed to meet specific objectives. Once objectives are established each tree or group of trees is placed on a pruning cycle. A **pruning cycle** is the elapsed time between one pruning and the next. The **utility tree-care** industry has adopted this approach to pruning trees near wires and other utilities to ensure reliable delivery of electricity to its customers.

If urban trees are allowed to grow for more than about five to ten years without pruning, significant structural problems can develop. These problems can be difficult to correct. If pruning is delayed another ten to fifteen years, the tree may develop poor form that is difficult or impossible to correct with only pruning. Tree removal or measures such as cabling and bracing may be recommended for these problem trees. This is the management spiral so many of our urban trees are in.

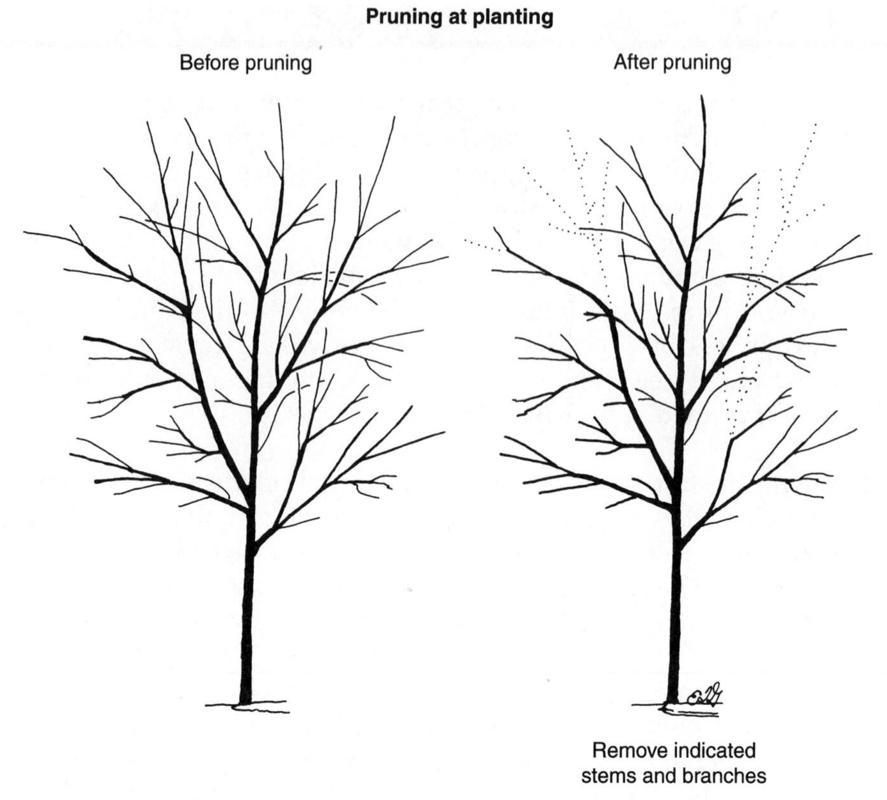

FIGURE 7-1. The tree at planting (left) has three stems all about the same diameter. If it will not be pruned in the next few years, prune it now to create good structure. Shorten stems that compete with the leader using reduction cuts (remove stems indicated with dotted lines).

MUNICIPAL PRUNING CYCLES

Pruning cycles vary tremendously around the world. Some communities have developed means to prune certain trees each year. For example Sarasota County, Florida prunes its crape-myrtle street trees annually. Many municipal tree care programs have the resources to prune shade trees on a five-year cycle. That is, each tree is pruned every five years. Other cities are on a twenty-five-year cycle. Still others do not know what a pruning cycle is, so their trees may have never been pruned. Political and professional skills are required to secure an appreciation in the community for the need for preventive pruning programs that include a defined pruning cycle.

It would be impractical to attempt to defend one pruning cycle for all communities and tree species. Species attributes, canopy forms, growth rates, soil conditions, wind patterns, and other factors play a role in developing appropriate cycles. For example, in colder climates trees grow much slower than in warm climates, so problems develop slower. It is true, trees grow slower in cold climates, but they recover from pruning slower too. If you were to create a void in the canopy in order to improve structure (Figure 7-1) in a cold climate, the void would take more time to fill in than in a warm climate. There is little research done on comparative shade tree pruning cycles for different climates although Miller (1981) suggests 5 years as optimum.

One way to determine how much time can elapse before the next pruning is to decide what is the maximum size pruning cut that should be made. The more time that elapses between prunings, the larger the pruning wound you will make when the tree is finally pruned. Larger pruning wounds have an increased likelihood of creating decay and cracks in the wood behind the cut. So it makes sense to keep pruning wounds as small as possible. This is accomplished by keeping the pruning cycle short.

Reduction cuts play a big role in accomplishing the objective of creating structurally strong tree architecture. It seems reasonable to keep reduction cuts less than 2 to 3 inches diameter on most trees. Pick a pruning cycle that keeps most reduction cuts less than your chosen maximum diameter while accomplishing your objectives. Young trees in the landscape require more frequent pruning than older trees to direct growth into desirable branches. See Chapter 15 for examples of pruning schedules.

TIME OF YEAR

Dead and diseased branches can be removed at any time. Live branches are best pruned in the dormant season or following a flush of growth after leaves harden and turn dark green. But if pruning is required to improve structure, in most instances go ahead and do it at other times as well with the following precautions. Root growth is slowed temporarily when live branches are removed in summer, and since most root growth occurs during the growing season removing large numbers of live branches during the summer is ill-advised unless your objective is to slow growth. However, **light pruning** (removing less than about 10 percent of the foliage) can be performed safely on most species at any time.

Many trees sprout excessively in response to moderate (10% of foliage) or **heavy pruning** (>25% of foliage) or pruning during a growth flush while leaves are expanding and shoots are elongating. Bark and cambium are easily damaged by pruning during a growth flush. Energy resources are usually low at this time as well. For these reasons it is best not to prune heavily during this time of year. Late-summer pruning may stimulate an additional flush of shoot growth on species that flush several times each year. These shoots may be damaged by an early frost. Some people think the worst time to prune is when trees are going into or breaking out of dormancy. More research is needed to confirm this.

Some recent research suggests that pruning during the summer results in less stem decay than dormant-season pruning (ISA, 1994). However, more research needs to be done before suggesting summer as the best time to prune. Pruning in the dormant season can help minimize undesirable sprouting.

Moderate to heavy pruning in late fall through mid-winter can stimulate new growth on a few trees such as crape-myrtle and some evergreens, particularly in the warmer climates during a warm winter. This is not a concern in tropical regions or in northern climates with native and adapted species. These succulent stems are not cold hardy and can be easily damaged, even by a light frost. Light pruning does not usually stimulate sprouting. Low winter temperatures can also cause cambium damage near pruning cuts made in late fall and winter in subtropical climates, even if growth is not stimulated by pruning. This is particularly true of plants that are marginally cold hardy. If in doubt about the cold hardiness of a tree planted in a subtropical climate, it is best to delay heavy pruning to just before buds begin to swell in the spring. Heavy pruning of live branches in any climate should be postponed if trees have been subjected to an unusually severe drought.

Growth Rate Control

Trees grow fastest without pruning. To encourage the greatest total growth when pruning deciduous and semievergreen plants, prune during the dormant period. Evergreen trees in tropical and subtropical climates may continue to grow during the cooler parts of the year. To encourage greatest overall growth, they can be pruned just prior to the first spring growth flush.

To retard growth, for a maximum dwarfing effect on all trees, prune just after each growth flush, when the leaves have fully expanded and turned dark green. Pruning at this time slows root growth and depletes energy reserves, which causes a dwarfing effect. Do this only on healthy, vigorous trees. Pruning live branches from unhealthy trees, including unhealthy trees impacted by construction activities, at a time of low energy reserves just after the growth flush could deplete them further of much-needed energy reserves and energy-producing tissue (i.e., leaves). This stress could lead to their eventual death.

Wound Closure Rate

The closure of pruning wounds on most trees should be rapid if pruning is conducted just before the spring growth flush (before buds swell). Wounds made immediately after the spring growth flush has ended and the leaves turn dark green also close rapidly. This is desirable because a closed wound is more aesthetically pleasing and because some diseases and decay organisms may be discouraged from entering the plant. Rapid closure cuts off oxygen to some decay organisms sooner.

Flowering Trees

Shaping and major structural pruning. Most people make too much fuss about pruning flowering trees at the correct time. In general, follow the guidelines discussed earlier in "Time of Year," but keep in mind the following: When existing twigs and branches are removed, the number of branch tips is reduced, which reduces the number of flower buds or potential flowers.

To minimize the reduction of next year's flower display, prune spring-flowering trees, such as magnolias, dogwoods, redbuds, lilacs, trumpet tree, and others, soon after the flower display (Table 7-2). These trees form flower buds in mid-summer for the following year's flower display. Pinching during the time between the end of the flower display and late spring will not reduce the number of flower buds set for next year's display because they will not have formed yet. In fact, pinching or heading the new growth may increase the number of branches and enhance the flower display next year (Figure 7-2). There is nothing wrong with pruning spring-flowering trees during other times of the year, but flower buds will be removed and the flower display reduced the following year. This should not affect flowering in subsequent years.

TABLE 7-2. Trees that flower from buds formed the year before. If these trees must be pruned, do it just after the leaves turn dark green and stiff in order to preserve the flower display for next year.

African Tulip Tree *(Spathodea)*	Camelia *(Camelia)*
Cherry *(Prunus)*	Crabapple *(Malus)*
Dogwood *(Cornus)*	Flowering Almond *(Prunus)*
Fringe Tree *(Chionanthus)*	Horsechestnut *(Aesculus)*
Lilac *(Syringa)*	Magnolia (Magnolia)
Orchid Tree *(Bauhinea)*	Pear *(Pyrus)*
Redbud *(Cercis)*	Serviceberry *(Amelanchier)*
Silver Bell *(Halesia)*	Trumpet Tree *(Tabebuia)*
Witch Hazel *(Hamamelis)*	Yellowwood *(Cladrastis)*

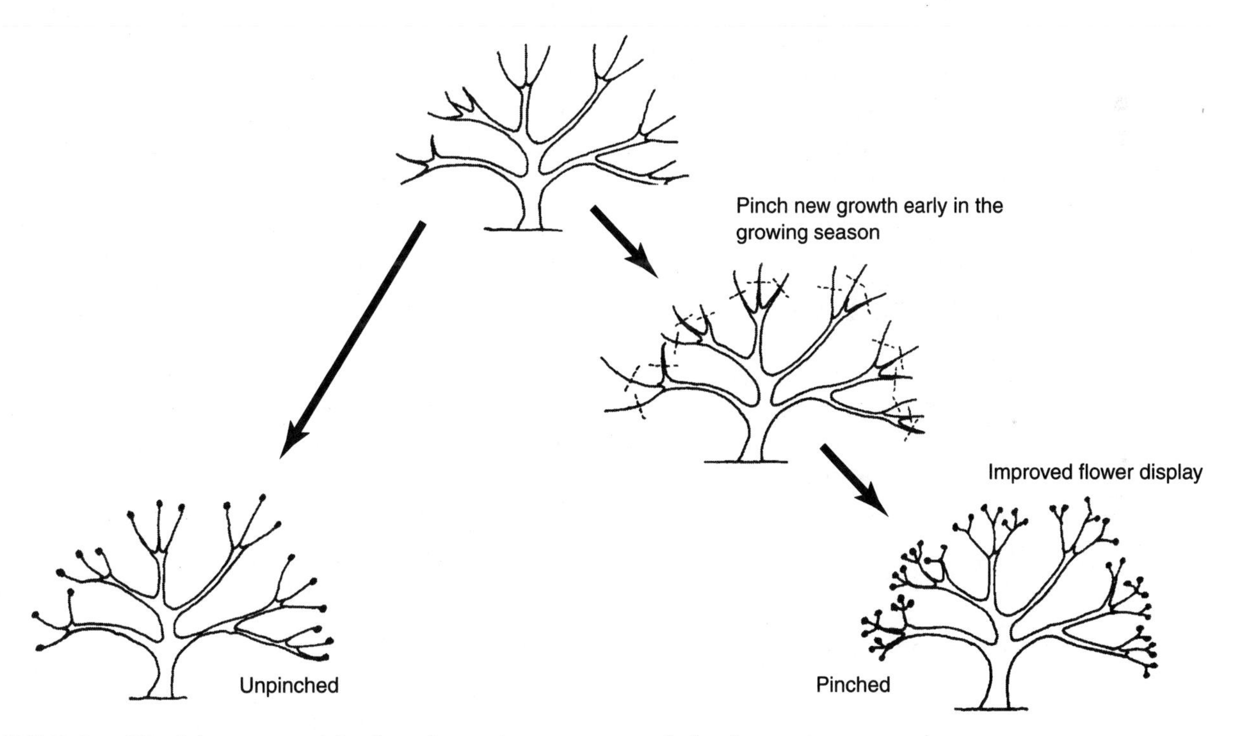

FIGURE 7-2. Pinching several inches from the new growth in the spring or early summer can generate more flower buds. On many trees that flower on the current year's growth, more (but perhaps smaller) flowers will be produced later in the summer. Pinching should be done before flower buds form.

Increase flower number. Pinching new shoots on summer-flowering trees such as crape-myrtle starting several weeks after they begin elongating through early summer will usually encourage lateral branching. Each of these laterals is likely to develop flower buds or a flower, depending on the tree species. Thus, the pinched plant produces more, but smaller, flowers than one that has not been pinched (Figure 7-2). This is time-consuming work and usually is not done except in the most highly maintained landscapes.

Larger flowers. Severe heading often encourages the production of fewer, but larger, blossom clusters on trees such as crape-myrtle, goldenraintree, and others. Crape-myrtle is often disfigured annually in this fashion. Consider pollarding (Chapter 10) instead. Those who perform this unnecessary task should know of the large variety of dwarf cultivars available that flower nicely without severe pruning.

Moderate to severe pruning reduces the number of shoots on a tree. The new shoots that develop in response to this hard pruning are vigorous, which often delays the flowering for a week or more, depending on when and how hard the tree was pruned. If certain trees (e.g., dogwood) are topped, flowering on the resulting vigorous shoots may not occur for a year or more after pruning. (Flower buds on existing, uncut branches usually develop and open at the normal time.)

Fruit Trees

Deciduous fruit trees are best pruned in the late dormant season just prior to spring bud swell. This minimizes the chances for cold damage, which can be aggravated by pruning in early to mid-winter. Light pruning (less than 10 percent of foliage removed) can be performed in summer, preferably following a growth flush and after the leaves turn dark green and have become firm. Pear, apricots, and other trees susceptible to bacterial canker are best pruned in late fall to avoid risk of infection through pruning cuts. It is best to remove dead branches in summer before the disease can spread.

Minimize Bleeding

Some trees ooze sap from pruning wounds made in late winter or early spring, just before or during active growth (Table 7-3). Called bleeding, this is not usually harmful to the tree. Some think it might help keep fungi from entering the wood (Lonsdale, 1999). The dripping sap can be objectionable and can stain the bark. To minimize bleeding on these trees, prune after full leaf development in summer. Some arborists also find that pruning in the fall minimizes bleeding. Removing only small-diameter branches by pruning frequently, beginning at an early age, can also reduce bleeding since smaller cuts are less likely to bleed heavily.

TABLE 7-3. These trees often bleed sap when pruned in late winter or early spring.

Birches (*Betula*)	Black Locust (*Robinia*)
Chinese Wingnut (*Pterocarpa*)	Dogwoods (*Cornus*)
Elms (*Ulmus*)	European Hornbeam (*Carpinus*)
Goldenchain Tree (*Laburnum*)	Hackberry (*Celtis*)
Honeylocust (*Gleditsia*)	Kentucky coffee tree (*Gymnocladus*)
Magnolias (*Magnolia*)	Maples (*Acer*)
Mulberry (*Morus*)	Mesquite (*Prosopis*)
Poplars (*Populus*)	Scholartree (*Sophora*)
Silk-oak (*Grevellia*)	Silverbell (*Halesia*)
Sumacs (*Rhus*)	Walnuts (*Juglans*)
Willow (*Salix*)	Yellowwood (*Cladrastis*)

Pest Control

Cytospora canker is a serious disease of spruce trees in certain regions that disfigures many trees by killing branches. Quick action in the early stages of infection can reduce the likelihood of the disease spreading to other portions of the tree. Branches with symptoms can be promptly removed back to the trunk. Some pines are susceptible to *Sphaeropsis* shoot blight which could kill the tree. The best time to prune both trees is when weather is dry. This could help retard infection of freshly exposed wood. Incidence of *Thyronectria* canker disease can increase if honeylocust (*Gleditsia*) is pruned in rainy spring weather, so it is best to avoid pruning at this time in regions where the disease is prevalent.

American elm (*Ulmus americana*) and other susceptible elms pruned in the dormant season may be less prone to infection from Dutch elm disease (DED) than those pruned at other times. Structural pruning and other maintenance pruning can be conducted at this time. However, removal of DED infected branches and limbs should be done as soon as symptoms are noticed.

Borers (also called longhorned beetles—*Phoracantha*) can lead to the demise of eucalyptus trees in the western United States, especially those under stress. In areas of heavy infestation, prune only in winter or very early spring to minimize infestation.

In areas where oak wilt is prevalent, susceptible oaks should not be pruned in the spring or very early summer because the insect vector that spreads the disease is especially active at this time. Some vectors can be found year-round in warm regions. If pruning is necessary during the time the beetles are in flight, wound dressings may help exclude them from the cut surface. This could reduce likelihood of infection.

Palms

Many palms are pruned in spring and at other times to remove flower stalks. This prevents the formation of fruit, which can make a mess or create a hazard in some landscape situations. For instance, flower removal on coconut palm eliminates formation of coconuts, which are hard and heavy and can become a liability in certain instances such as near roads, parking lots, and patios. Flowers or developing fruit can also be removed from recently transplanted palms to help them regain vigor. Some arborists find that it is best to wait until fruit is well developed to remove it. Removing it too early could induce more flowers and fruit. Removal of dead fronds can be done at any time of year.

Small Ornamental Trees

Young, multi-stemmed ornamental trees planted in a nursery may not need pruning for a couple years after planting depending on species and objectives. Letting them grow for two or three years before pruning promotes rapid growth. In the third or fourth year, some branches can be removed to produce the traditional multi-stemmed look and the canopy can be shaped. Removal of lower branches too soon results in slower and leggy growth. The result can be a poor crop that needs to be topped to fill out the canopy. Topping slows growth further.

CHECK YOUR KNOWLEDGE

1) The most important reason to prune a balled-and-burlapped (B&B) tree when transplanting is:
   a. to promote vigor.
   b. to develop good structure.
   c. to compensate for root loss.
   d. to slow top growth so roots grow.

2) Recently transplanted nursery trees should be pruned:
   a. to compensate for root loss.
   b. to eliminate foliage on the interior of the canopy.
   c. hard to help them regain vigor.
   d. sparingly so they can regain vigor.

3) If your objective is to slow growth, prune:
   a. just after bud break.
   b. immediately after leaves fall from the tree in fall or early winter.
   c. during the dormant period.
   d. just after each growth flush.

4) For the fastest tree growth:
   a. do not prune at all.
   b. prune during the dormant period.
   c. prune only outer branches.
   d. prune in early fall.

5) The closure of pruning wounds on most trees should be MOST rapid if:
   a. pruning is conducted just before or just after the spring growth flush.
   b. branches are removed using collar cuts.
   c. pruning paint is applied to the wound in a thin coat.
   d. pruning is conducted in the dormant season.

6) To minimize the reduction of next year's flower display on spring-flowering plants, prune:
   a. in the fall just before dormancy.
   b. lightly any time making collar cuts.
   c. just after the spring growth flush.
   d. soon after the flower display.

7) To minimize bleeding from pruning cuts:
   a. prune only small-diameter branches.
   b. prune in the late dormant season just before leaves emerge.
   c. remove less than 10 percent of the foliage.
   d. prune in the summer after twigs stop growing.

8) The most appropriate pruning cycle for community trees is:
   a. three years.
   b. ten years.
   c. fifteen years.
   d. depends on age, tree species, and other factors

Answers: b, d, d, a, a, d, d, d

CHALLENGE QUESTIONS

1) Write a recommendation for an oak tree planted from a B&B nursery in the spring that has a double leader in the bottom half of the canopy. Assume the tree will not be pruned for about five years.

SUGGESTED EXERCISES

1) Remove a few branches from the trunk in mid-winter. Remove another set of branches of the same size just after the growth flush in spring. Remove another set in mid-summer. Record the number of months required to close the wound on all cuts.
2) Pinch the emerging spring growth on a summer-flowering plant. Do not pinch new growth on another plant of the same species. Watch which one develops the best flower display.
3) Remove a branch in spring from a tree that bleeds after pruning. Remove another branch in summer from the same tree. Compare what happens.
4) Find a set of young trees in your community. Find a set of the same species or cultivar planted about fifteen years earlier. Use what you see in the older trees to decide on a pruning cycle for the young trees. Present this program to the city arborist.

CHAPTER 8

NURSERY SHADE TREE PRODUCTION PRUNING: DEVELOPING THE TRUNK AND LEADER

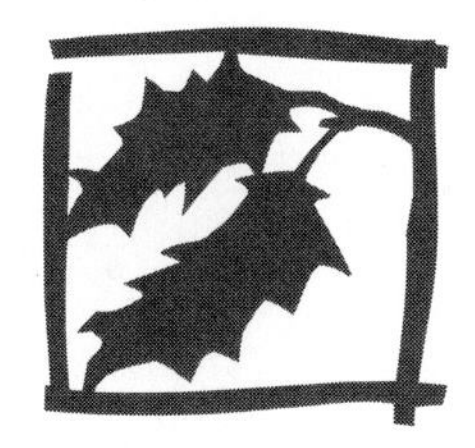

OBJECTIVES

1) Describe methods for crafting trees with a dominant trunk and a straight leader.
2) Develop taper and a strong trunk.
3) Show the importance of beginning with quality liners.
4) Describe the role of tree spacing strategies in canopy development.
5) Determine when and how to top trees.
6) Control canopy growth rate with lower branch management.
7) Show how to create branches where you want them.

KEY WORDS

Caliper
Canopy
Central leader
Clear trunk
Dominant trunk
Leader training process
Liners
Permanent nursery canopy
Single leader
Splints
Subordination pruning
Tapered trunk
Temporary branches
Topped
Tree shelters

OBJECTIVE OF NURSERY LEADER PRUNING

The main objective of pruning shade trees in the nursery is to develop one dominant leader in the shortest period of time inside a pleasing **permanent nursery canopy** (Figure 8-1). The permanent nursery canopy is made up of the branches that will be on the tree when it is sold. A dominant leader is the one stem that grows much larger than all other stems and branches on the tree. It eventually develops into a **dominant trunk.** The dominant leader structure is the most durable. The term dominant leader is often used synonymously with dominant trunk, **central leader,** or **single leader.**

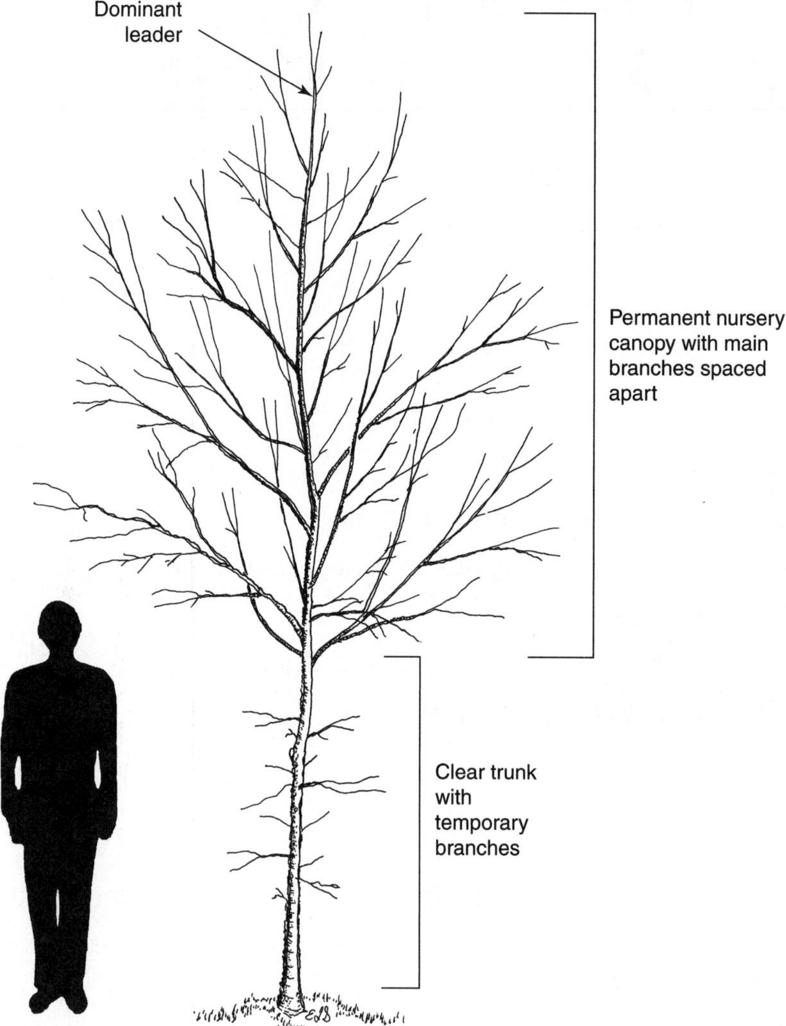

FIGURE 8-1. The objective of growing shade trees is to produce a nice canopy on main branches that are spaced along a dominant trunk. Small diameter branches can be left along the lower trunk of the finished crop to help protect it from injury.

The diameter of branches and stems should be no more than half to two-thirds the trunk diameter. Some growers prune to prevent branches from growing more than half the trunk diameter because this helps keep the leader dominant. The largest diameter branches should be spaced along the dominant trunk. A weak structure could develop if they originate too close to each other, or if they are clustered together in one spot. The leader may loose dominance and vigor if they are clustered together. Smaller diameter branches can grow anywhere between the larger ones. The result is a strong, aesthetically pleasing mix of large and small-diameter branches along a single, dominant trunk.

There should be no large-diameter branches growing below the permanent nursery canopy. It is a waste of resources to allow a large branch develop here because it will be removed. This portion of the lower trunk is referred to as the **clear trunk.** The amount of clear trunk depends on tree size. Typically, 2- to 3-inch-**caliper** shade trees that are ready to ship to a customer have about 4.5 to 5 feet of clear trunk. Caliper is trunk diameter measured 6 inches from the ground level on trees up to 4 inches diameter, or 12 inches from ground level on larger trees. Lower branches, called **temporary branches** (Figure 8-1), on shade trees should not be removed too early in the nursery because poor growth and form could result.

PURCHASING QUALITY LINERS

Liners is the term used to describe the young seedlings planted in a container nursery or a field nursery for growing on to landscape-sized trees. They can be containerized or bare root. *Short liners with abundant lower branches* (Figure 8-15, bottom) *are usually a better buy than tall liners with few low branches* (Figure 8-15, top). Short liners have adequate trunk caliper to hold themselves erect, whereas tall liners often require stakes. This chapter will show you that it is easier to craft quality trees from shorter liners with plenty of low branches. If liners are container-grown, they are less likely to be root bound if they are small. Trees oversized for their containers are usually poor quality.

SUBORDINATION OF COMPETING STEMS

Subordination is the cornerstone of production shade tree pruning (Figure 8-2). It essentially mimics what naturally occurs in the forest. In the forest, shading of lower branches subordinates or suppresses them and forces the tree to grow one leader on many trees. Pruning subordinates those branches that would grow too aggressively in a nursery if left unpruned. This aggressive growth results because trees in a nursery have access to unlimited sunlight. In most instances, if you are not subordinating codominant stems in the nursery, you will not be producing quality shade trees with good structure.

Subordination pruning essentially allows you to control two things: sunlight and growth rate. When you prune a stem or branch with a reduction cut or heading cut, growth rate on that stem or branch is slowed, subordinated, or suppressed (all mean the same thing) because there is less foliage on it. As a result, several months from now the cut stem or branch will be smaller, compared to the trunk, than it was before pruning. Growth rate increases on branches higher in the canopy that were previously shaded by the removed portion of the stem (Figure 8-2, right) because now they have access to more sunlight. Before pruning, these branches were suppressed because of shading by the removed portion of the stem. The net result is a faster growing trunk and a slower growing branch.

Making these subtle cuts has a dramatic impact on future tree structure. Instead of developing into a triple-trunk tree, with included bark in the unions, the tree in Figure 8-2 will have one dominant leader, at least along the existing portion of the trunk. The two cut stems (Figure 8-2, center) will grow slower than before they were cut but will remain dominant enough to become major limbs on the tree. The two cuts made on the right-hand tree in Figure 8-2 removed one stem entirely and also removed a large portion of the other one. This cut subordinates stem A even more than the cuts made in Figure 8-2 center. This cut could slow growth on stem A enough to prevent it from developing into a major dominant branch. Both pruning strategies are acceptable. The choice is governed by the position on the trunk, size of the branches, objectives of the grower, tree cultivar or species, and growth rate. Without such pruning, codominant

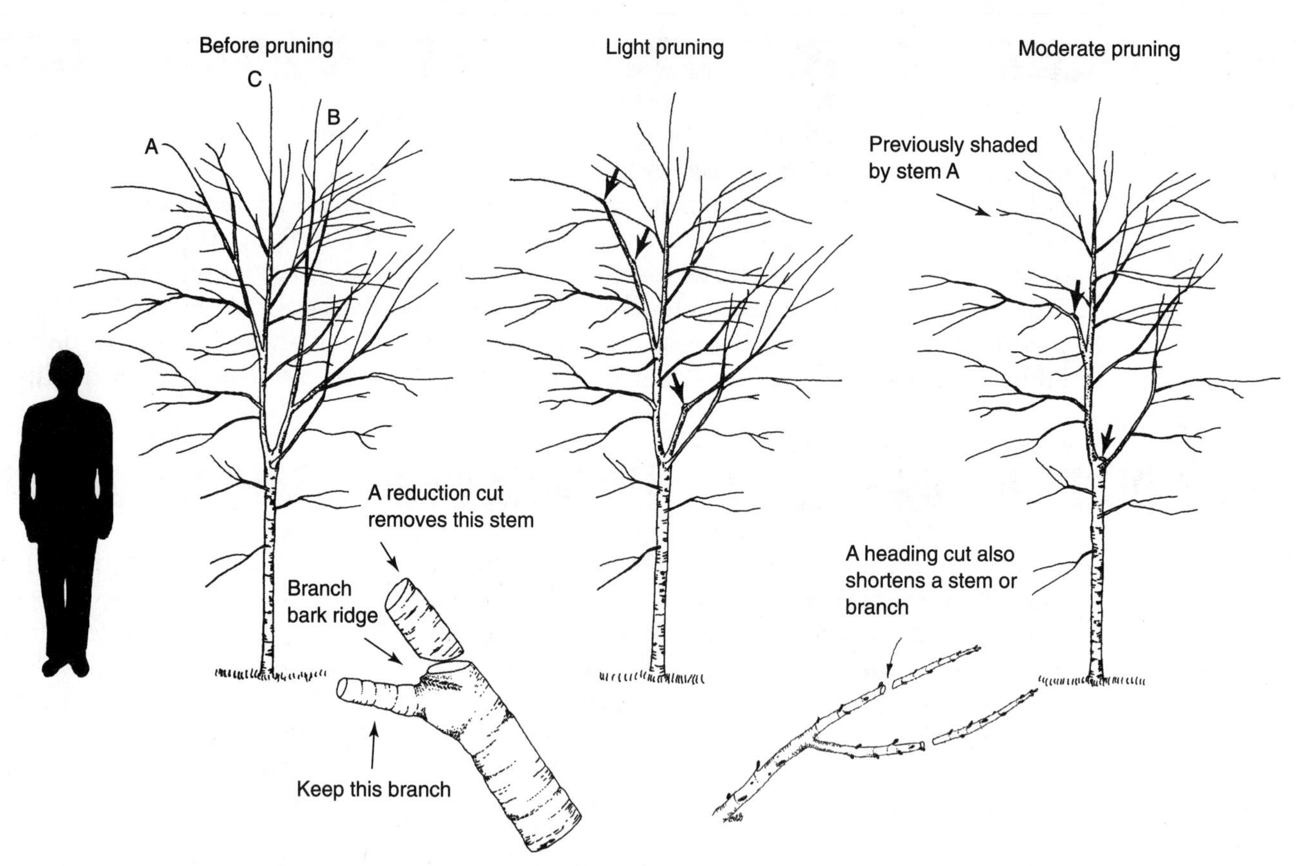

FIGURE 8-2. Three codominant stems are present: A, B, and C. All (A and B) but one stem (C) is subordinated with heading cuts (young liners) or reduction cuts (older trees). This allows the portion of the tree not cut (C) to grow faster. Remove the upright portion of stems A and B back to a more horizontal branch (see arrows center and right). This creates a small void in the canopy above the cut. As a result, more light reaches branches above the cut that were previously shaded by the removed stems. This stimulates growth on suppressed, shaded branches higher in the tree.

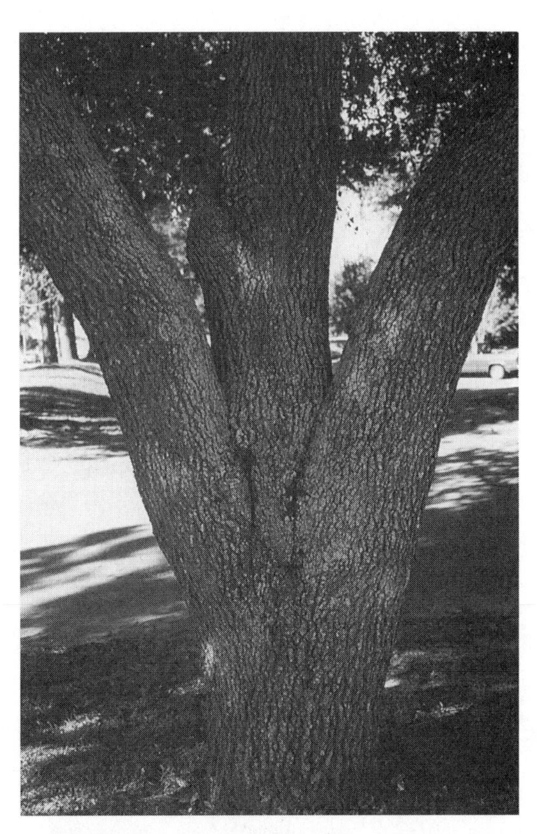

FIGURE 8-3. Without pruning, the young tree (left) would develop the structure shown at the right in which both side branches are poorly attached to the trunk due to the included bark in the crotches.

stems will form on many trees (Figure 8-3). Sometimes subordinating to a small lateral branch slows growth more than subordinating to a more vigorous lateral. Other times, reducing to a small lateral results in more sprouting. You will learn by practice how small a branch to cut back to.

Young trees less than about 18 months to 2 years old can be subordinated with heading cuts. Heading cuts are quicker and easier to make than reduction cuts and they are very useful for pruning on young trees in the nursery. Heading cuts are acceptable because most cuts are made on branches that will be removed from the tree prior to marketing it (Figure 8-4, top). Reduction cuts are used on older trees (18 months and older) in the nursery to subordinate stems and branches in the permanent nursery canopy (Figure 8-4, bottom).

Often, the top of the leader is hidden by surrounding stems and branches before a tree is subordinated (Figure 8-5, left). After a tree is subordinated, you should be able to see the top of the leader clearly from all around the tree when standing away from it a distance equal to the tree's height (Figure 8-5, right). This assures that enough competing foliage has been removed. Caution pruners against isolating the leader by removing too many branches and stems near the top of the leader (Figure 8-5, center). This overpruning often result in the leader growing too fast and extending too far beyond the rest of the canopy. Growers that overprune in this fashion often need to return to top the tree to bring the leader back into scale with the rest of the canopy. This slows growth, interrupts leader development, and adds to the time required to produce the tree for sale.

Subordination as a technique to produce quality shade trees

Heading cuts on young nursery tree about five feet tall

Before

After

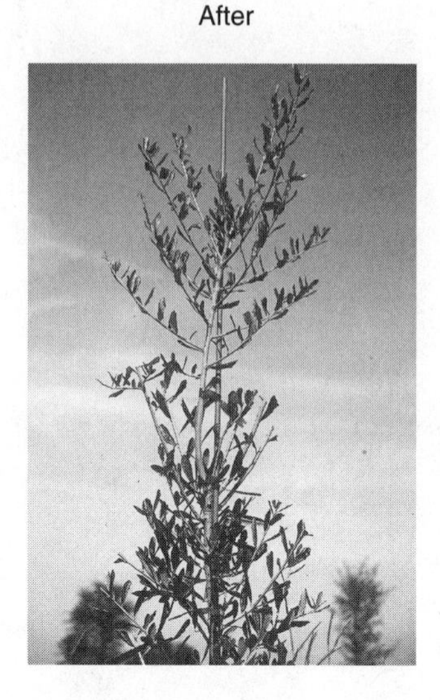

Reduction cuts on older nursery tree about twelve feet tall

Before

After

FIGURE 8-4. Branches and stems on young nursery trees can be headed to subordinate them (top). If they are removed instead of headed, too much growth may be forced into the leader. Branches and stems in the permanent nursery canopy on old trees should be subordinated using reduction cuts (bottom). Reduction cuts give the tree a more finished appearance and results in a more professionally grown tree.

Tip: *Shorten all stems and branches blocking your view of the tip of the leader as you walk completely around the tree.*

An easy method of training employees to perform subordination pruning is to teach them to grow a cone, teardrop, or oval on top of a cylinder (Figure 8-5). The base of the cone is located at the bottom of the developing permanent nursery canopy. The tip of

Allow light to reach the leader, but do not overprune

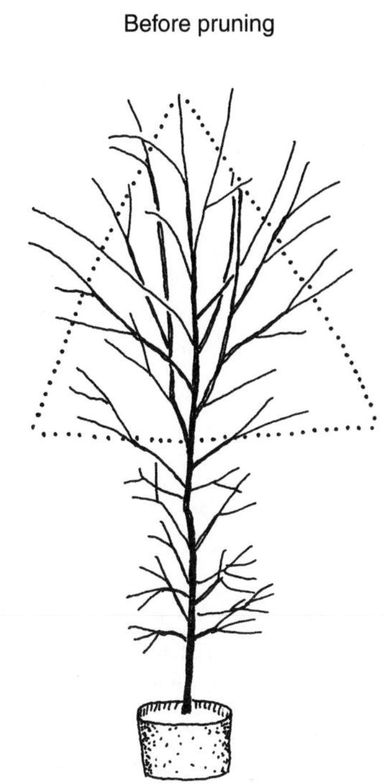

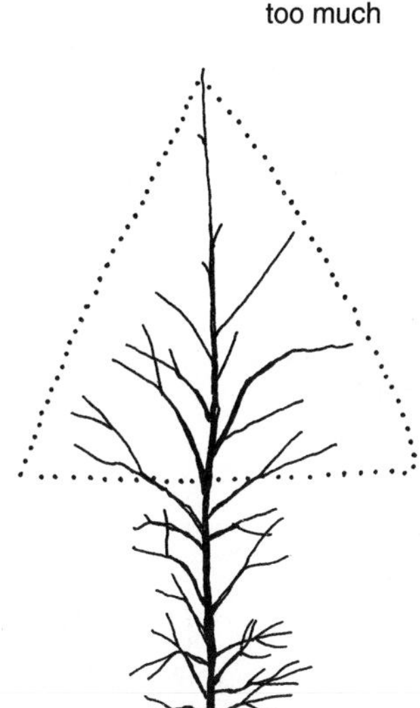

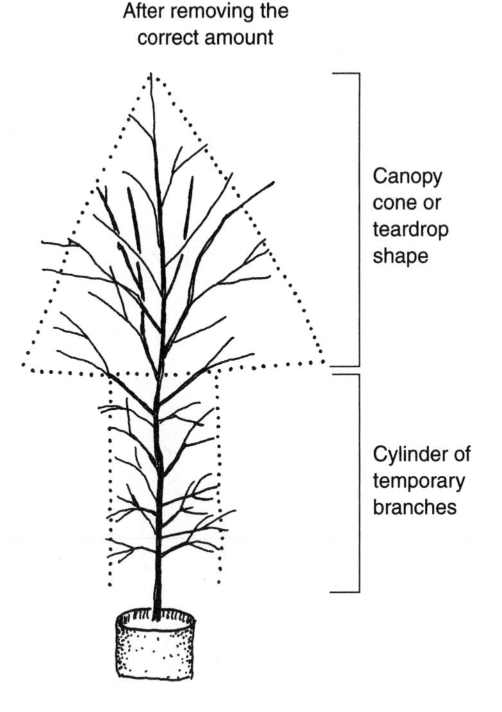

- Several leaders are competing for dominance
- Competing leaders are blocking sunlight from reaching the main leader
- Branches are growing beyond an imaginary teardrop or cone with its apex at the top of the leader

- Too much was removed from around the leader
- The leader will grow too fast and is likely to flop over

- Sunlight can reach the leader from all sides
- The leader will remain dominant because competing stems were subordinated with reduction and some heading cuts
- Branches in the permanent nursery canopy are pruned back to an imaginary cone or teardrop
- A cylinder of temporary branches is left along the lower trunk

FIGURE 8-5. Several stems and branches should be subordinated to help develop one dominant leader (left). Over-pruning isolates the leader and forces too much growth into the leader (center). A more modest approach reduces the length of competing stems but not so much as to force over-extension of the leader (right). One way to think of subordination pruning is to grow a cone, teardrop, on top of a cylinder. Reduce the length of branches that grow outside the cone. Making heading cuts creates a sheared look; using reduction cuts creates a less formal appearance and results in a more professionally grown tree.

the cone is more or less at the tip of the leader. The cylinder is the group of temporary branches along the lower trunk that will be removed from the tree before the tree is sold.

Some have misunderstood the role of subordination. Instead of reducing the length of competing stems some growers have removed competing stems entirely, back to the trunk. While this technique can produce a tree with a single dominant leader, it sure looks terrible. In most instances shorten the competition instead of removing it

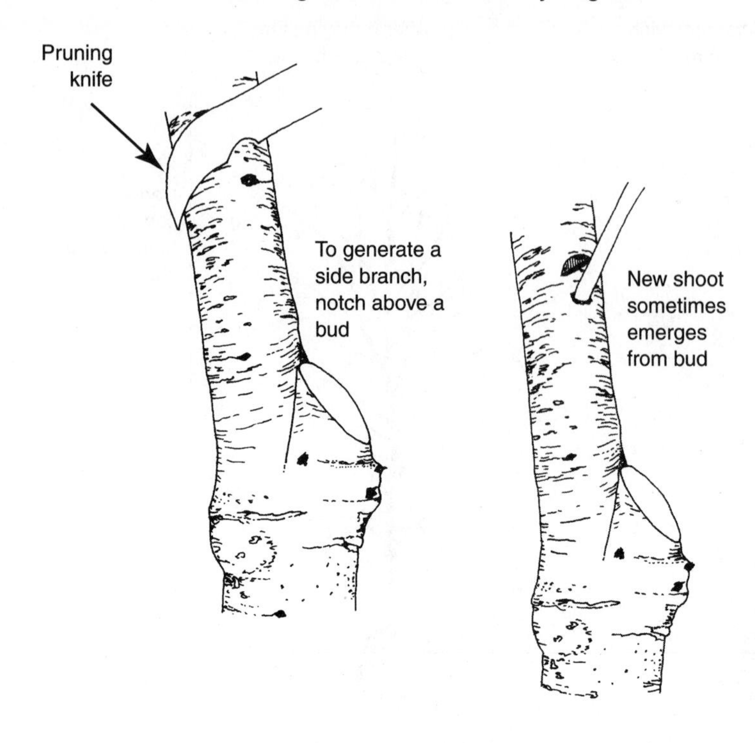

FIGURE 8-6. Notching the trunk above a bud on small-diameter trunks and branches may stimulate the formation of a sprout from that bud. The sprout will be weakly attached for a short while but will gradually become sturdier.

especially in the permanent nursery canopy. There are some instances where removal of a codominant leader less than 2 years old is appropriate, *but only if there are plenty of branches close to the pruning cut.* These branches will help retard the remaining leader from overextending.

Sometimes there is a void in the canopy following removal of a codominant stem. Branches can be generated along the trunk of some young trees by notching the trunk just *above* a bud at a point where a branch is needed (Figure 8-6). This is most effective on 1- and 2-year-old wood when performed as buds swell in the spring. A new lateral branch will sometimes emerge from the bud. This technique seems to work better on some species than others. It can be used to fill in the canopy on some trees after stems are removed.

CREATING AN UPRIGHT DOMINANT TRUNK

The tree training process should begin in the year of planting. On many decurrent trees, annual pruning in the first few years in the life of the tree makes it easy to develop a suitable dominant leader. In the first year, one or two cuts are all that is needed on some trees, requiring perhaps ten to twenty seconds per tree. In the second through the fourth year, one (cool climates), two (warmer climates), or three (subtropical and tropical) annual prunings result in the best-looking, fastest growing trees. Pruning less often can result in removing too much foliage at one time in order to accomplish the goal of developing a dominant leader. This slows growth, creates large voids in the canopy, and creates bigger pruning wounds, which close slowly and can

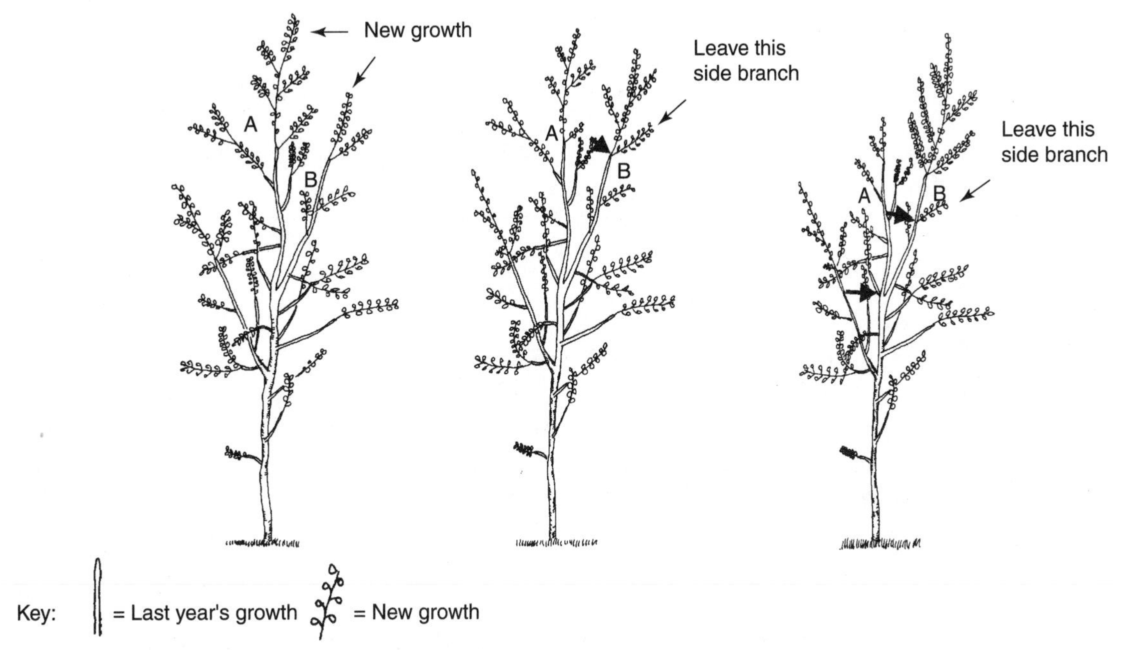

FIGURE 8-7. No pruning is needed when new growth on the leader (A) is much more vigorous than on a potentially competing stem (left). Note the greater amount of branching on new growth on A, indicating a more aggressive, dominant stem. Competing stems may need to be reduced (see arrow on B) in cases where they are equally dominant with the leader (center). Depending on your preference and the relative vigor of the competing stem, either the leader (A) or the competing stem (B) can be reduced (see arrows) when the competing stem is more dominant than the leader (right).

cause cracks and discoloration in the trunk. Slow-growing trees or those growing in extremely cold climates may need pruning less often when they are young. In tropical and subtropical climates several prunings each year are needed to develop a dominant trunk in the early years because of the rapid growth rate. When the dominant leader has been selected and is growing vigorously, by about year three, the tree will be well on its way to developing good structure. Less frequent pruning may be needed in years four and five.

Less pruning is needed to develop a leader on some trees that have lateral branches on current year's growth extending at nearly right angles to the trunk (excurrent growth habit). These species typically form a strong dominant leader and are somewhat conically shaped for the first twenty or more years. Examples include many conifers, birches (*Betula*), sweet gum (*Liquidambar styraciflua*), black gum (*Nyssa sylvatica*), lindens (*Tilia*), tropical almond (*Terminalia catappa*), Norfolk Island pine (*Araucaria heterophylla*), and sycamore (*Platanus*). However, there are instances when double leaders and codominant stems form on these trees. When this occurs, one should be removed or cut back with a reduction cut as outlined above.

Trees provide you with information on how aggressive each branch is growing and this can guide your pruning plan. For example, note that the three stems in Figure 8-7 have identical structures. However, as new growth emerges in spring, clues develop that help you determine how to prune.

TABLE 8-1. Developing a leader in shade trees can be reduced to three simple steps.

1. Locate the best stem to develop into the leader. This is usually the one located close to the center of the canopy.
2. Locate the stems and branches that are competing with the leader.
3. Cut back competing stems and branches to allow the leader to dominate the structure.

The **leader training process** that develops a leader is easier to learn and teach if it is broken into three simple steps (Table 8-1): (1) choose one stem to be the leader; (2) locate those stems and branches competing with the leader; and (3) decide where to cut these competing stems back to. Practice this on a variety of tree species and cultivars. On young one-year-old seedlings, you can hold the leader with one hand while making heading cuts on competing stems. This is a good way to ensure the leader will not get cut because most people will not cut their hand.

Figure 8-8 provides many examples of different situations that trees present when they are very young. Each is unique requiring slightly different pruning strategies for developing a leader. Go through the three-step leader training process for every tree. Heading, reduction, and removal cuts are utilized, in that order of frequency, on young seedling trees. In other words, it is more appropriate to shorten branches or stems than to remove them on seedlings.

Maintaining a leader on slightly older trees (greater than about 2 years old) requires similar pruning cuts, except that heading cuts are used much less frequently (Figure 8-9). These are instances on elms, maples, and other trees where heading cuts are appropriate because no lateral branches are in place to cut back to. If large branches (greater than half the trunk diameter) were removed, visit the tree a month or two later or early in the next growing season to check for sprouts. Remove these sprouts with your hand before they become woody, or, to hasten wound closure on a large cut, shorten sprouts after they become woody. Remove them several months later. If you do not remove the sprouts, they may fill in the canopy with undesirable, weakly attached branches. *If trees are regularly pruned, large branches such as this will not have to be cut from the tree and sprouting will be minimized.*

CREATING A STRAIGHT TRUNK

Staking

The trunk can be tied to a stake that is inserted into the ground to produce a straight trunk (Figure 8-10, option A). Many materials can be used to stake or secure trees so trunks grow straight (Table 8-2). Stakes that allow the trunk to move in the wind may result in slightly stronger trunks and shorter trees that are more in balance with the caliper. Ridged stakes could result in more trunk breakage in strong winds. If the tree has two leaders about equal size, one can be headed to about 6 inches long and temporarily

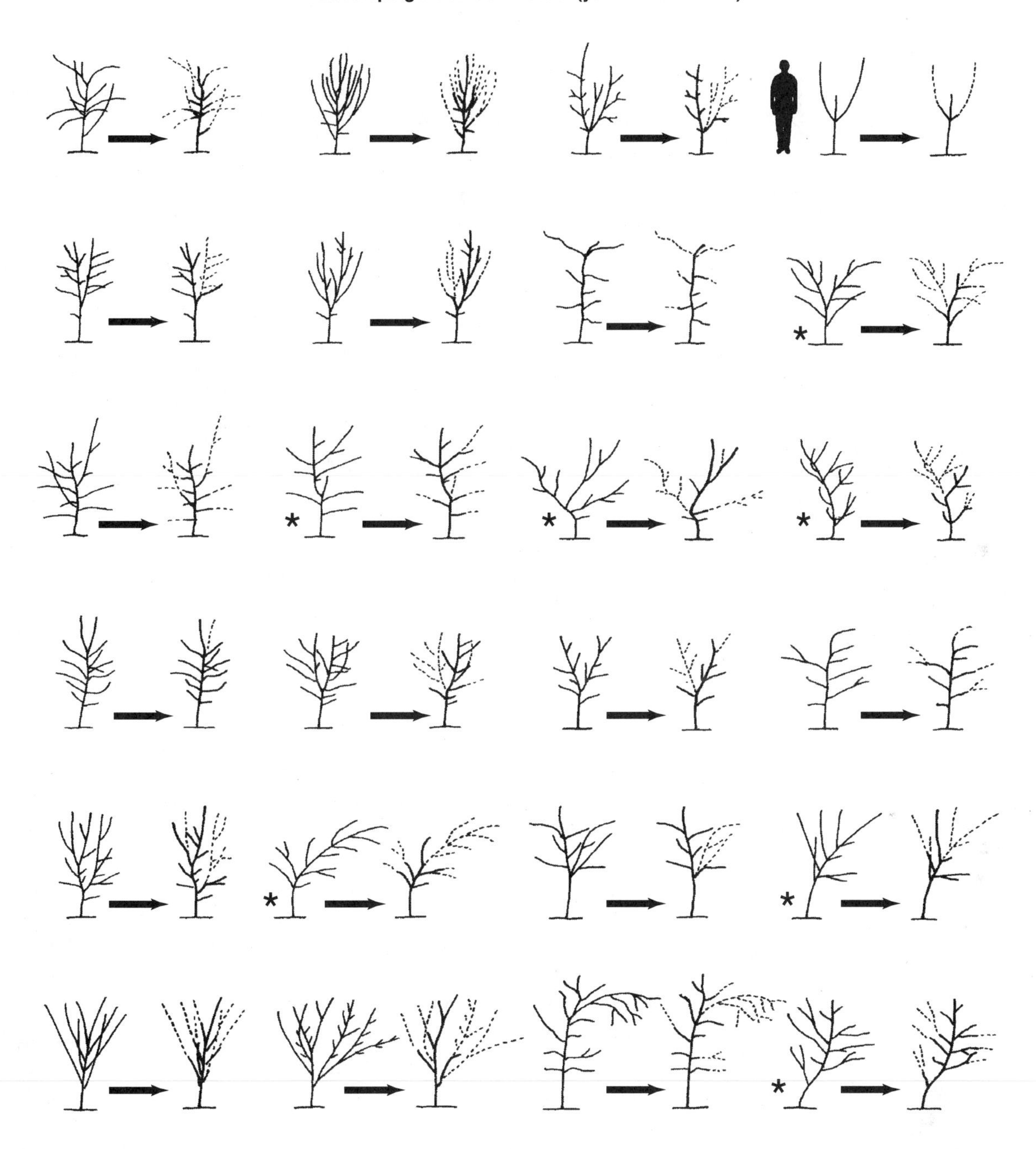

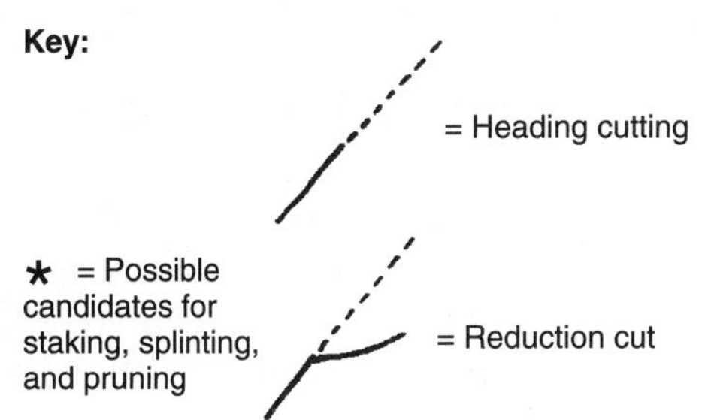

FIGURE 8-8. Developing a dominant leader on different young trees with reduction, removal, and heading cuts. Branches marked with dashed lines should be removed. Trees marked with an asterisk might also benefit from staking. Few of the branches presently on these trees (except for one or two at the top) should be on the trees when they are sold. Be sure to *shorten any that are growing upright* into the permanent canopy to ensure they are removed before sale.

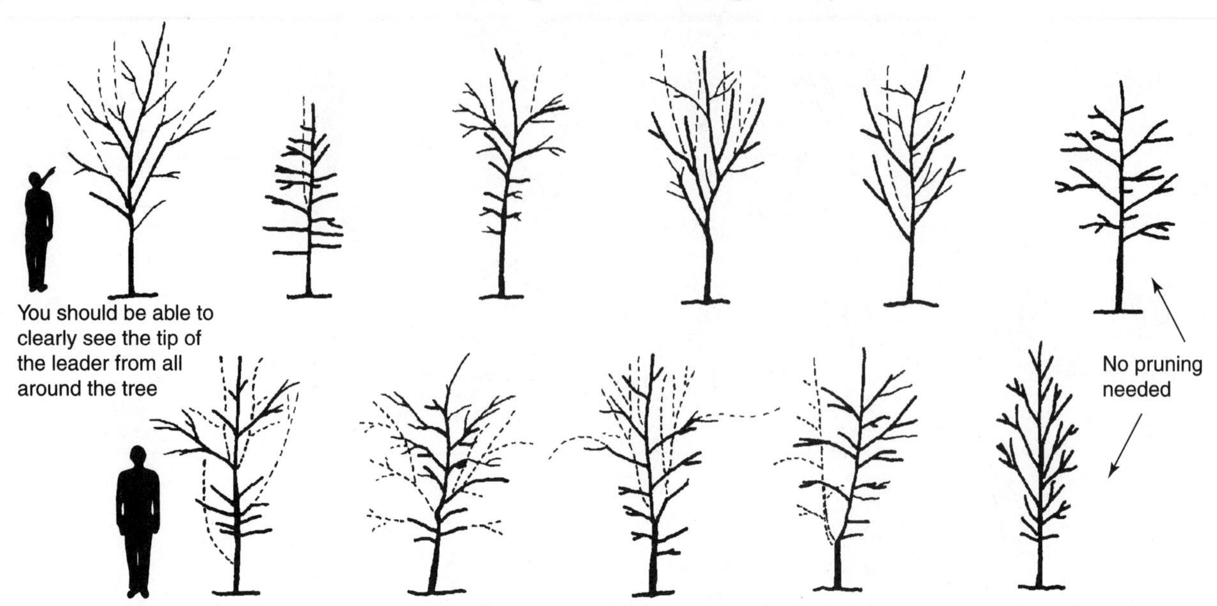

FIGURE 8-9. Remove or shorten branches such as those marked with a dashed line in order to encourage the continued development of a dominant leader. Two trees on the right do not need pruning because the leader is well developed. On some trees, the original leader should be removed so an aggressive branch can become the new leader.

tied to the other leader, which is not cut, acting like a short splint (Figure 8-10, option B). This pulls the uncut leader toward the center of the tree; it is only appropriate for young trees. Secure the trunk to the stake or headed leader with stretchable plastic ties or other mechanism to allow it to increase in diameter. Remove the stub later.

Do not remove the lower branches from the trunk. The trunk may not develop adequate strength if the lower branches are removed too soon (Figure 8-11). Lower branches increase trunk diameter, which helps the tree hold itself up when the stake is removed. It is essential to allow these low branches to develop adequate length. This is covered in depth later in the chapter. While it is staked, make sure to check the tree during the growing season to ensure that the ties are not choking the trunk, and the trunk or root system are not growing into the stake.

One drawback of using stakes or a securing mechanism comes when they are left on the tree for an extended period. The roots may grow around a portion of the stake in the ground and it could become stuck. The trunk could be damaged by the ties used to secure the stake to the trunk. In addition, trunk diameter could increase slower when trees are ridgedly staked for too long, and they may not develop the taper needed to hold the tree erect. They also grow taller and could produce less of a root system. Occasionally, the tree will bend away from the stake and not stand up on its own. The stake should be removed as soon as the trunk is sturdy enough to support the tree or is straight. Check this by periodically unfastening some stakes.

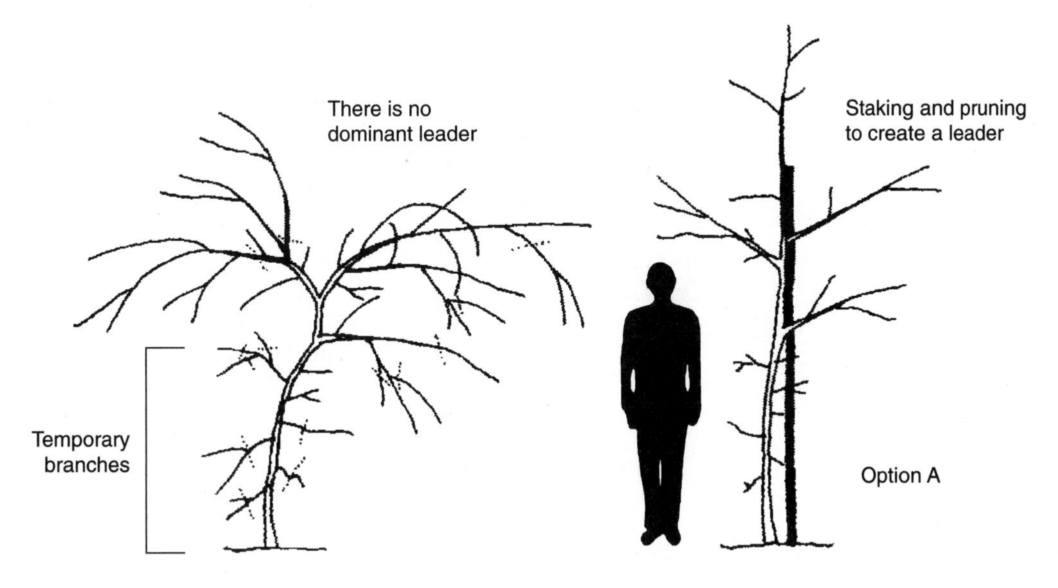

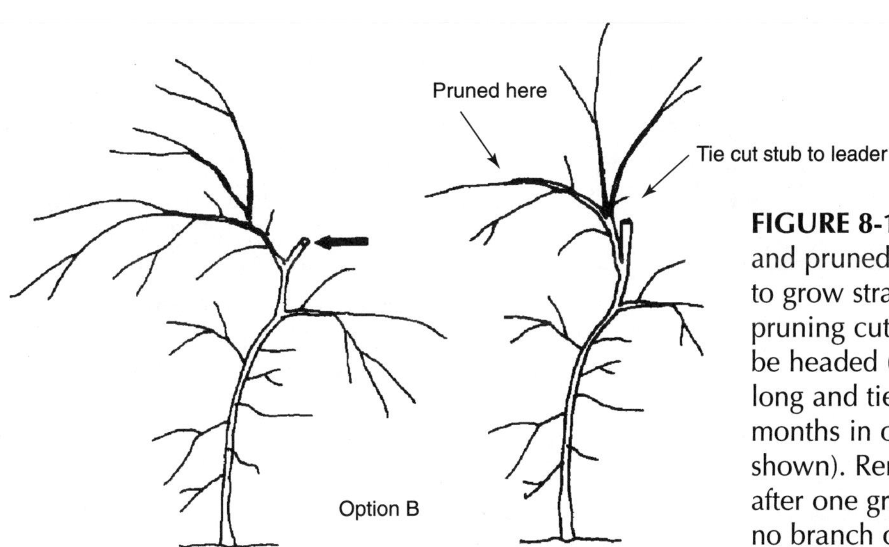

FIGURE 8-10. A tree can be tied to a stake and pruned in order to train a dominant leader to grow straight (top, dotted lines indicate pruning cuts). One side of a double leader can be headed (see arrow, left) to about 6 inches long and tied to the other leader for several months in order to straighten it (ties not shown). Remove the 6-inch stub and the ties after one growing season. A small tree that has no branch or stem suitable for training to a leader can be cut to the ground or cut back to just above the graft union (bottom). One shoot can then be trained into the trunk.

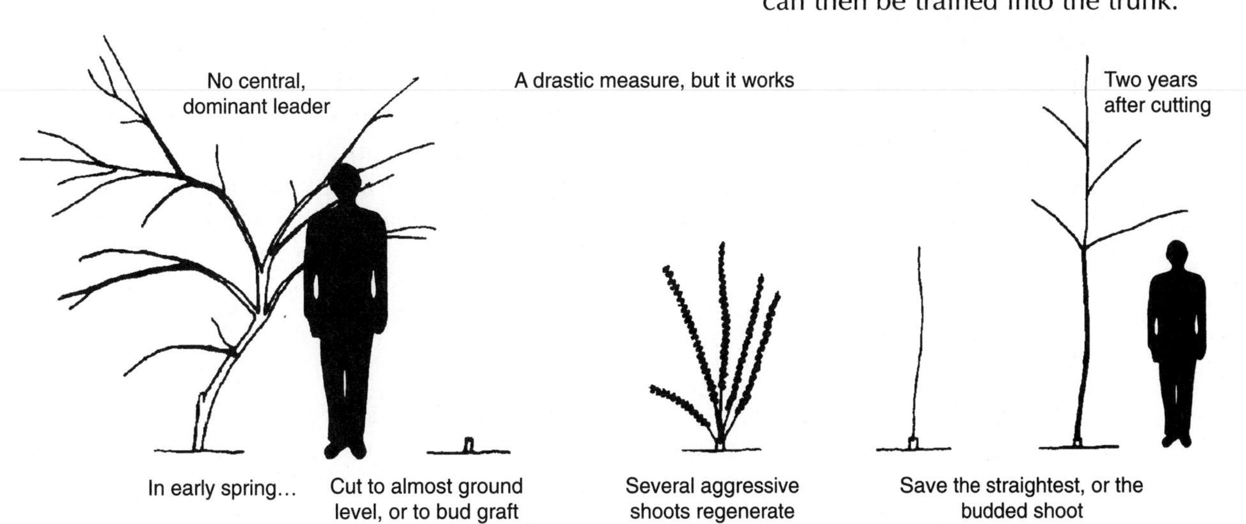

TABLE 8-2. Materials commonly used to stake trees in a nursery.

Staking Material	Advantages	Disadvantages
Bamboo stems	Strong	Short life
pvc pipe	Inexpensive, flexible	Needs trellis for stiffness
Galvanized rods	Long life, flexible, strong	Thin gage bends in storms
Electrical conduit	Strong, durable	Too stiff
2 × 2 wooden lumber	Strong	Damages trunk, too stiff
Trellis system (wire stretched tight between posts in the ground)	Flexible	Expensive, doesn't necessarily produce straight trunk

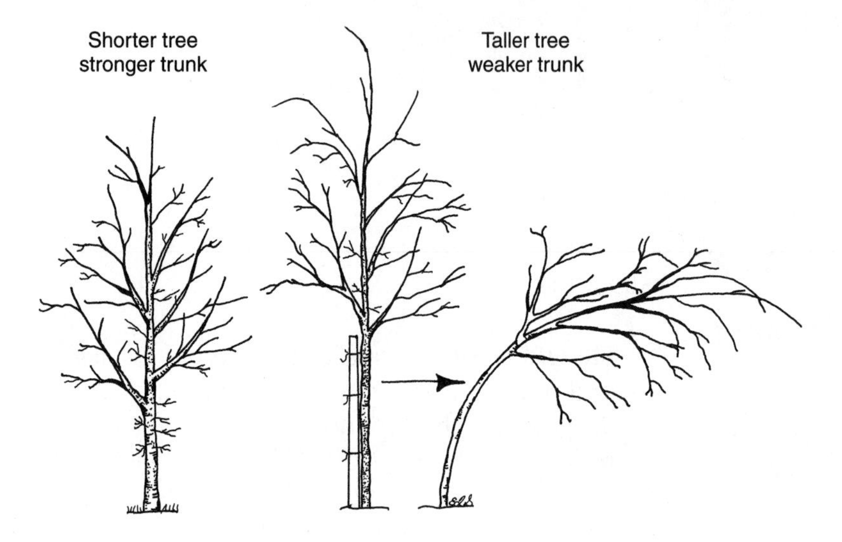

FIGURE 8-11. Trees without stakes and those with lower branches (left) are shorter and sturdier than staked trees with no lower branches (right). Be sure to leave shortened branches on the lower trunk, especially on trees that are staked.

Tip: *Be sure to leave plenty of low temporary branches on the trunk, especially when using stakes.*

Another disadvantage with using stakes is trunk injury. The trunk can be injured as it is blown in the wind against the side edge, and especially against the top edge, of the stake. The tip of the stake can be beveled or bent to minimize injury. Galvanized metal stakes have a small diameter and may cause less injury than other types. To help prevent injury secure the tree firmly to the stake. The other option is to not be concerned about developing an arrow straight trunk. A perfectly straight trunk is not necessary for the structural strength of the tree. A single, dominant trunk is far more important than a straight trunk. Trees may grow slower without stakes since more of the canopy may have to be removed to produce a dominant leader.

Topping

Some nurseries routinely cut trees nearly to the ground in their first or second year. Cut the tree off several inches from the ground (or just above the graft union) just as buds are beginning to swell in spring, or early in the growing season (Figure 8-10, bottom). Some horticulturists slant the cut away from the sun (i.e., facing north in the northern

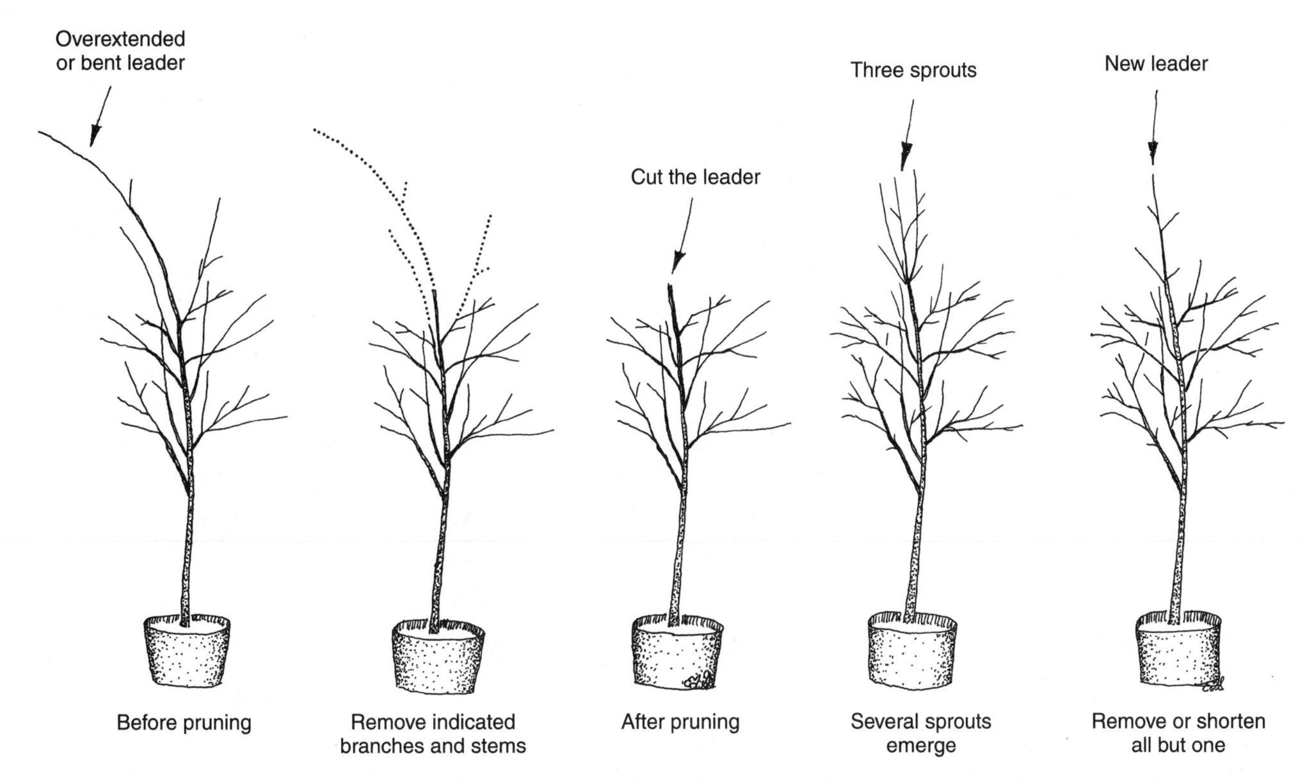

FIGURE 8-12. Topping can be used as a means of producing quality nursery stock on trees that have overextended leaders with few branches (left). However, it is essential that follow-up pruning be performed to reduce all but one of the sprouts that result (right). It is better to solve the problem causing an overextended leader than to regularly top trees in the nursery.

TABLE 8-3. Causes of overextension of the leader.

Trees spaced too close together
Lower temporary branches shortened too much
Removal of branches along lower trunk
Reducing canopy width too much
Removing too many branches from around the leader
Too much nitrogen fertilizer

hemisphere). Allow the numerous shoots that grow from the cut to reach about 10 to 20 inches long. Leave the straightest shoot, head back two or three, and remove the others. Remove headed shoots at the end of the growing season or when the pruning wound closes.

The leader can be **topped** if it has overextended and grown too tall and out of scale with the rest of the canopy, or if it has a severe bend (Figure 8-12). Many production problems can cause too much growth in the leader (Table 8-3). You must follow up soon after topping to reduce the length of or eliminate many of the sprouts that result. Your goal is to end up with only one stem as the leader. This treatment corrects the symptom of an overextended leader but does not result in a solution to the production problem that caused the overextension. To produce quality trees in the shortest possible time period, find the cause of the overextended leader and correct it. This might eliminate the need for topping.

Splinting

Splints are stakes that have no contact with the ground. Bamboo or galvanized rods are commonly used as splints. Splints are used to train the middle and top portion of the leader to be straight (Figure 8-13). They can be attached to a drooping leader of one that has a bend in it. The splint is placed along the portion of the trunk that is bent and enough of the straight portion below to straighten it. The stake inserted into the ground to support the young seedling can be pulled from the ground and moved up the tree as a splint. The bottom portion is no longer in the ground. Or, add a second stake as a splint and leave the original one into ground if needed. All trees in a field may not need to be splinted. Trees that are staked and splinted often reach marketable size quickest because less of the tree needs to be pruned off at each pruning to construct a dominant leader. Stake some trees, and leave others unstaked as a small test. Which ones reach marketable size quickest?

Tip: *Stakes and splints can reduce pruning needs and reduce the time required to grow a quality tree.*

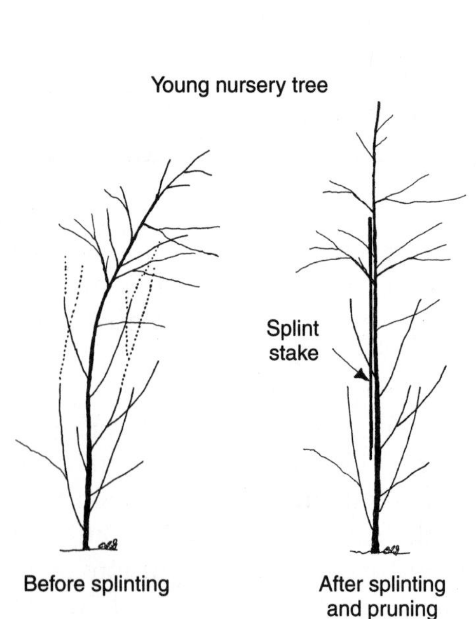

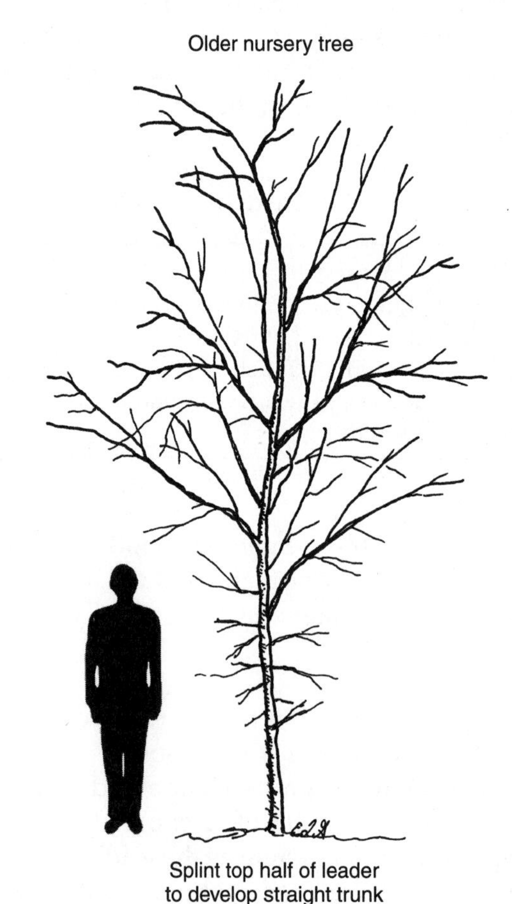

FIGURE 8-13. Young trees with a bend or lean in the trunk can be splinted with a stake to develop a straight trunk (left). Dotted lines indicate removed branches. Older nursery stock can also be splinted to keep the leader straight and in the center of the canopy (right, splint not shown). The splint is secured to the trunk below the point where the trunk begins to bend. It is attached to the trunk in several points to straighten it. Remove when the trunk has developed the strength to remain straight.

Tree shelters

Tree shelters are tubes 2 to 4 inches in diameter about 3 to 4 feet long made from various materials that are placed around the young trunk. They were designed to protect young seedlings in reforestation efforts from rodent and other animal damage. They have been suggested as a nursery production tool for aiding in the development of a leader. There are numerous downsides to using tree shelters (Table 8-4) that probably far outweigh any advantages. They are not commonly used in the nursery industry.

TREES WITH AN UNBRANCHED TRUNK

Trees such as maples (*Acer*), Chinese pistache (*Pistacia chinensis*), goldenraintree (*Koelreuteria*), royal poinciana (*Delonix regia*), jacaranda (*Jacaranda*), Kentucky coffeetree (*Gymnocladus dioica*), Japanese scholar tree (*Sophora japonica*) and some elms may not branch effectively on new growth. The young tree can become lanky because of sparse branching. Branching can be enhanced by a simple technique designed for wood less than about two 2 years old. It is not appropriate to cut older wood in this manner.

On a young (1- to 3-year-old) plant, the leader may be headed at a point (existing lateral bud) where the first lateral branch is needed. Several weeks to a month or two later (depending on growth rate), select the most vigorous, topmost sprout from the cut as the main leader and another slightly lower for a side branch, and severely head the others leaving only two to four buds (Figure 8-14). If one sprout is not trained into the leader, a weak, multi-stemmed tree will often develop. The young leader can be pruned several times in this manner (sometimes in the same growing season) at any place a lateral branch is needed. Lateral branches can be pruned in the same manner to increase secondary branching. The thickest, most vigorous stems will produce the greatest number of branches or sprouts. Only one sprout may form on less vigorous stems. If all new shoots grow upright, try heading earlier in the season, when the leader is younger. Vigorous plants respond best to this treatment; trees lacking vigor may only generate one shoot at each cut.

TABLE 8-4. Advantages and disadvantages of using tree shelters as a nursery production tool.

Advantages	Disadvantages
Helps create upright trunk	Trunk could be injured at the top of the tube
	Tall, weak trunk is produced
	Trunk could be damaged by birds
	High temperatures in the tube
	Increased trunk and stem damage from ice and snow loads
	Trees enter dormancy later
	Ants, bats, and other animals nest in the tubes
	Reduced root system

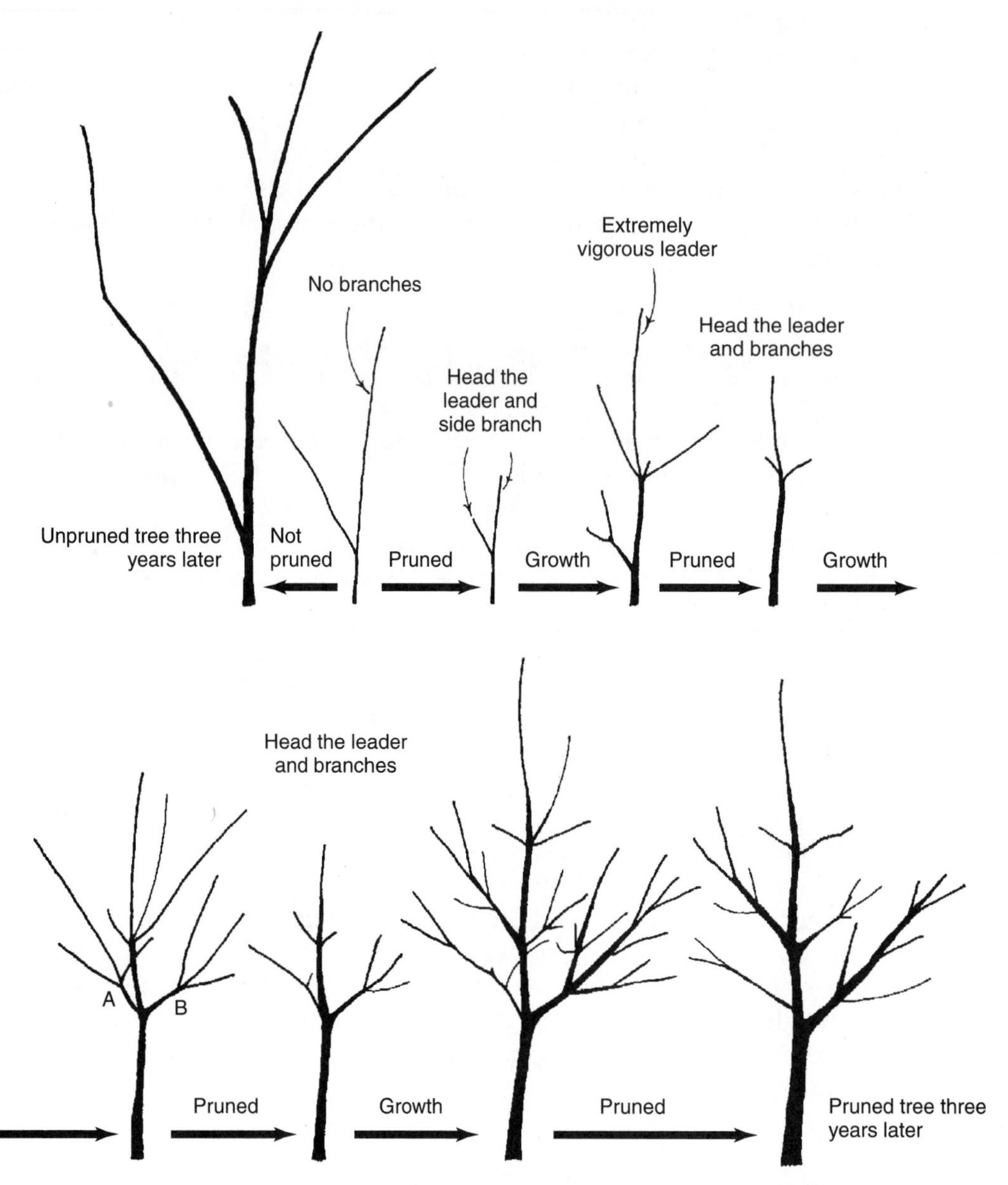

FIGURE 8-14. Creating branches along an unbranched trunk forms a more balanced and denser tree. On some trees, two or three sets of branches can be created and spaced in a single year by pinching two or three times during the growing season. Cut branch A more than branch B so only one large branch (B) develops at one point on the trunk (bottom left).

TREE SPACING STRATEGIES

Tree spacing in the nursery has a dramatic impact on tree structure and leader development (Figure 8-15). Close spacing, especially in the first two years, can spoil an otherwise good nursery production system. Trees spaced too close together will grow taller than those spaced apart because they are shaded from the side by adjacent trees. This encourages upright growth and development of codominant stems as trees stretch upward toward the sunlight. Close spacing discourages lateral expansion of the canopy, creat-

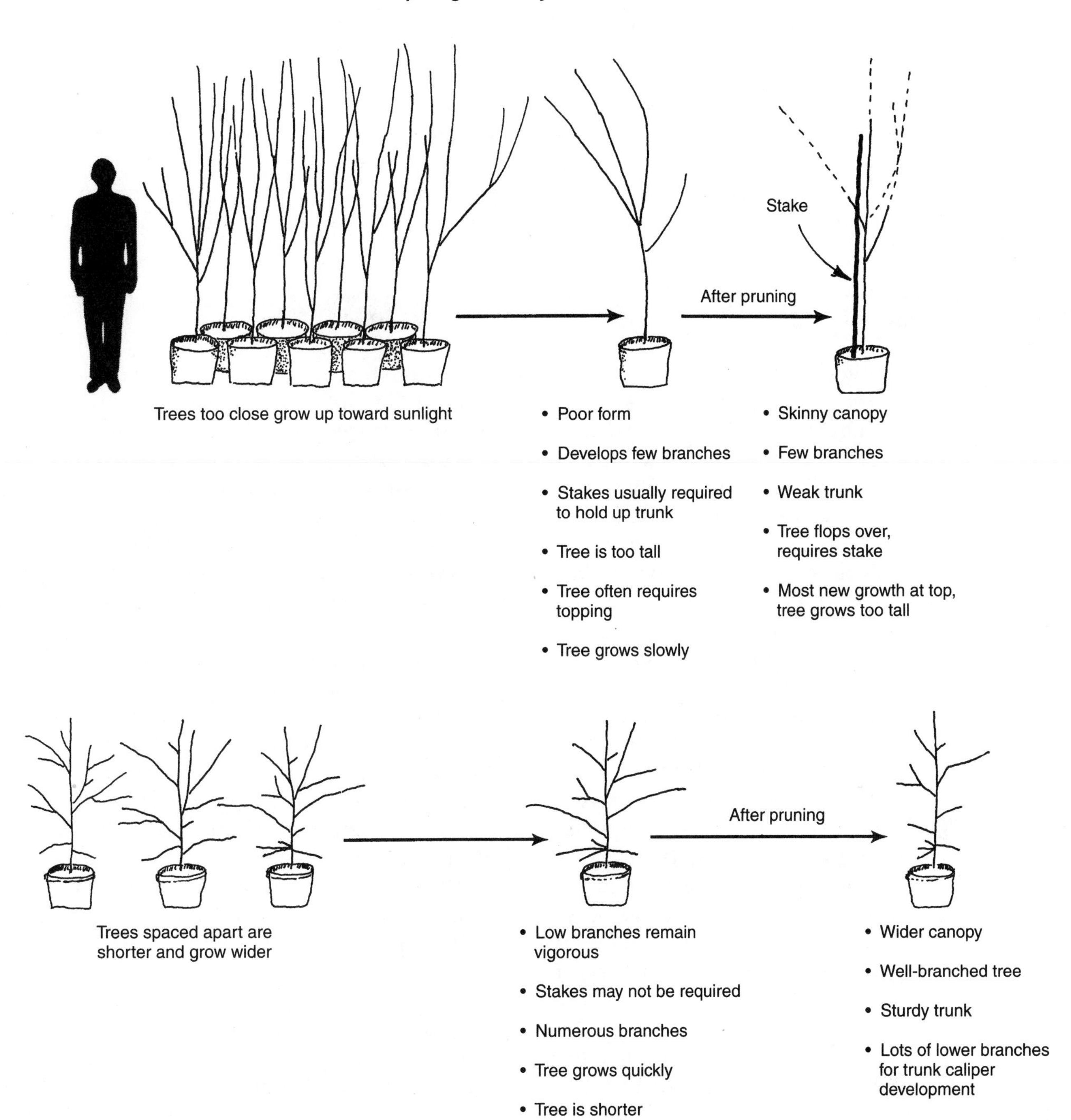

FIGURE 8-15. Saplings spaced too closely in the nursery will develop poor form and weak trunks (top). The side branches are starved for sunlight, so they grow straight up toward the sun. The leader stretches too tall. You should space trees apart to retain low, more horizontal branches and develop a sturdy tree (bottom).

ing a tall narrow tree. When codominant stems are pruned to subordinate them, the tree looks terrible and is poor quality because there are too few side branches and little spread. Some nurseries choose not to subordinate resulting in a weak tree with upright codominant stems. In addition, *lower branches and foliage are shaded and die when trees are close to each other,* creating a skinny, weak trunk that requires staking. These trees can be very difficult to train into quality trees because there are few lower branches to help build caliper. Lack of low branches also can cause trees to grow too tall.

Seedlings up to about 2 feet tall in containers (smaller than the 2-gallon size) can be placed nearly pot-to-pot (with pots touching each other) in the nursery. Seedlings in a liner field can be spaced closely as well because this encourages development of an upright leader. However, it is essential to space apart larger trees to encourage growth in horizontal lateral branches. When trees are adequately spaced, lower branches grow well because they receive plenty of sunlight. This is the key to producing quality trees!

DEVELOPING TRUNK CALIPER QUICKLY

Trees with **tapered trunks** (thicker at the bottom than at the top) are able to hold themselves erect without stakes and can withstand greater stress from wind and vandals than those with little or no taper (Table 8-5). A tapered trunk is often a sign that caliper developed quickly. Open-grown trees produce trunks that are more tapered than trees in the forest, and widely spaced trees in a nursery develop a more tapered trunk than their closely spaced counterparts. Trees that are not staked or are secured with flexible stakes usually taper more and may be stronger than ridgedly staked trees (Figure 8-11). A slightly shorter, stronger, thicker tree is produced when lower branches are kept on the trunk of a young tree. More caliper means a greater return on investment.

Temporary branches should be left along the trunk below the lowest branch in the permanent nursery canopy to strengthen the trunk and protect it from sun and mechanical injury (Figure 8-16). Trees with temporary branches grow quicker than those without them. On young trees, treat all branches below the lowest desirable permanent branch as temporary. At each pruning, shorten only the most aggressive temporary branches and leave those with weak to moderate vigor. Cut them to 12 to 36 inches long depending on species or cultivar. Temporary branches can be left long on trees that grow tall too rapidly; *this will help moderate overaggressive leader growth.* Remove any that grow bigger than half the trunk diameter or about half-inch diameter. Remove or shorten those that grow upright into the permanent nursery canopy, or where the permanent canopy will be. All temporary branches will be removed in the nursery as the tree grows larger.

As the trunk gains strength, temporary branches can be gradually removed from the tree. When trunks are 2 to 3 inches caliper the number of temporaries can be reduced. Most 3- to 4-inch caliper trees do not need temporary branches on the lower trunk

TABLE 8-5. Factors affecting trunk strength.

Encourages Taper and Sturdy Trunk	Encourages Weak Trunk and Stretched Trees
Well-spaced trees	Close spacing
Retaining lower branches	Removing lower branches
No staking	Inappropriate staking
Moderate nitrogen fertilization rate	Too much nitrogen fertilizer

because their contribution to growth is small compared to the rest of the canopy. Bark on many trees is thick enough to provide some protection for the cambium from sun injury. On thin-barked trees such as lindens (*Tilia*) and maples (*Acer*), consider leaving temporary branches for a longer period of time to provide protection from the sun and mechanical injury. They could reduce incidence of sunscald in regions where this is a problem.

Some young trees in nurseries are grown void of temporary branches on the lower trunk. Because of a lack of trunk taper caused by removing these lower branches too soon and growing trees close together, the trees are often staked to hold them erect, which ironically further increases the height and restricts caliper growth.

Two questions often arise: How long should temporary branches grow (what length should they be) before reducing their length? How much of their length should be removed at each pruning? Trees respond to pruning in a unique manner and you must learn, by experimentation, how each species and cultivar reacts to temporary branch management. Trunk caliper growth rate will be most rapid if all temporary branches are left to grow without pruning them. However, growth in the permanent nursery canopy might be restricted too much because the aggressive low branches sap growth from the permanent canopy. On the other hand, permanent canopy growth rate might be too fast causing overextension of the leader if temporary branches are cut too short (Figure 8-17). There is a compromise that results in the best-looking, fastest growing crop for your operation. You must find this compromise.

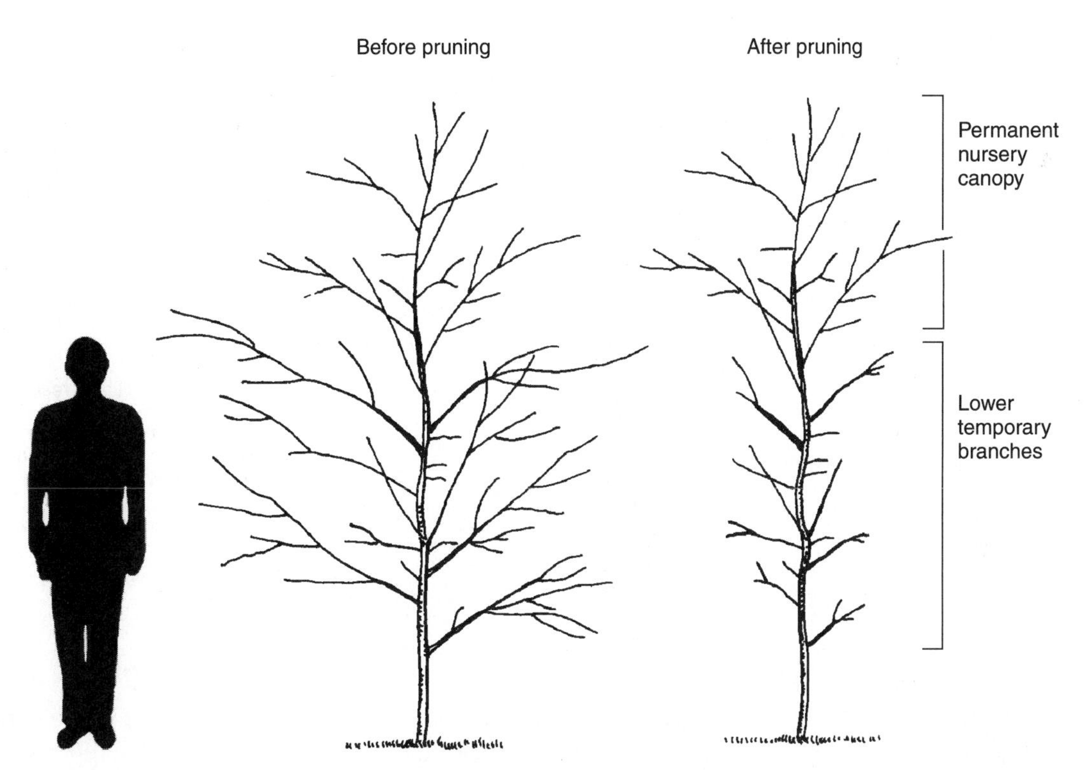

FIGURE 8-16. Shorten the lower branches with heading cuts to encourage more aggressive growth in the permanent nursery canopy. Leaving those lower temporary branches greatly increases trunk caliper and overall tree growth rate compared to removing them.

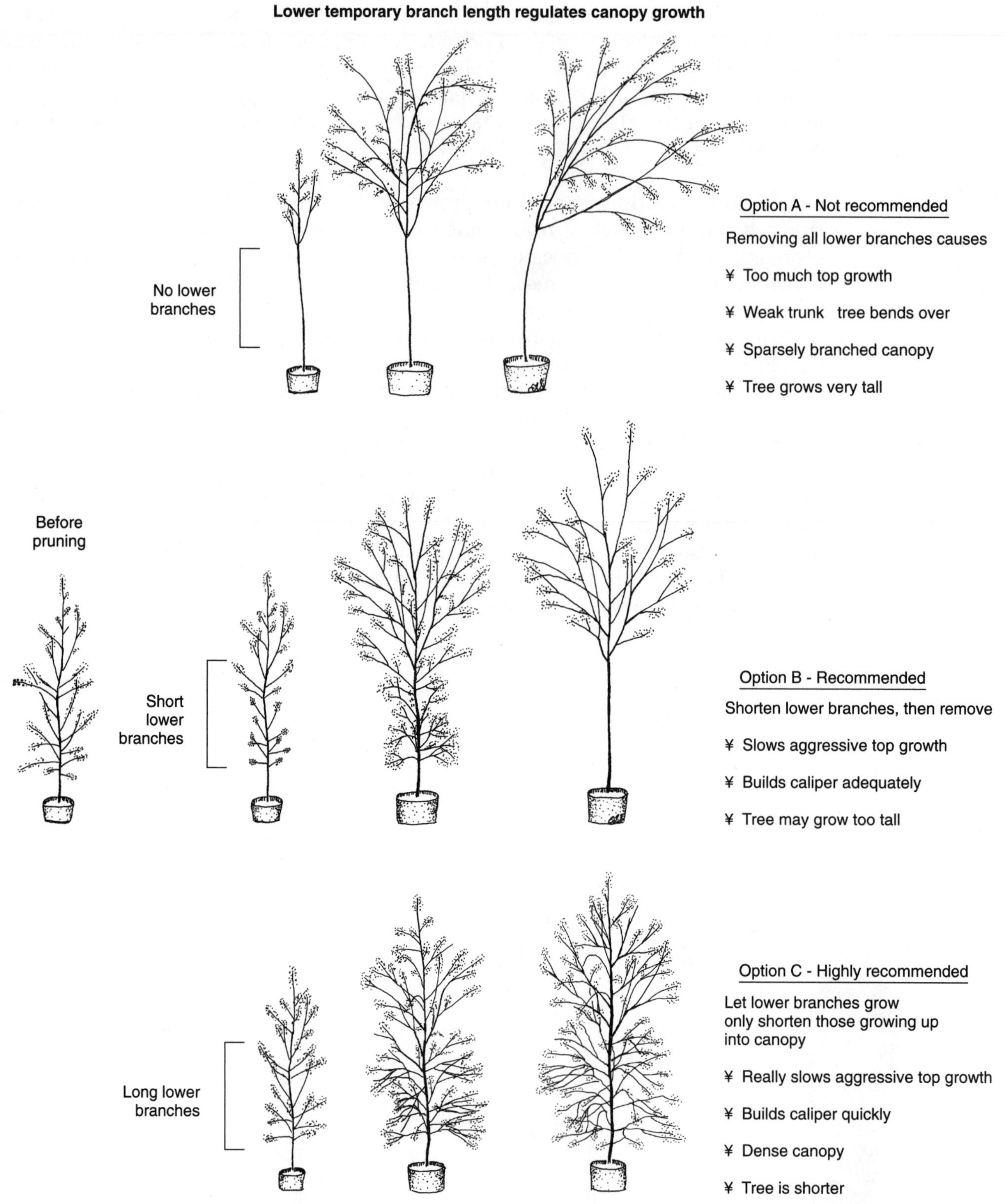

FIGURE 8-17. Length of lower branches affects growth rate of the permanent nursery canopy. The canopy grows too quickly causing over-extension of the leader and a tall tree if all lower branches are removed too soon. Trunk caliper growth is also slowed causing the tree to bend over (top). Growth rate of the canopy can be slowed so it develops more in balance with caliper growth rate by leaving lower branches on the tree (center). If canopy growth rate is too rapid, leave lower branches extremely long. This will slow the leader and canopy growth rates and prevent the canopy from becoming too tall too fast. Caliper growth rate is extremely rapid (bottom).

This compromise is found by experimenting with your own trees. Set up a test in your nursery on about sixty trees of one type. Remove lower branches from the first tree, keep lower branches shortened on the next one, and leave most of the lower branches untouched on the third—only shorten those that grow up into the canopy. Trees may look something like those in Figure 8-17; try other temporary branch lengths if you wish. Alternate these pruning methods down the row of sixty trees. Keep the trees pruned in this fashion for two to three years and watch what happens.

In a sense, the lower temporary branches can be used to regulate growth rate of the permanent nursery canopy. Shortening them forces more growth into the canopy. Keeping them long slows growth in the canopy. Use this relationship to produce the tree you desire.

CHECK YOUR KNOWLEDGE

1) It is appropriate to head or top large-maturing young trees in the nursery:
   a. if the trees are spaced close in the nursery.
   b. if you follow up with appropriate techniques to maintain a dominant leader.
   c. if the trees will be planted under power lines.
   d. if the trees have strong wood with a high specific gravity.

2) When subordinating a stem in the permanent nursery canopy, what type of pruning cut is most effective and appropriate?
   a. heading cut
   b. thinning cut
   c. jump cut
   d. reduction cut

3) When reducing the length of temporary branches, which cut is most efficient?
   a. heading cut
   b. thinning cut
   c. jump cut
   d. reduction cut

4) Each of the following can be used to develop an upright trunk on a young sapling tree EXCEPT:
   a. the trunk can be tied to a stake driven into the ground next to the trunk.
   b. make regular removal cuts along the trunk to push growth into the leader.
   c. cut the tree nearly to the ground and train the straightest sprout into the main trunk.
   d. head the leader where it makes a bend.

5) One danger in using stakes to hold a young tree erect for too long a period of time is:
   a. the tree might produce less of a root system.
   b. top growth can be slowed.
   c. staked trees remain shorter than trees that are not staked.
   d. lower branches lose vigor and die.

6) How do you encourage more aggressive growth in the upper canopy of a young sapling?
   a. remove or subordinate a codominant leader.
   b. remove or shorten branches on the lower portion of the tree.
   c. make as few pruning cuts as possible.
   d. tip or head as many branch ends as possible.
7) What is the BEST method of building caliper in the lower portion of the trunk on a young sapling?
   a. remove branches on the lower portion of the trunk.
   b. secure it with flexible stakes.
   c. a prune in the dormant season.
   d. space trees apart in the nursery.
8) Which type of pruning cut would you use in the nursery to create branches along a stem or branch that has few lateral branches?
   a. reduction
   b. thinning
   c. pollarding
   d. heading
9) Which is NOT used to help create a dominant leader in young nursery trees?
   a. subordination
   b. heading cuts
   c. keeping lower temporary branches
   d. root pruning
10) What would be the best method to straighten the trunk of a 2-inch caliper tree if there was a bend in it toward the top of the canopy?
   a. keeping lower temporary branches
   b. securing the trunk to a stake inserted into the ground
   c. placing a splint in the middle and top portion of the canopy
   d. making a reduction cut back to a lateral pointed upward

Answers: b, d, a, b, a, b, d, d, c or d

CHALLENGE QUESTIONS

1) Why do topped trees take longer to reach marketable size than trees that are not topped?
2) What impact would cutting lower temporary branches very short have on the permanent nursery canopy? Compare the tree's response to leaving these branches very long.

SUGGESTED EXERCISES

1) Set up a test on about nine trees of one type. Remove lower branches from the first one, keep lower branches shortened on the next one, and leave most of the lower branches untouched on the third and only shorten those that grow up into the canopy. Alternate these three pruning methods down the row of nine trees. Keep the trees pruned in this fashion for two to three years and learn what happens. Measure the caliper, height, and canopy spread on each tree at the end of each growing season. Which pruning method produces the desired tree?
2) Grow trees with splinting and staking and some without. Prune both to a dominant leader. Which ones produce a crop the soonest?
3) Space a group of liner trees pot-to-pot in a container nursery or on very close spacing in a field nursery for a year and a half to two years. Space another group about a pot's width or two apart. Upsize the trees to a larger container when appropriate and space about 4 feet apart. After a year and a half to two years, which group of plants has the best quality?
4) Divide a class into groups of five. Have each group prune a row of young trees while the instructor circulates among groups. Watch, demonstrate, and react to the pruning you see taking place.

CHAPTER 9

NURSERY SHADE TREE PRODUCTION PRUNING: DEVELOPING THE CANOPY

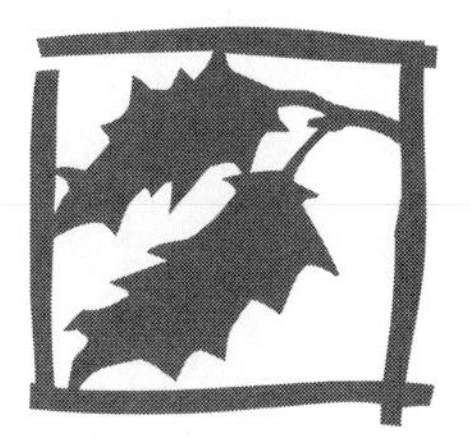

OBJECTIVES

1) Develop a uniform full canopy around a dominant trunk.
2) Create adequate space between main branches.
3) Choose the lowest branch in the canopy.
4) Select strategies for excurrent evergreens.
5) Craft a production protocol for small caliper and large shade trees.

KEY WORDS

Branch arrangement
Large caliper trees
Main branches
Matching trees
Permanent branches
Production protocol
Secondary branches
Tipping
Upright-growing tree cultivars
Vigorous branches

OBJECTIVES OF DEVELOPING A NURSERY TREE CANOPY

The objective of growing a quality shade tree with a good **branch arrangement** is to space main branches about 6 to 12 inches or more along a dominant trunk (Figure 9-1). **Main branches** are those that are the largest on the tree. They should be kept smaller than about half to two-thirds the trunk diameter. Main branches are spaced apart by reducing the growth rate on competing branches nearby. Small branches should be growing from the trunk between the larger diameter ones forming a dense permanent nursery canopy when the tree leaves the nursery. All branches are not the same length and diameter on quality shade trees. Decurrent trees with all branches the same diameter may require extra work to train into strong trees once planted in the landscape because branches may be too crowded. Excurrent trees with all branches the same diameter are normal.

Branch spacing is more crucial on large caliper nursery trees than on small ones because some of the branches on large trees may be permanent branches. **Permanent branches** are those that will remain on the tree for a number of years in the landscape. Young nursery trees with a 2- to 3-inch trunk caliper will not have any permanent branches on the trees when they leave the nursery. Since the top of the tree may be at or below the point where the first permanent branch will be located, all the branches on the tree will be removed as the tree grows (Figure 3-2). Therefore, spacing is not terribly critical on young trees.

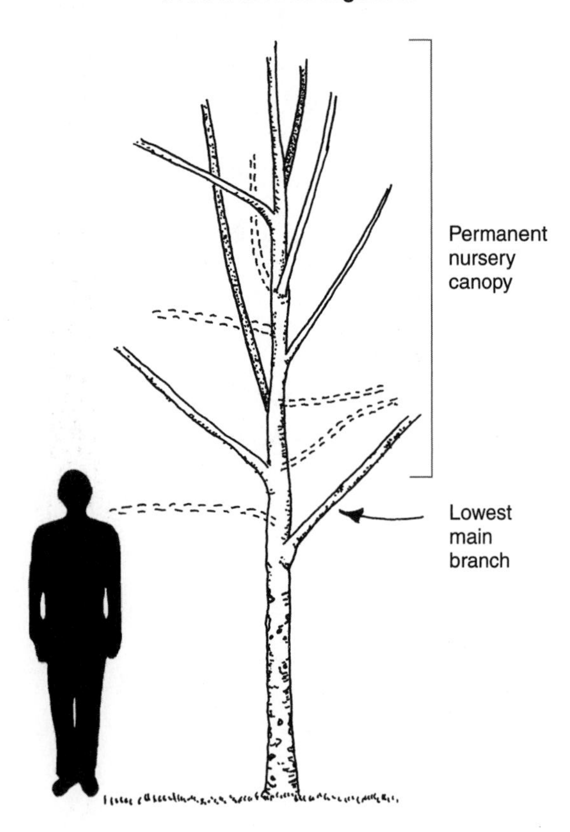

FIGURE 9-1. Main branches should be spaced apart along a dominant trunk. This is especially important on large caliper nursery trees because some branches could be on the tree for many years. Shorten branches (indicated with a dashed line) between those you want to become main branches. There is no need to remove these branches entirely back to the trunk as indicated in the drawing. Shortening branches will encourage main branches that were not shortened to grow faster. In addition to the lowest main branch indicated with the arrow, there are six other main branches in the permanent nursery canopy.

CONTROLLING VIGOROUS GROWING BRANCHES

Vigorous branches (also called aggressive branches) in the permanent nursery canopy that are growing taller than the leader can be headed (on branches 2 years old and younger) or subordinated with a reduction cut (Figure 9-2, top left). This may allow the original leader to develop a more aggressive growth rate. Another option is to remove the original leader if it is growing much slower than a vigorous branch that appears to be capable of taking over the role of the leader (Figure 9-2, bottom).

Ashes (*Fraxinus*), maples (*Acer*), dogwoods (*Cornus*), and other trees that have buds opposite each other on the twigs often develop aggressive lateral branches opposite each other on the trunk. Trees with alternate buds can develop aggressive branches nearly opposite each other also. When this happens, the growth rate of the leader often

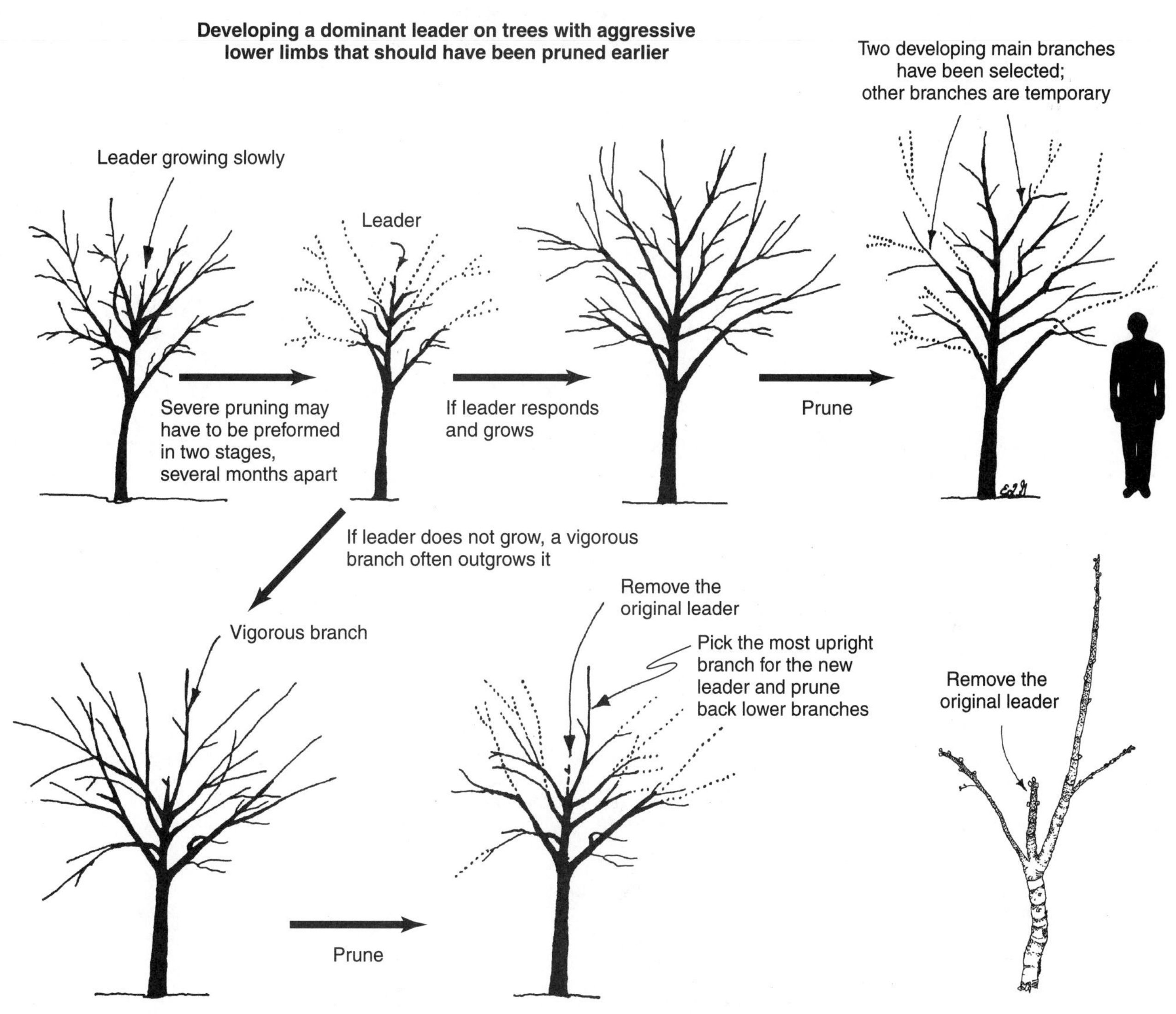

FIGURE 9-2. Trees with fast-growing lower branches often require several prunings to develop and maintain a dominant leader. Dotted lines indicate removed branches. Sometimes the original leader should be removed to allow an aggressive branch to assume the role of the leader (lower right).

slows down and the leader will be outgrown by the two lateral branches. This discourages development of the leader and can lead to formation of codominant stems and/or included bark (Figure 9-3, left). On young nursery trees, remove one of the vigorous lateral branches and reduce the length of the other with a reduction cut (or heading cut on current year growth). The other option is to reduce the length of one branch so it is well below the tip of the leader and cut the other branch even harder. Both strategies allow the original leader to develop. As an alternative, remove the original leader, select the most upright lateral branch closest to the top of the tree as the new leader, and cut back the other aggressive branch (Figure 9-2, bottom). This is especially useful when branches grow substantially taller and are larger in diameter than the original leader.

UPRIGHT TREES

Branches and stems on the 'Bradford' Callery pear and some other **upright-growing tree cultivars** form narrow angles with the trunk. The branches are clustered together (Figure 9-4, left). Included bark regularly forms in the crotches. Some horticulturists

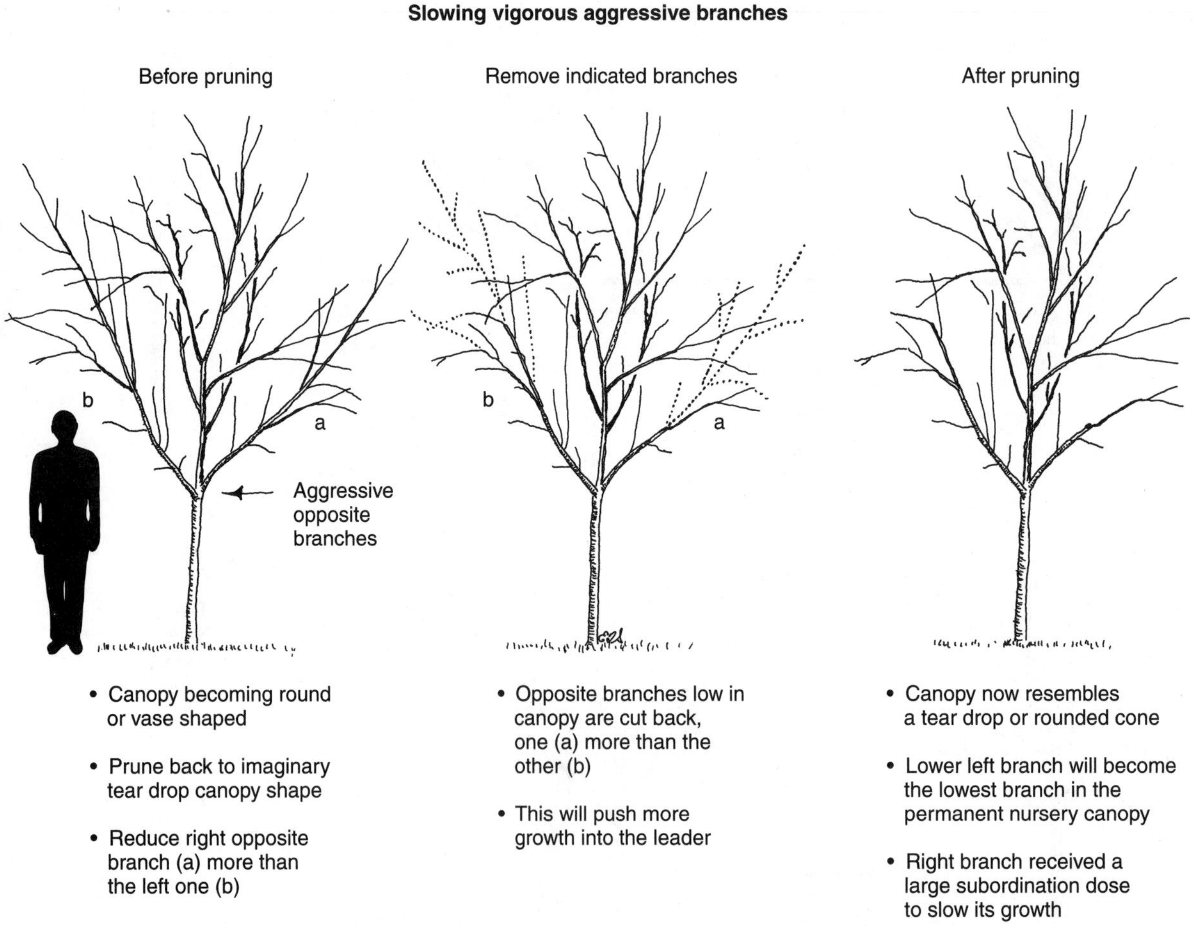

FIGURE 9-3. Shorten vigorous, aggressive branches in the permanent nursery canopy using reduction cuts. If two vigorous branches are growing opposite one another, shorten one (lower right branch 'a') more than the other (lower left branch 'b'). The one shortened the most will grow slowest. The other one will become the one main branch at this point.

recommend heading to an outside bud or cutting back to an outward-pointing twig to create wider branch angles and less aggressive limbs when the tree is a young sapling. This keeps lateral branches small in diameter and may improve structural strength on many trees, but it requires a great deal of labor on Bradford pear (Figure 9-4). If these branches or stems remain small relative to the trunk and do not form included bark, they will remain on the tree longer than if included bark forms in the crotches. More often than not, included bark develops. Most nurseries consider the 'Bradford' Callery pear too vigorous and time consuming to train properly in this manner. In the case of the Callery pear, it would probably be more efficient to grow a cultivar, such as 'Aristocrat', 'Capital', 'Chanticleer', or 'Redspire', that has the form you desired than to prune 'Bradford' into a form that it does not develop naturally. Unfortunately, 'Aristocrat' and 'Capital' are very susceptible to fire blight disease.

To increase the strength and improve structure of young trees, space branches along the trunk by reducing or removing the largest upright branches with the tightest angles or those with the worst included bark (Figure 9-5). Prevent remaining branches from getting larger than half the trunk diameter with regular reduction cuts. This reduces branch crowding on the trunk, allowing the remaining branches to better secure themselves to the tree. The canopy will be thinner, and the smooth canopy outline will be slightly spoiled for a while because a number of branches will be removed back to the trunk. However, the canopy will thicken back to its original density and structural strength should be greatly improved. If the tree is pruned regularly in this manner beginning when it is very young, small amounts of foliage will be removed at each pruning and the effects on canopy density and shape will be hardly noticeable.

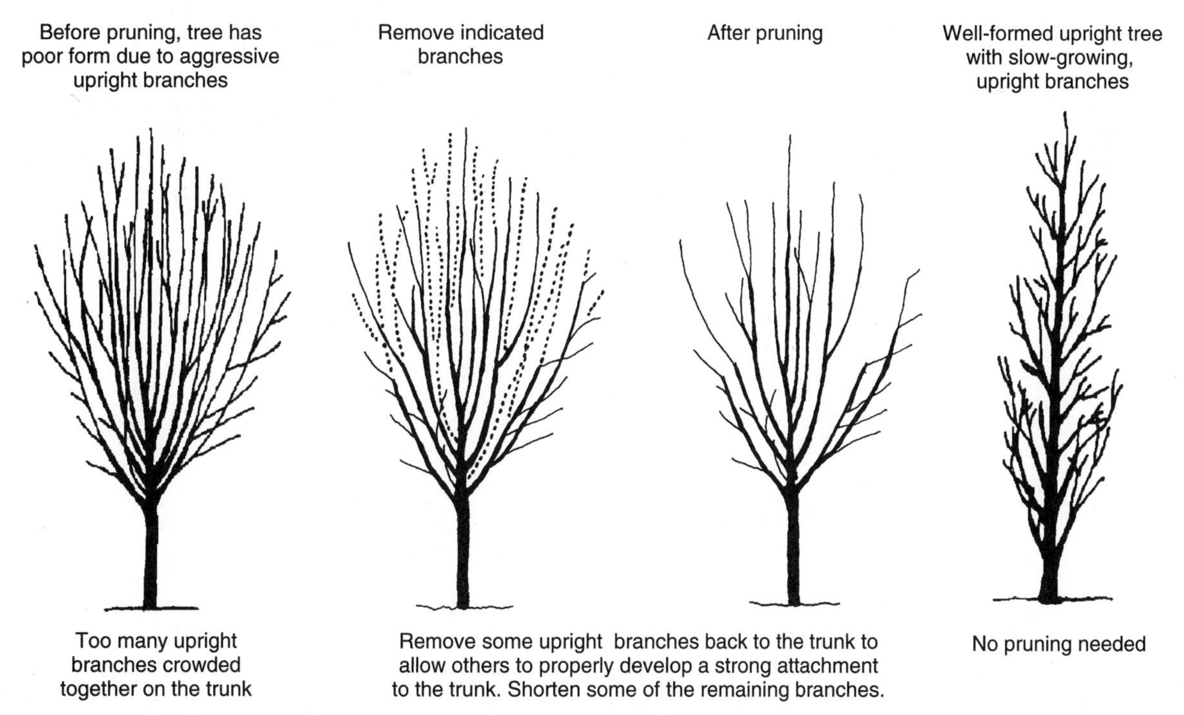

FIGURE 9-4. You can reduce or remove some of the aggressive branches on upright trees (left and center). Remove some all the way back to the trunk. Shorten remaining limbs that are large in diameter to keep them smaller than half the trunk diameter. This will help reduce damage from ice and snow loads. Other upright trees have well formed branches and require little pruning because they are less aggressive (right).

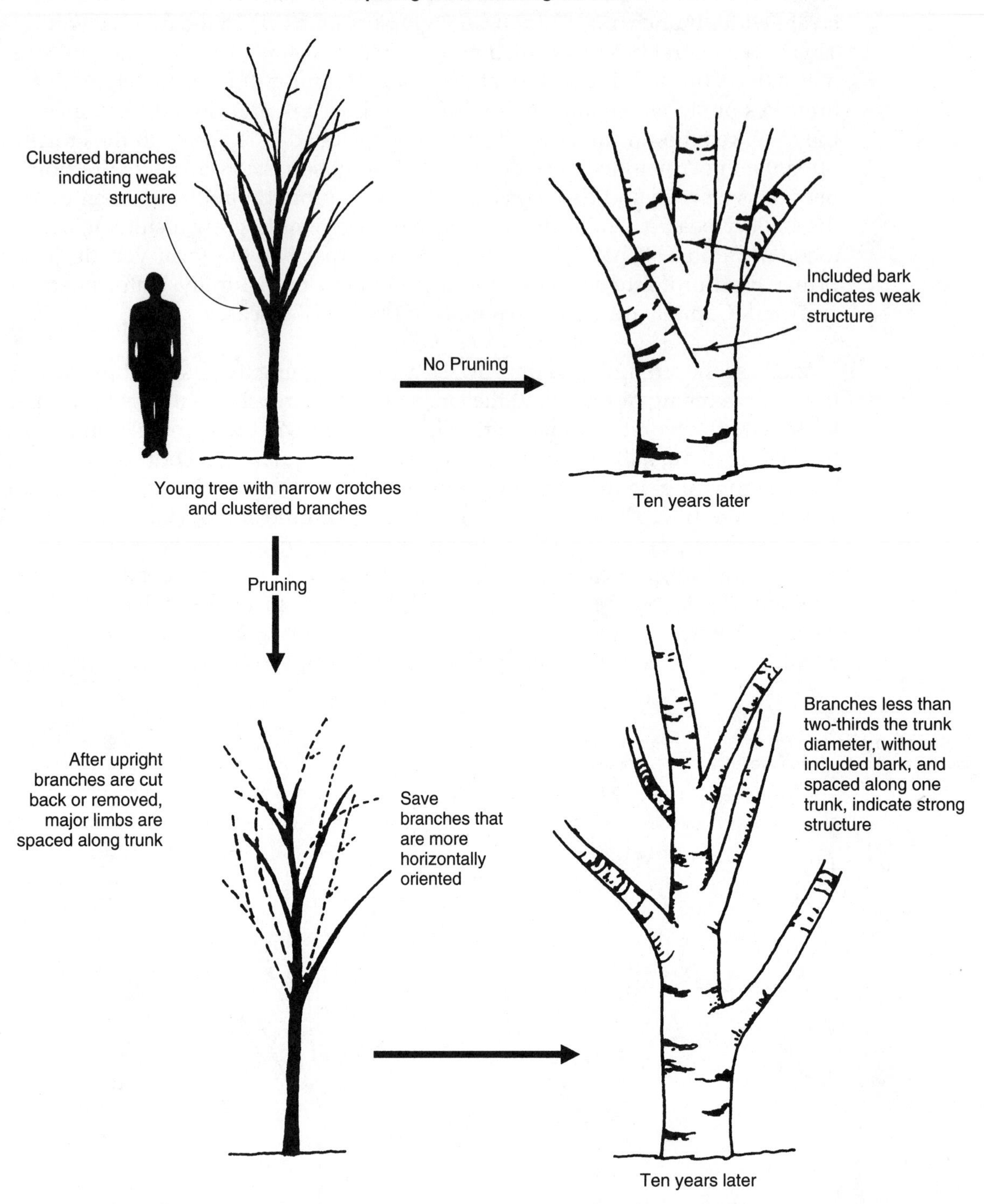

FIGURE 9-5. Rapidly growing limbs clustered together on the trunk often form included bark and are poorly attached to the tree (top). The crowded limbs develop few side branches and so taper poorly, which makes them weak. You should prune the young trees to remove or shorten (removal is shown) some branches so that the remaining limbs are spaced apart (bottom). Remaining limbs on nursery trees should be at least 6 inches apart, preferably 12. Shortening branches instead of removing them results in a denser canopy and faster tree growth.

TABLE 9-1. Upright, non-coniferous trees with small branches.

Acer x *freemanii* 'Armstrong'	*Acer saccharum* 'Temple's Upright'	*Ilex* x *attenuata* 'Foster #4'
Acer x *freemanii* Scarlet Sentenial™	*Agathis robusta*	*Ilex* x *attenuata* 'Fosteri #2'
Acer platanoides 'Columnare'	*Alnus glutinosa* 'Pyramidalis'	*Liquidambar styraciflua* 'Festival'
Acer platanoides 'Erectum'	*Araucaria* sp	*Magnolia grandiflora*
Acer platanoides 'Olmsted'	*Betula* sp	*Podocarpus nagi*
Acer rubrum 'Bowhall'	*Carpinus betulus* 'Fastigiata'	*Pyrus calleryana* 'Stonehill'
Acer rubrum 'Columnare'	*Franklinia lasianthus*	*Pyrus calleryana* 'Whitehouse'
Acer rubrum 'Doric'	*Ginkgo biloba* 'Fastigiata'	*Quercus petraea* 'Columna'
Acer saccharum 'Arrowhead'	*Ginkgo biloba* 'Magyar'	*Quercus robur* 'Fastigiata'
Acer saccharum 'Coleman'	*Ginkgo biloba* Princeton Sentry™	*Quercus robur* Skymaster™
Acer saccharum 'Endowment'	*Ilex* x 'Nellie R. Stevens'	*Ulmus* x 'Regal'
Acer saccharum 'Goldspire'	*Ilex* x *attenuata* 'East Palatka'	*Ulmus* x 'Urban'
Acer saccharum 'Skybound'		

Not all upright trees are poorly formed with included bark in branch unions. The lateral branches on some, such as 'Fastigiata' English oak (*Ouercus robur* 'Fastigiata'), 'Fastigiata' European hornbeam (*Carpinus betulus* 'Fasigiata'), and others stay small relative to the trunk (Table 9-1). Small upright branches with narrow crotch angles are relatively well secured to the trunk (Figure 9-4, right).

CLUSTERED STEMS OR BRANCHES

The form on trees with a cluster of deformed young (1-year-old) shoots at the tip of the leader can be corrected one of three ways. The first removes the cluster of shoots at the top of the tree just before new growth emerges in the spring (Figure 9-6). After cutting, remove the top bud and apply a piece of wide masking tape loosely around the second bud to just beyond the top of the cut shoot. This will provide a straight channel for the emerging shoot, causing it to grow straight. You might have to apply tape after the tree flushes so you can choose an appropriate sprout to tape as the leader. Remove any other emerging shoots that grow upright. This technique is especially useful on young trees when the cluster of branches or stems is closer to the ground. The second method removes all but two of the shoots in the cluster. Train one into the new leader and the other one into a branch by heading it back by about ½. The third technique removes the cluster back to an existing lateral branch that could become the leader. Staking or splinting may be required to straighten out the crook in the stem at this point. This really amounts to a reduction cut and is most suitable on stems less than 2 years old. Terminal bud cluster pruning can also be used to prevent formation of clustered, multiple leaders (see Chapter 5).

CHOOSING THE LOWEST BRANCH IN THE CANOPY

The lowest branch in the permanent nursery canopy can often be identified the second (warm climates) or third (coldest climates) year after planting from liners in the nursery. It is important to identify this branch as soon as possible. The lowest permanent branch is located just above the highest temporary branch. Unlike the lower temporary

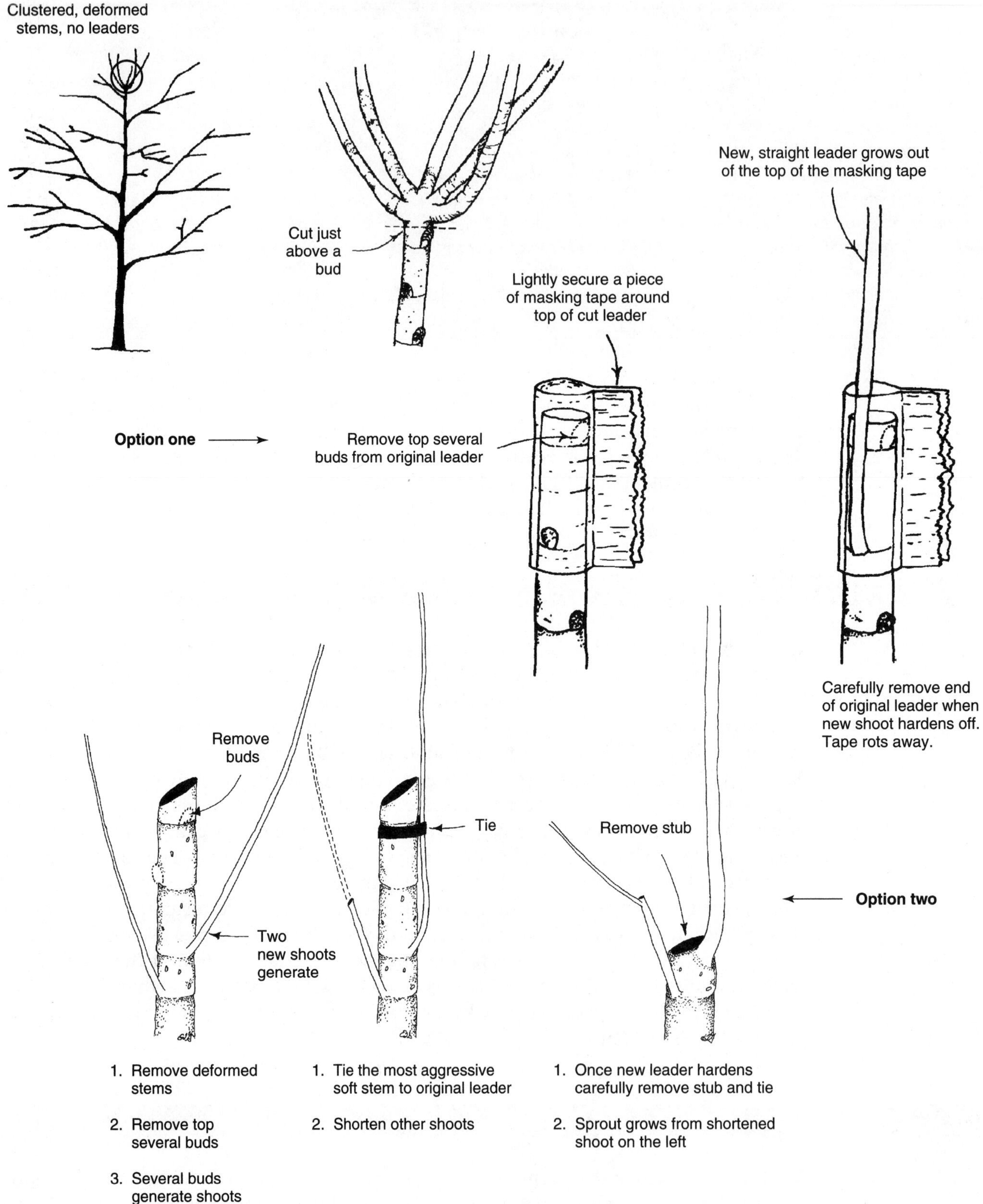

FIGURE 9-6. A tuft (cluster) of stems may form at the top of the leader. The leader can be reestablished using these simple techniques, which also work quite well on young trees with a cluster of trunks originating close to the ground. Option one uses tape or another device to deflect a new shoot to the upright position. Option two allows the new shoots to grow, then one is secured to a short stub made from the original leader.

branches, the permanent branches will be allowed to grow to a wider spread. The lowest branch in the permanent nursery canopy is usually the longest. Do not cut these back like the temporary branches.

Due to the importance of selecting these branches, some growers send the most highly trained crews into the field to identify the lowest permanent branch. Several branches just below the lowest permanent branch can be removed back to the trunk with removal cuts creating an 8-inch-long section of clear trunk (Figure 9-7). This makes it easy for other members of the crew to identify the lowest permanent branch later. Less skilled crew members can then follow to manage lower temporary branches, if necessary. They can be instructed to shorten only those branches below the clear section on the trunk. By the end of the year in which the first permanent branch was identified, it will be easy to see where the permanent canopy begins because the canopy will typically be widest at this point (Figure 9-7). Some species or cultivars may have temporary branches longer than branches in the permanent nursery canopy at the end of the first year or in the second year.

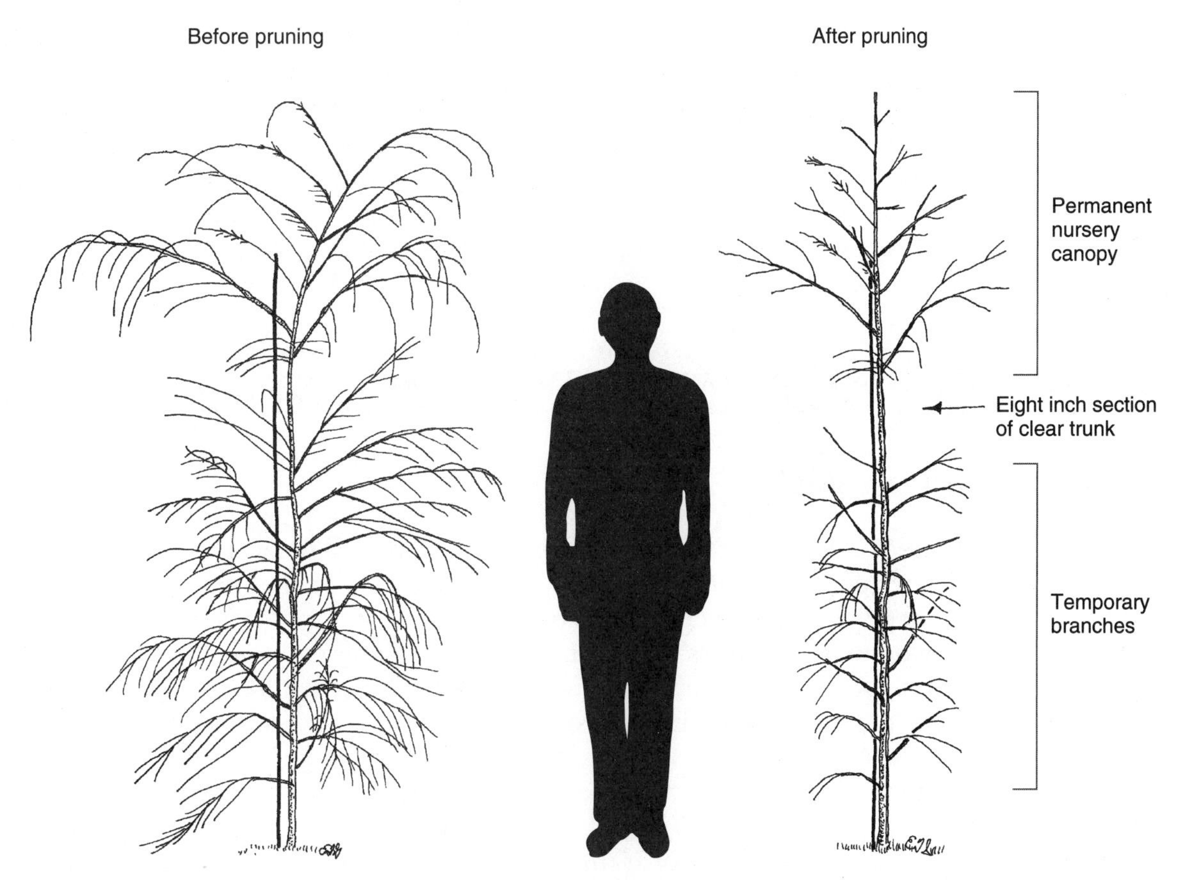

FIGURE 9-7. Clear a short section of the trunk to delineate the bottom of the permanent nursery canopy. Lower branches remain to build caliper and prevent over-extension of the leader. Only 2 or 3 branches may have to be removed from the trunk to form a clear zone immediately below the lowest branch in the permanent nursery canopy. Be sure to shorten any temporary branches growing up into the permanent nursery canopy.

The amount of clear trunk, and hence the position of the lowest permanent branch is dictated by the tree type, size, and the market. Typically, 2- to 3-inch caliper trees have a clear trunk of about 4.5 to 5 feet, whereas larger trees have a clear trunk of 6 to 7 feet. Exceptionally large nursery trees may have more clear trunk.

SPACING BRANCHES ALONG THE TRUNK

The largest branches on most shade trees should be spaced apart along the trunk, not crowded and touching (Figure 9-5). If left too close together, leader development on some trees could be restricted. If too crowded, branches on certain trees do not develop enough strength to hold themselves erect so they droop or sag. Canopy growth rate can sometimes be sparked by removing a number of branches back to the trunk on very slow-growing trees with crowded branches. This can cause the remaining branches and leader to grow faster and can bring a runt back into the ball game. Smaller branches can (and should) be left between the larger ones. Branch spacing is usually not an issue on baldcypress (*Taxodium*), pines (*Pinus*), and other excurrent trees.

Tip: *Space main branches along the trunk by shortening nearby branches.*

The best method of spacing main branches along the trunk is to shorten other branches nearby. Your goal should be to finish the crop with the largest branches at least 6 to 12 inches apart. Keep these smaller than about half the trunk diameter measured directly above the branch union. This is accomplished by removing the ends of the main branches with a reduction cut or by pyramidal shearing, cutting through only current year's shoots. Keeping main branches less than half the trunk diameter encourages the leader to dominate the structure. This creates a strong structure and a high-quality nursery tree.

CREATING A UNIFORM CROP

Cultivar selection

Many tree buyers want all trees in a group to look identical. These identical trees are often referred to as **matching trees.** They have the same amount of clear trunk, the same shape, the same height, the same foliage color, and so on. One of the best ways to develop matching trees is to plant a cultivar. Many regions of the world plant only cultivars. Other regions are just catching on to the advantages of cultivars so most trees are planted as seedlings. Besides the advantage of easily producing matching, nearly identical trees, cultivars are much easier to manage in the nursery because they are more predictable. Whereas there may be five or more different canopy forms and types in a field of seedlings, a field of cultivars has just one form. Crews can be trained to prune just one form instead of several. This is likely to result in increased crew efficiency.

Topping

Topping in the nursery should cut through wood no more than about one year old. Some nursery operators routinely top trees to create predictable forms, sizes, and so on. This practice appears efficient because all crews do precisely the same thing to each tree. It is easy to train crews to perform this task. But the structure is usually unacceptable on shade trees that are routinely topped through wood older than one year. Growth rates are much slower resulting in an increase in the time required to grow a crop for market. I believe that if these operations would crunch the numbers, they would find that the increased time required to grow trees that are routinely topped through older wood results in a very expensive, inefficient protocol. Cutting the leader through 1-year-old wood to help develop good structure is appropriate and necessary on a number of trees.

Shearing and rounding over

Shearing makes heading cuts through wood less than about one year old. Once a strong structure has been developed with main branches spaced along a dominant trunk, the sides of the canopy can be sheared to create uniform, dense canopies. This often occurs in years three or four, depending on the growth rate and climate. Some growers shear one or more times each year as a method of shortening aggressive branches and creating uniform canopies. If you do this, do not forget to prune for a dominant leader and space branches apart. Many nurseries that shear regularly create trees with too many branches close together. Close spacing is usually not a problem on hollies and other excurrent trees.

Christmas trees are often pruned in this fashion. Because interior branches sometimes die from shading, foliage is essentially redistributed from the inside of the canopy to the outer edge. This creates a tree with a smooth silhouette that appears denser. Growers are able to produce identical looking canopy forms using this technique. Trees pruned excessively in this manner can appear odd when they grow out into the landscape unless shearing is continued (Figure 12-5e and f). They can look like a hedge that needs to be clipped. Left unclipped and with some training to create good branch structure, they regain their more natural appearance in several years.

INCREASING CANOPY SPREAD

The need for this type of pruning can be minimized by selecting a tree species or cultivar that naturally has the desired spread and shape. Removing selected branches can have some impact on tree form (Figure 9-8). The crown on young trees can be widened by pruning vertically oriented branches with reduction cuts and leaving the more horizontal ones. Removing horizontal branches creates a more vertical crown. Trees pruned to a more vertical form often develop vigorous lateral branches due to the vertical scaffold branch orientation. Be sure these branches do not outgrow the leader or develop included bark. If they do, make reduction cuts to slow their growth.

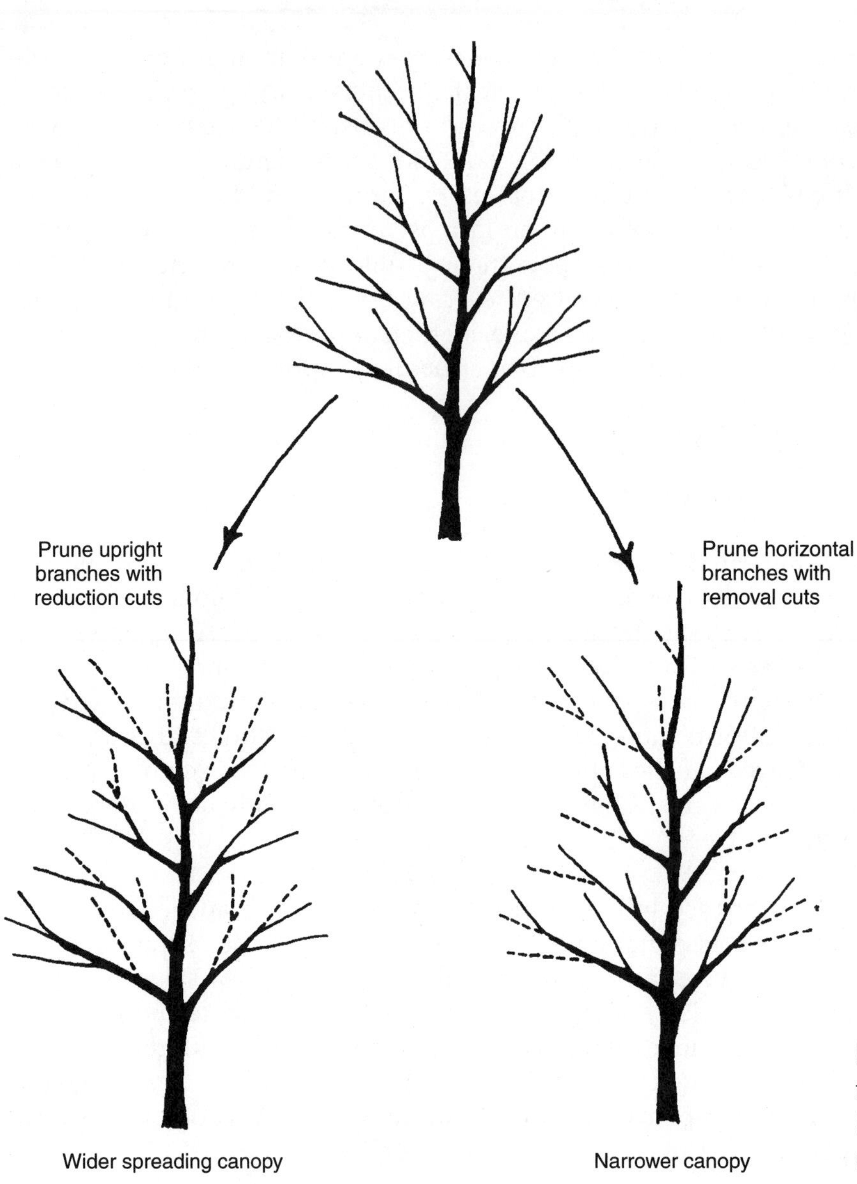

FIGURE 9-8. Prune upright branches to create a more spreading canopy. Prune horizontal branches to form an upright canopy.

DEVELOPING LARGE CALIPER TREES

Some branches in the upper half of the canopy of **large caliper trees** (more than about 5 inches caliper) will remain on the tree for a long period of time, perhaps decades. For this reason, it is more important to appropriately space apart the largest diameter main branches in the middle and upper portion of the tree. Be sure none have included bark in the branch union. Spacing is not as much a concern for branches in the lower portion of the canopy because these will eventually be removed. However, trees look best if branches have more or less the same spacing throughout the canopy. Spacing branches apart keeps the interior **secondary branches** alive which helps prevent the main branches from overextending and drooping. In addition close spacing on low branches could result in reduced leader vigor, especially if these low branches get large in diameter.

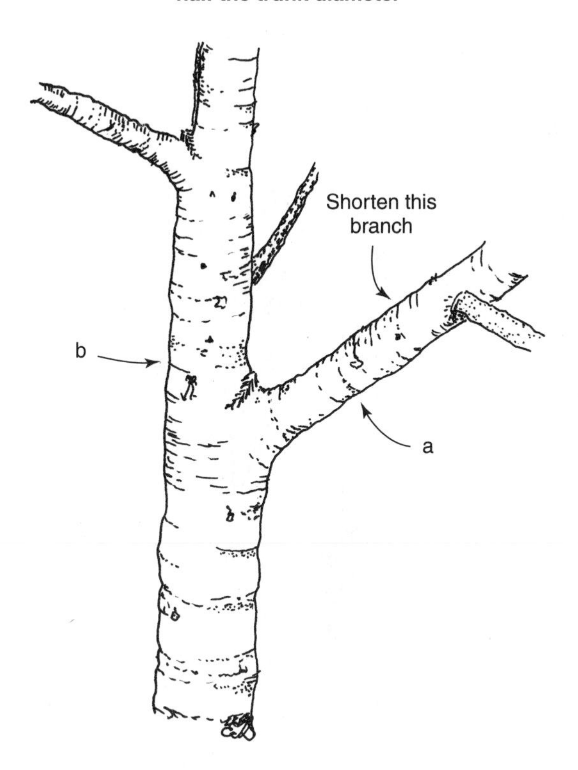

FIGURE 9-9. Keep branches less than about half the trunk diameter by shortening them with reduction cuts or heading cuts. Keep the diameter of 'a' less than half the diameter of 'b'.

It would be best to space main branches at least 12 to 18 inches apart on large caliper trees (Figure 9-1). Greater space between branches is fine and encouraged on some species and cultivars. Shorten branches that are between those you want as main branches. Keep all branches less than half the trunk diameter by shortening them (Figure 9-9). Branches that are small compared to the trunk help encourage leader development, are better attached to the tree, and more likely to develop a branch protection zone. Specialized equipment such as pruning platforms, carts, and extension pruners may aid in reaching higher portions of the canopy (see Chapter 6).

Subordinate (using a reduction cut) branches that grow from the bottom half of the tree that have reached into the top third of the canopy (Figure 9-10). This will help ensure that the leader remains dominant and that branches on the upper part of the leader have access to enough sunlight from the side to develop appropriate spread. Adequate spread on these upper branches is very important because some of these are likely to be on the tree for a long time. If they are not allowed to grow laterally and are forced to grow upright due to shading from upright lower branches the resulting upright orientation could cause poor tree structure. Subordination of these low branches also helps ensure that they will be removed from the tree as the tree grows larger. They are too low on the tree to be part of the permanent landscape canopy. Their more horizontal growth as a result of subordination keeps them from becoming a permanent part of the canopy. If trees are on a regular annual pruning program, each time you subordinate, you will make only small voids in the canopy.

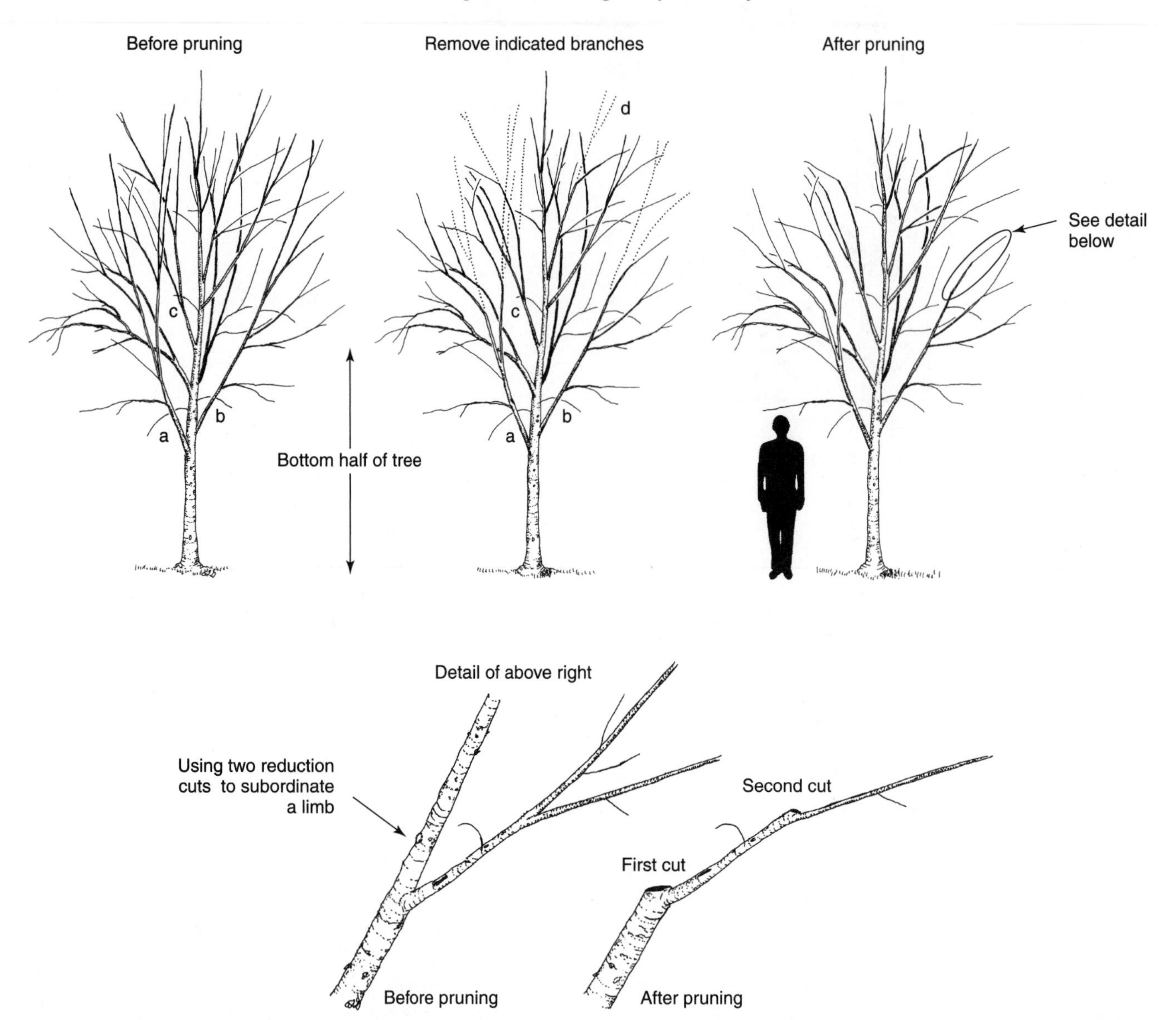

FIGURE 9-10. Shorten limbs a, b, c, and d because they are growing into the upper third of the canopy and they originate from the bottom half of the tree (left). Most, if not all, of these shortened limbs will eventually be removed from the tree as it grows in the landscape. They were shortened using reduction cuts (center). If the lateral branch left after making a reduction cut is too long, also shorten the lateral branch with a reduction cut (bottom detail). This essentially results in making two reduction cuts to accomplish subordination of a limb. Branch d was shortened because it was forming a codominant stem in the upper canopy (center).

Remember to splint the leader if it begins to grow crooked or bend over (see Chapter 8) if you want the trunk straight. Make as few heading cuts as possible in the permanent nursery canopy of large caliper trees. Heading cuts make the tree look pruned and could present an unprofessional image. Use reduction cuts instead where possible.

Tip: *Subordinate branches or stems that grow from the bottom half of the tree that have reached into the top third of the canopy.*

It would be best to identify sometime before the third year after planting from liners whether the trees will be grown to the 2- to 3-inch caliper size, or to larger trees. If not, you could waste energy and lengthen the time to produce the trees. Let's say, for example, in the third or fourth year after planting you take trees originally destined for the 2- to 3-inch market and decide to grow them on to the 6-inch caliper size. The lowest permanent branch for the 2- to 3-inch tree was identified more than a year ago and it has been allowed to develop a nice spread and a large diameter (half the trunk diameter). It is probably about 5 feet off the ground. The tree is beginning to commit significant resources to its development. Now it must be removed because it may be too low on the trunk for the 6-inch crop. A large trunk wound is created and you wasted time forcing the tree to grow this branch large. You would have been better off forcing growth into branches higher in the tree earlier by reducing the length of this branch and keeping it small from the outset.

Place large caliper trees on a regular root-pruning program. This prevents roots in the top portion of the soil from growing to a large size. In certain instances, roots unable to be reached with digging spades or balling spades can be undercut with a blade. Root-pruning fabrics placed under the liner at planting can also help restrict downward root growth. Some root-pruned trees will grow slightly slower than trees not root-pruned; others will not be impacted provided irrigation and fertilizer are managed appropriately. Survival can be improved and you might be able to dig root-pruned trees successfully in summer and in other nontraditional times.

DEVELOPING STREET TREES

Nursery managers in some European countries use different methods of training for specimen and street tree planting. There might be certain nurseries in the United States doing this as well (Figure 9-11) but it is not common. It should be considered. Trees are grown as columns when young to keep low branches in check. As the tree reaches a height of about 15 to 18 feet, one or two of the lowest permanent branches can be selected about 12 feet off the ground. *Be sure to leave smaller branches below the permanent canopy* until the tree is of adequate size to not need these lower branches. In addition to the previously mentioned benefits, these branches shade the trunk. This is especially important on thin-barked trees such as lindens (*Tilia*), some maples (*Acer*), and golden chaintree (*laburnum*) that could be subject to injury from direct exposure to the sun. These low branches also provide some protection from mechanical injury.

EXCURRENT EVERGREENS

Many pines, spruces, firs, podocarpus, large-growing magnolias, cedars, cypress, some hollies and other evergreens develop a central, dominant leader without pruning. Little pruning is needed to develop strong structure. About the only leader pruning needed is when two or more sprouts develop to take the place of a damaged leader. One may have to be splinted up into the vertical position so it can become the leader (Figure 9-12). If a double leader is present, shorten or remove one of the stems entirely as soon as it is recognized. This will allow the remaining leader to produce branches on the side of the leader formerly shaded by the removed stem. Some of these plants are

Training trees for specimen and street tree use

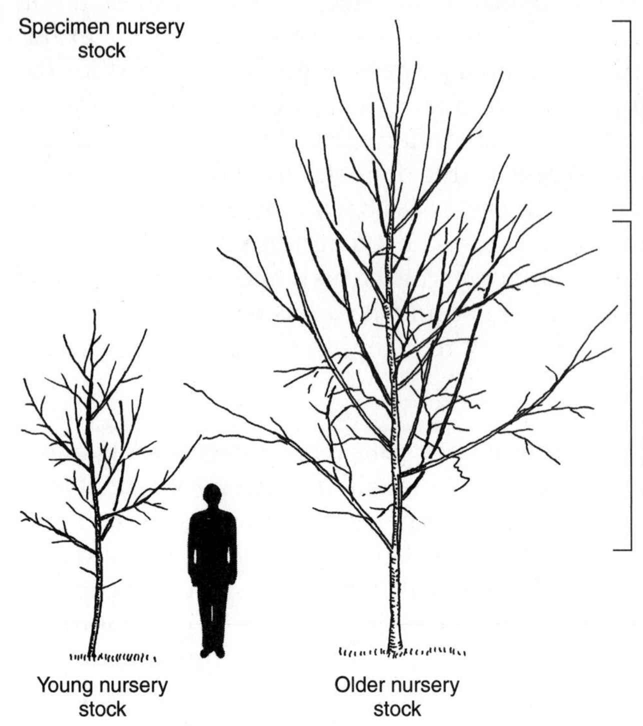

Permanent canopy

- Poor candidate for street tree planting because low temporary branches are too aggressive and large
- Suitable candidate for lawn and park planting

All these branches eventually come off the tree in most landscapes

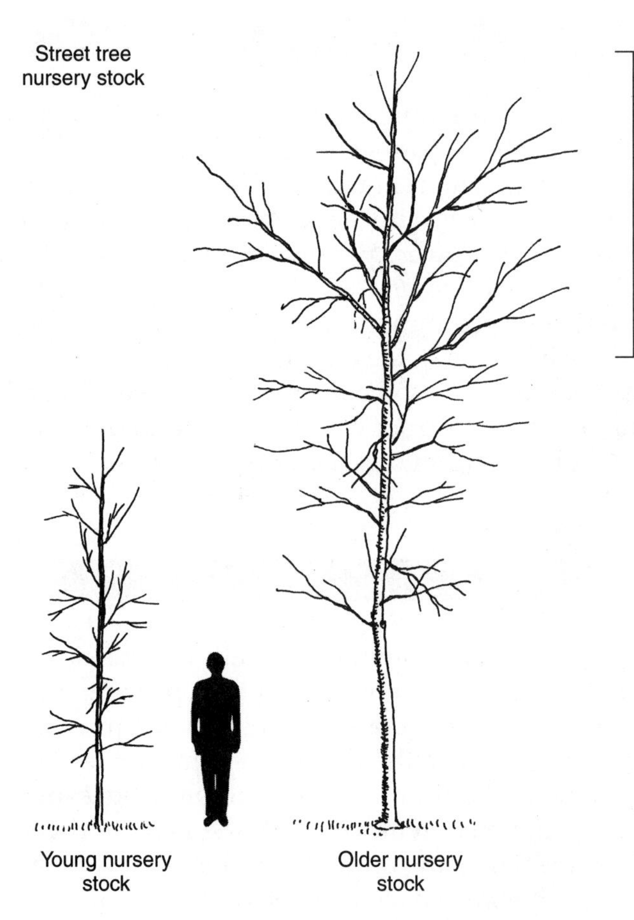

Permanent canopy

- Good candidate for street tree planting because low temporary branches are small
- Suitable candidate for lawn and park planting

FIGURE 9-11. Growing trees for street tree planting that have only small diameter branches up to about 15 feet makes it easy to train branches to grow well over the traffic (bottom). The first big branch is 15 feet from the ground. Trees planted in open landscapes such as in a lawn area can have the lowest scaffold branch closer to the ground (top). Trees trained in the nursery as specimen trees (top) that are planted along streets require more work to prune because large low branches are in the way.

Splinting a damaged leader

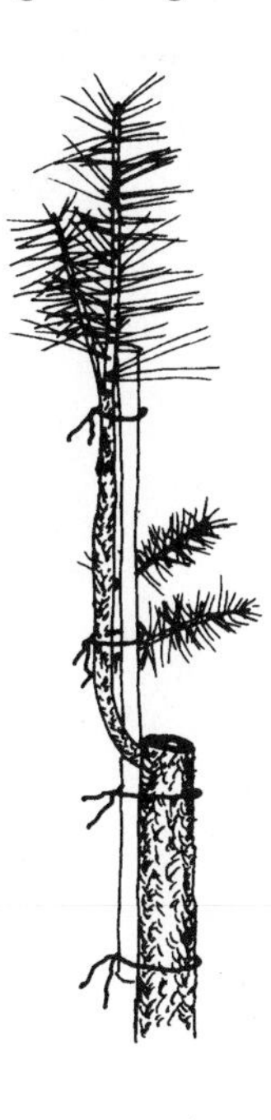

FIGURE 9-12. Fixing a damaged or missing leader on a conifer is easy. Simply attach the upper-most branch to a short splint which will help direct this limb to become the new central leader.

sheared regularly to develop a tight, dense canopy. This is conducted just prior to bud break and at other times during the growing season. If the leader is cut, be sure to return to train the plant to one straight main trunk if needed.

PRUNING FOR SALE IN THE NURSERY

In shaping the canopy in the weeks before trees leave the nursery, only remove small twigs or make small heading cuts (not on the leader) to help in wrapping or tying the plants. This is not the time to remove large limbs or perform major pruning to improve structure. Fine pruning of small dead and live twigs is best conducted at this time to present a finished, clean-looking, professionally grown tree. Subordinate branches and stems that might appear to be competing with the leader, if necessary.

DEVELOPING A PRODUCTION PROTOCOL

Putting together all that was presented in Chapters 8 and 9 into a production protocol may appear to be a daunting task. A **production protocol** is a written plan detailing what is to be performed and when. With some practice, you can become proficient and efficient at pruning shade trees using a protocol. Once you have taught yourself how to prune, then it is time to teach members of your crew. It might take you several years to hone these skills to a point where you feel you can teach them to others. Read, speak to your colleagues, attend continuing education classes, try new methods, and practice.

Making a written record of your protocol and procedures is likely to help develop efficiency in your operation. This can also help ensure that you produce a consistent, high-quality product each year. Specialized consultants can perhaps help develop these, but

TABLE 9-2. Items to include in a production protocol.

- Detailed drawings of what needs to be done each year
- The number of structural canopy prunings required each year; the number of temporary branch prunings
- Time of year competing leaders should be sudordinated
- How long temporary branches should be allowed to grow before reducing their length
- What length temporary branches should be cut back to
- Whether only upright temporary branches should be reduced in length, or shorten all branches
- When all temporary branches should be removed
- If stakes are needed; how long they should be left on the trunk; the diameter and length they should be; if splints are required
- The shape the permanent canopy should be pruned to: tear drop, oval, cone
- How far from the ground the lowest permanent branch is in the canopy
- How far apart to space main branches

be sure they have the drawing skills and the knowledge of pruning to communicate what needs to be shown. Along with detailed notes on when certain types of pruning are best performed for your crops and climate, a written protocol can act as a safeguard against personnel changes in the organization. It would be a shame if the one person in the operation who had the knowledge of the protocols suddenly left without leaving anything in writing.

A protocol should have notes on many aspects of pruning (Table 9-2). It should include drawings of what to do each year. Most people cannot learn how to prune trees from reading written words. In addition to hands-on training, they need at least a drawing, and perhaps photographs. Be sure the photographs silhouette the tree canopy so it is easy to see what was done. Notes should be made on what dose to apply; in other words, how much should be removed when performing each type of pruning. Dose is the most difficult aspect of tree care to learn and teach, yet it is one of the most important.

You might find it helpful to begin with a protocol such as the one in Figure 9-13 for a spreading tree. This protocol brings you through the first two years of production. In year one, trunks on liners can be staked (ties are not shown in drawing) when planted or sometime in the first year to make them grow straight. *Prune as little from the tree as is needed to meet objectives for the most rapid growth.* Do not shorten or remove *any* branches or stems early in the first year unless they compete with the leader—some can be shortened later in the dormant season, if needed. The tree essentially appears as a column or tall cone. All branches on the tree at this time will be removed before the tree leaves the nursery.

In year two, the trunk usually grows taller than the stake and could begin to weep or bend. The canopy might spread out on long poorly branched limbs. The lower portion of the permanent nursery canopy begins to appear in the second year. Slide the stake out of the ground and up the trunk at the appropriate time to act as a splint if needed. Be sure to pull the stake from the ground before roots bind it tightly. Begin to shape the permanent nursery canopy into a teardrop or rounded cone using reduction and heading cuts. Leave the lowest branches in the permanent nursery canopy the longest. Maintain low temporary branches to increase trunk caliper and to moderate canopy growth rate. Cut back any that grow up into the permanent nursery canopy. Some of

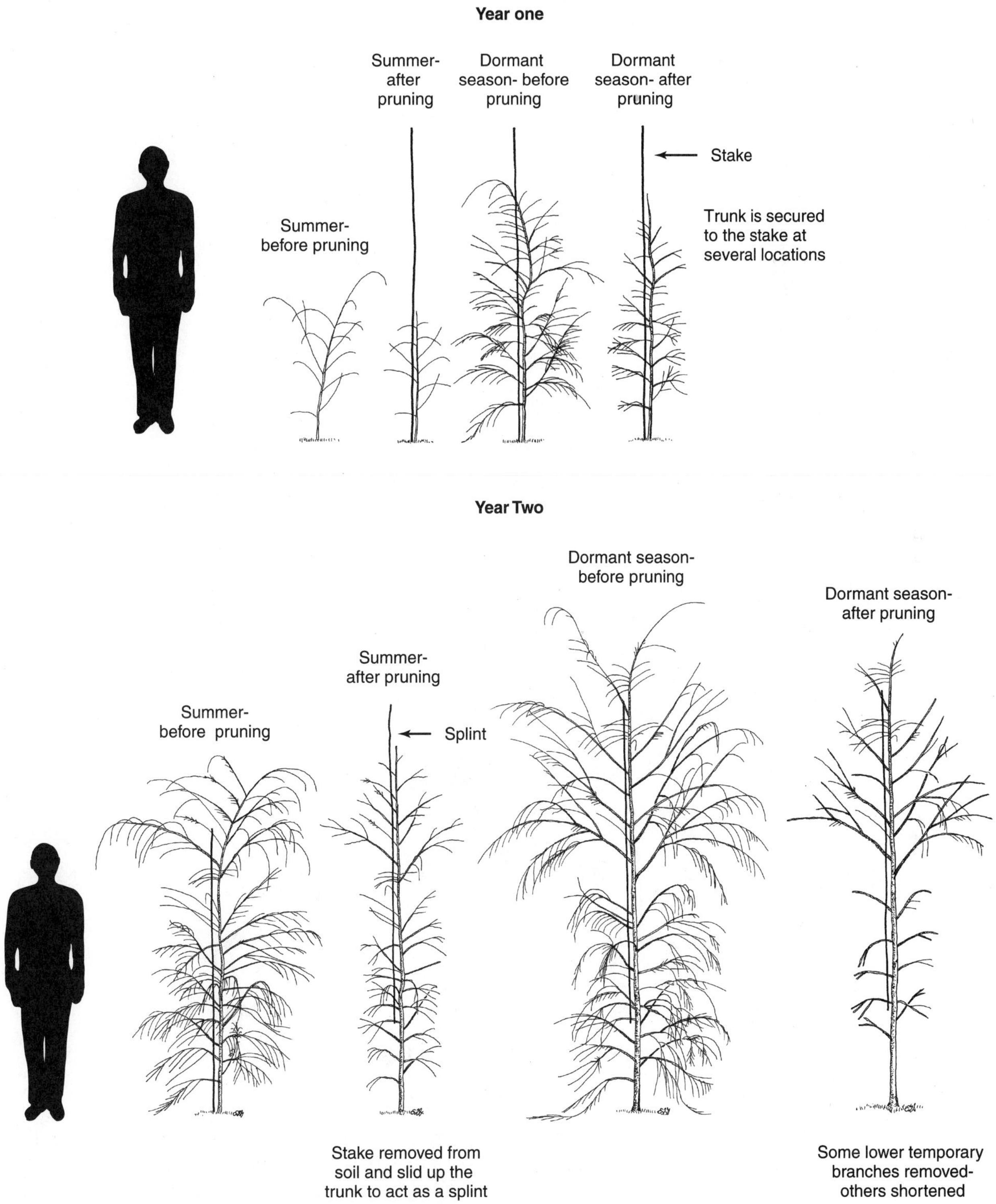

Year one

Summer-
before pruning

Summer-
after
pruning

Dormant
season- before
pruning

Dormant
season- after
pruning

Stake

Trunk is secured
to the stake at
several locations

Year Two

Summer-
before pruning

Summer-
after pruning

Splint

Stake removed from
soil and slid up the
trunk to act as a splint

Dormant season-
before pruning

Dormant season-
after pruning

Some lower temporary
branches removed-
others shortened

FIGURE 9-13. Production protocol for producing a spreading shade tree.

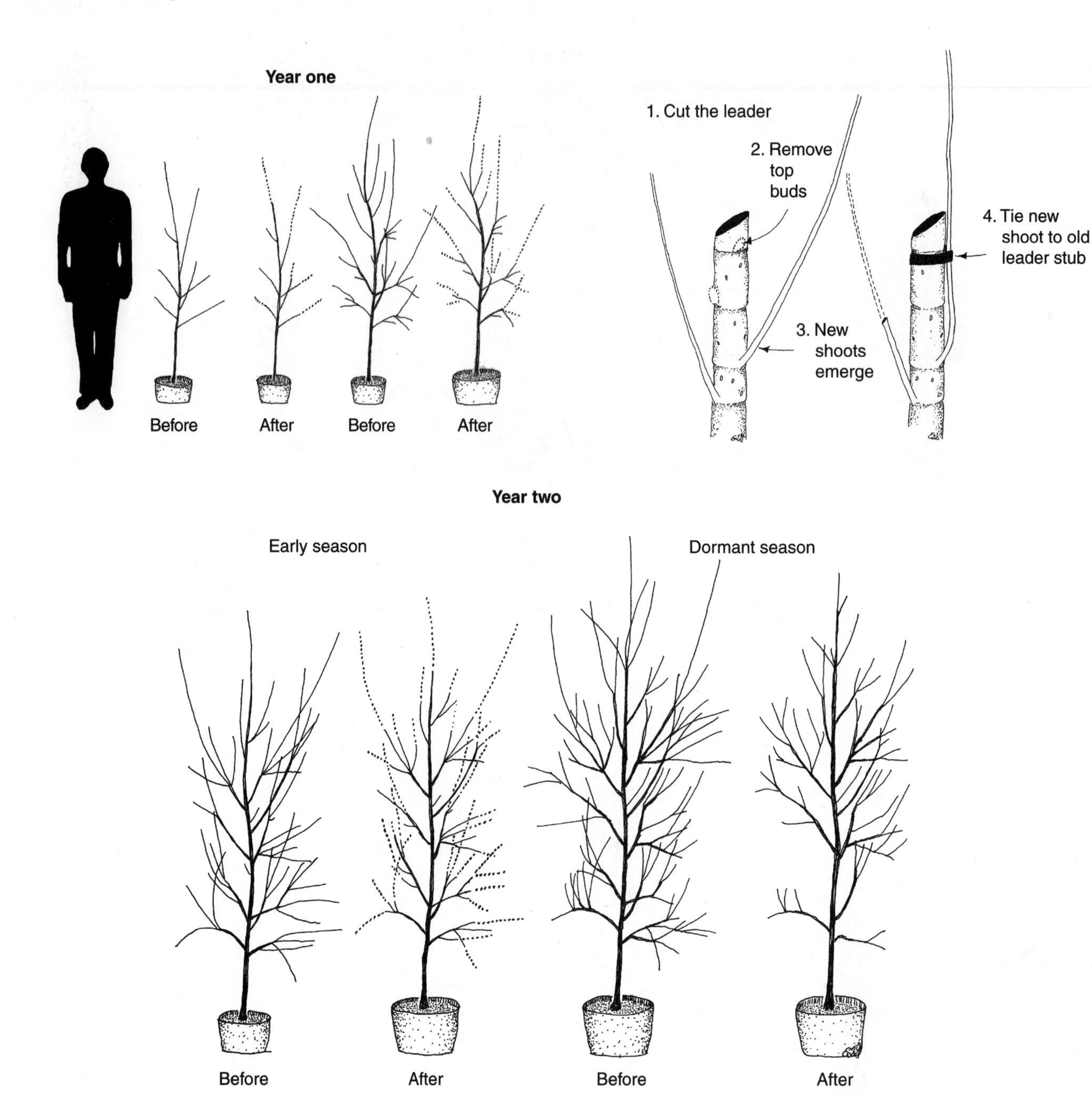

FIGURE 9-14. Production protocol for producing trees using heading cuts.

the largest temporary branches may be removed at the end of the second year if the tree will be dug and marketed at the end of year three. Appendix 7 shows protocol for producing a three-year and four-year crop on a more upright tree form.

Figure 9-14 shows how to use heading cuts to create a compact tree with a full canopy. Without pruning, the tree would stretch and produce several leaders or one very tall poorly branched leader and poorly branched laterals. Be sure to follow heading of the leader with a program to generate just one dominant leader if needed (upper right). Many genera respond to this technique including *Acer* (maples), *Gleditsia* (honeylocust), *Ilex* (holly), *Magnolia* (magnolia), *Sophora* (scholar tree), *Tilia* (linden), *Ulmus* (elms), and many others.

CHECK YOUR KNOWLEDGE

1) One way to teach shade tree production pruning is to train trees so they resemble a:

   a. teardrop on a cone.
   b. teardrop on a stick.
   c. cone of a cylinder.
   d. round ball on a stick.

2) How would you describe good branch arrangement on a quality nursery shade tree?

   a. several branches growing from the same position on the trunk
   b. all branches about the same diameter
   c. branches arranged in layers or nodes
   d. large branches spaced along a dominant trunk

3) If there are two large, equal sized branches growing opposite one another in the permanent nursery canopy:

   a. shorten one more than the other.
   b. encourage them to grow by removing several lateral branches.
   c. remove one so the other becomes the main branch at that point.
   d. shorten both of them so neither dominates.

4) On young trees that have many upright aggressive branches such as 'Bradford' Callery pear:

   a. shorten the longest branches.
   b. shorten all branches to improve strength.
   c. remove some branches back to the trunk and shorten the remaining large branches.
   d. these trees should not be structurally pruned.

5) What is the best method of spacing branches along the trunk?

   a. shorten some branches so others can become dominant.
   b. shorten all branches.
   c. remove some branches so others can become dominant.
   d. it is best to keep all branches close to each other on the trunk.

6) Topping or heading can be an important and useful tool in nursery tree production.

   a. true
   b. false

7) If a branch or stem is growing aggressively in the permanent nursery canopy and it is larger than about half the trunk diameter:

   a. remove it so it does not become a codominant stem.
   b. remove branches nearby to encourage it to grow faster.
   c. shorten it to a branch pointed away from the trunk.
   d. encourage another branch to also grow to this size to balance the canopy.

8) Why would you consider shortening stems and branches growing into the upper third of the nursery canopy that originate from the lower half of the tree?
   a. there should be few branches in the upper third of the canopy.
   b. it would help prevent them from becoming permanent branches on the tree.
   c. there should be few if any upright branches on trees.
   d. it is OK to allow these to grow, so they do not need to be shortened.

9) The best method of producing a consistent-looking crop is to:
   a. plant a cultivar with a known form.
   b. shear plants on a regular basis.
   c. top the leader and main lateral branches.
   d. prune often.

10) The last pruning before the plant is finished and ready for sale should:
   a. reduce the length of competing codominant leaders.
   b. not attempt to make major changes in the canopy so it should be light.
   c. shorten long branches to bring their ends back into the rest of the canopy.
   d. all of the above

Answers: c, d, a, c, a, a, c, b, a, d

CHALLENGE QUESTIONS

1) Describe what steps you would take to produce a consistent crop of red maple (hardiness zones 2–8) or Jacaranda (zones 9–11) trees up to the 3-inch caliper size.
2) Draw on a piece of paper how you would prune a young tree that has a vase shape with several codominant stems into a well-structured form with a dominant leader.

SUGGESTED EXERCISES

1) Draft a production protocol for the first two years in the nursery for a common tree grown in your area.
2) In a field of nursery trees, locate trees with large aggressive branches opposite each other. Prune so only one of them will dominate at that point on the trunk.
3) Over a two- to three-year period, have the same group come to a field of shade trees to prune them to a dominant leader with well-spaced branches. Discuss with the group how the trees have changed in response to the previous pruning.

CHAPTER 10

DEVELOPING SPECIAL FORMS ON YOUNG PLANTS

OBJECTIVES

1) Contrast options for developing multi-trunked trees.
2) Train multi-trunked trees to a single leader.
3) Describe the special forms trees can be pruned to and how to create them.
4) Present nursery production and landscape practices for developing special forms.

KEY WORDS

Architectural pruning
Bonsai
Crotch spreading
Curtain
Espalier
Modified central leader
Open-center system
Open-vase system
Pleaching
Pollard head
Pollarding
Scaffold branches
Standard
Topiary

DEVELOPING A SINGLE LEADER ON A SMALL ORNAMENTAL TREE OR SHRUB

Trees that stay small at maturity can be trained to a single trunk or left to grow into a multi-trunked habit. Most will grow into a multi-trunked habit if left unpruned. The strongest form resulting in the least likelihood of breakage is the single trunk with stems and branches less than half the trunk diameter (Figure 10-1). To train a plant to a single trunk, allow a recently planted liner to grow for about a year. Reduce the length of the stems that compete with one leader in the center of the canopy (Figure 10-2, center left). Cut back aggressive branches to smaller lateral branches using reduction cuts, or simply use heading cuts on young plants (Figure 8-2). Continue this procedure regularly until the desired form is achieved. If needed, remove lateral branches at appropriate times to push more growing into the main leader. You might need to use a stake or splint if you wish to grow a straight trunk on certain trees (see Chapters 8 and 9).

DEVELOPING MULTIPLE TRUNKS

Large-maturing trees are usually not suited for this form. A number of small-maturing trees and shrubs have showy bark and trunk structure that can be displayed nicely by creating a multi-trunked or low-branched tree (Appendix 2). Many develop a multi-stemmed form without pruning. Multi-stemmed shrubs and trees develop one of several forms. Stems can emerge naturally or be developed from a short, single trunk pruned to encourage multiple leader development. For the strongest tree, develop 4 to 6 inches of vertical spacing between multiple trunks arising from a short stem (Figure 10-1, top left, center, and center left). When the tree is young, keep the branch diameter less than one-half the diameter of the main trunk through regular subordination pruning. This allows trunk tissue to develop a collar around the base of the branch and will increase the strength of the attachment. The least desirable form is present when all stems originate from the same position on the trunk, or when included bark forms (Figure 10-1, right). One trunk could split from the plant due to this weak structure.

There are several methods of developing multi-trunked plants. One is to plant two or three seedlings 4 to 12 inches apart, forming a multi-stemmed effect (Figure 10-3, option B). It is crucially important to leave temporary branches along the trunks to help them develop a thick, tapered strong trunk. As the trees grow older in the landscape, the trunks may eventually touch; this may cause them to push each other apart because they are unlikely to fuse. Root systems can also develop asymmetrically, causing a trunk to fall over. Because this is a relatively new planting technique, time will reveal the merits of this practice. It is probably fine for shrubs and small-growing trees such as lilacs, crape-myrtle, ligustrum, and bottlebrush, because if one of the trunks falls over in the future, little damage to property will occur due to its small size. However, this technique should not be practiced on trees maturing at greater than 30 to 40 feet tall due to the potential for injuring a person or property if one of the large trunks should split from the tree. One trunk may grow slowly, spoiling the multi-trunk effect.

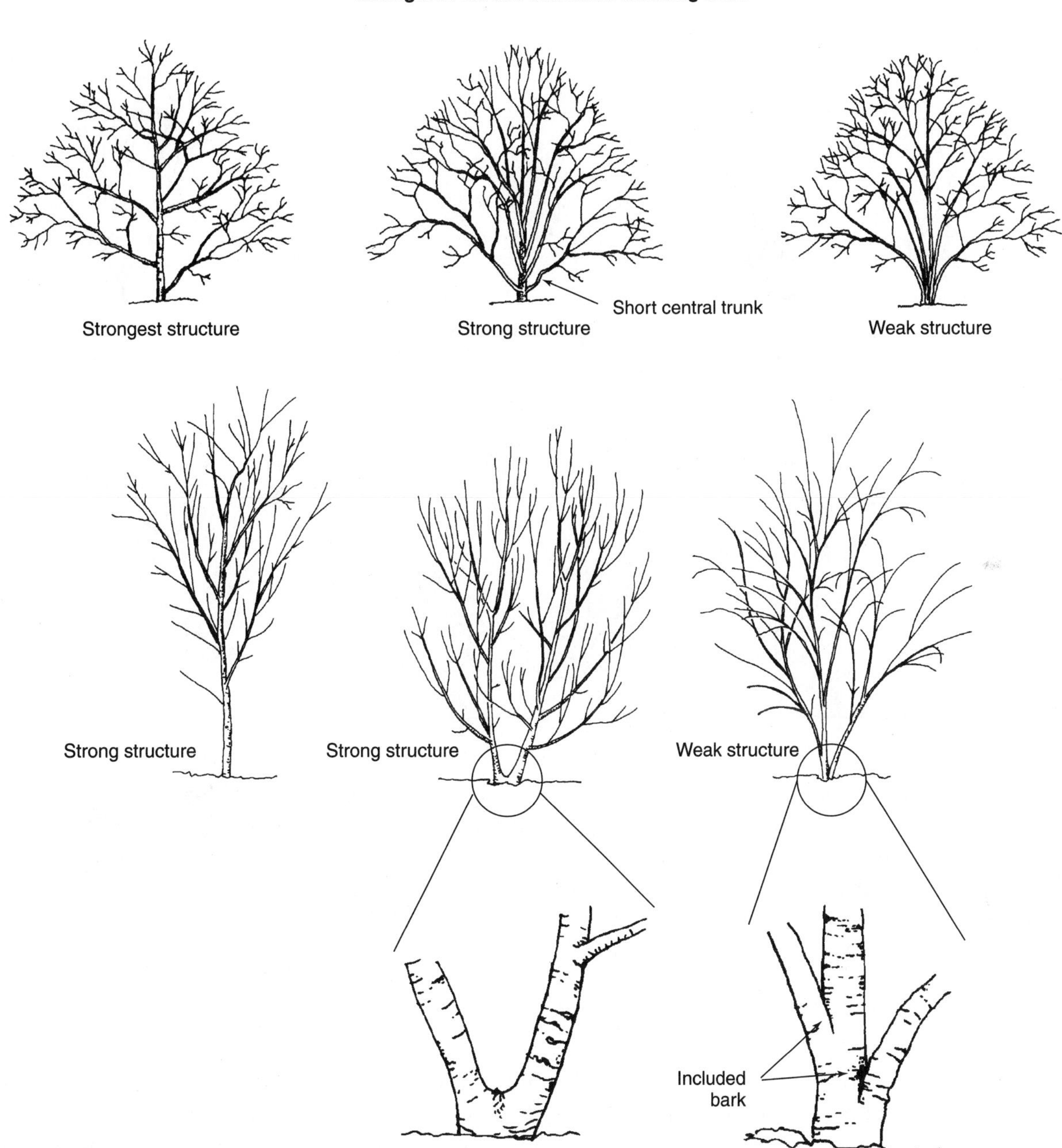

FIGURE 10-1. These three specimens of the same small-maturing tree were trained to be similar in size and canopy shape. However, the trunks and branch structures are entirely different, providing contrasting effects, especially in the dormant season (top). Small trees can also be pruned into other strong structures (center left and center).

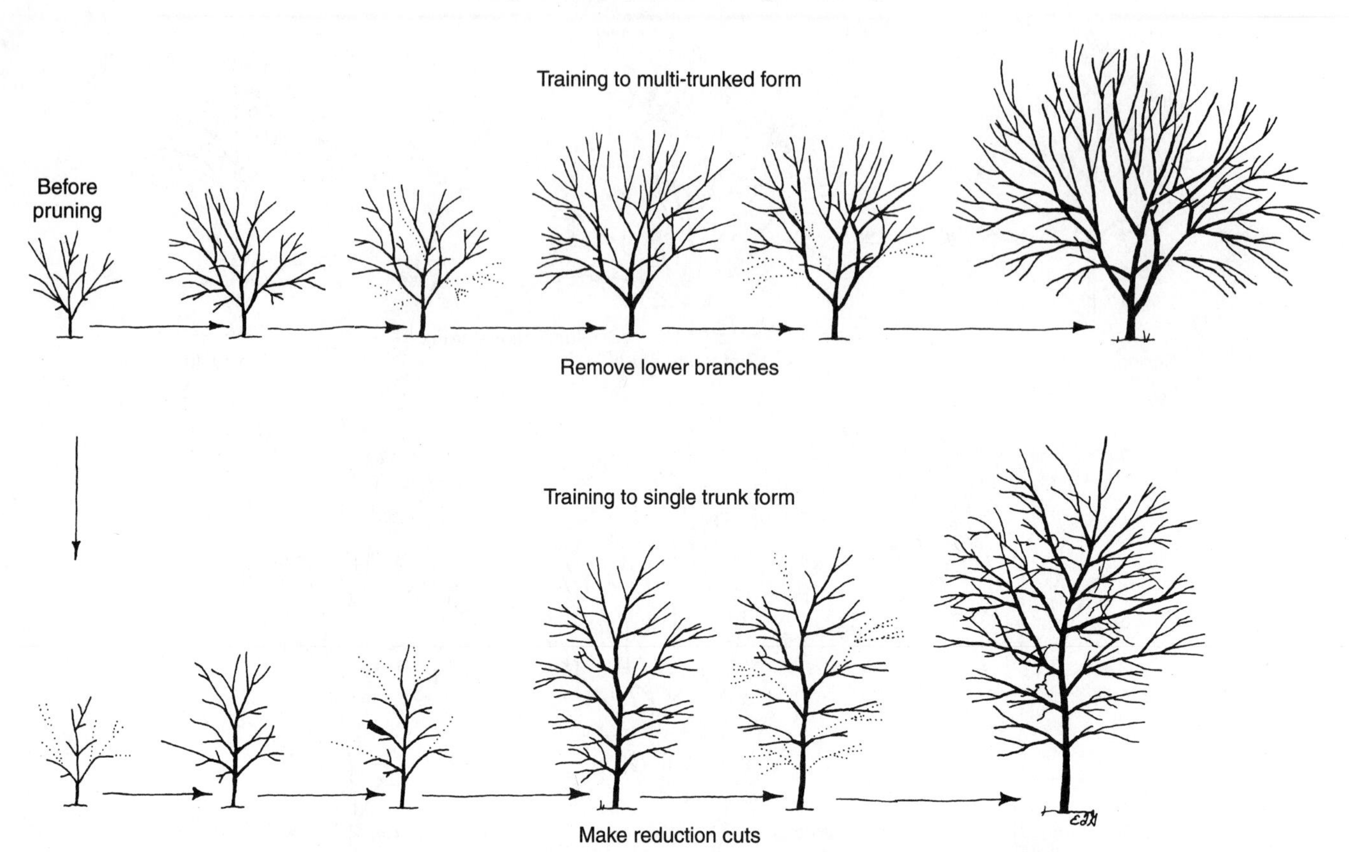

FIGURE 10-2. Small maturing trees can be trained and pruned to a single central leader form or to multi-trunked form. Remove low branches as the tree grows to encourage formation of multiple trunks (top). Slow the growth rate on stems that grow much faster than others by shortening them with reduction cuts. Make subordination cuts on all stems except the one you want to form a single trunk (bottom). This technique is similar to generating a dominant leader on a shade tree (Chapter 8). Repeat at regular intervals as needed. Stakes may be required to generate a straight or upright trunk on certain species.

A second method is to make a heading cut on the leader of a young sapling (Figure 10-3, option A). This will stimulate the development of sprouts and lateral shoots. One shoot will become the leader and the others the major branches to make up the structural framework of the tree. All sprouts from this cut are headed, each one to a different length (see Figure 10-3, E). The shoot cut back the least will become the leader; the others will be branches. This provides several stems of different lengths. Each sprout from these cuts is also pruned to a different length. Continue heading sprouts in this fashion until the tree takes on the desired form. Each time you head the sprouts, cut one of them back only slightly. This will become the leader stem. Training a tree in this fashion can create a strong, multi-trunked tree. Apples and crabapples are pruned in this fashion to create strong trees capable of holding fruit. This form is sometimes referred to as the modified central leader system.

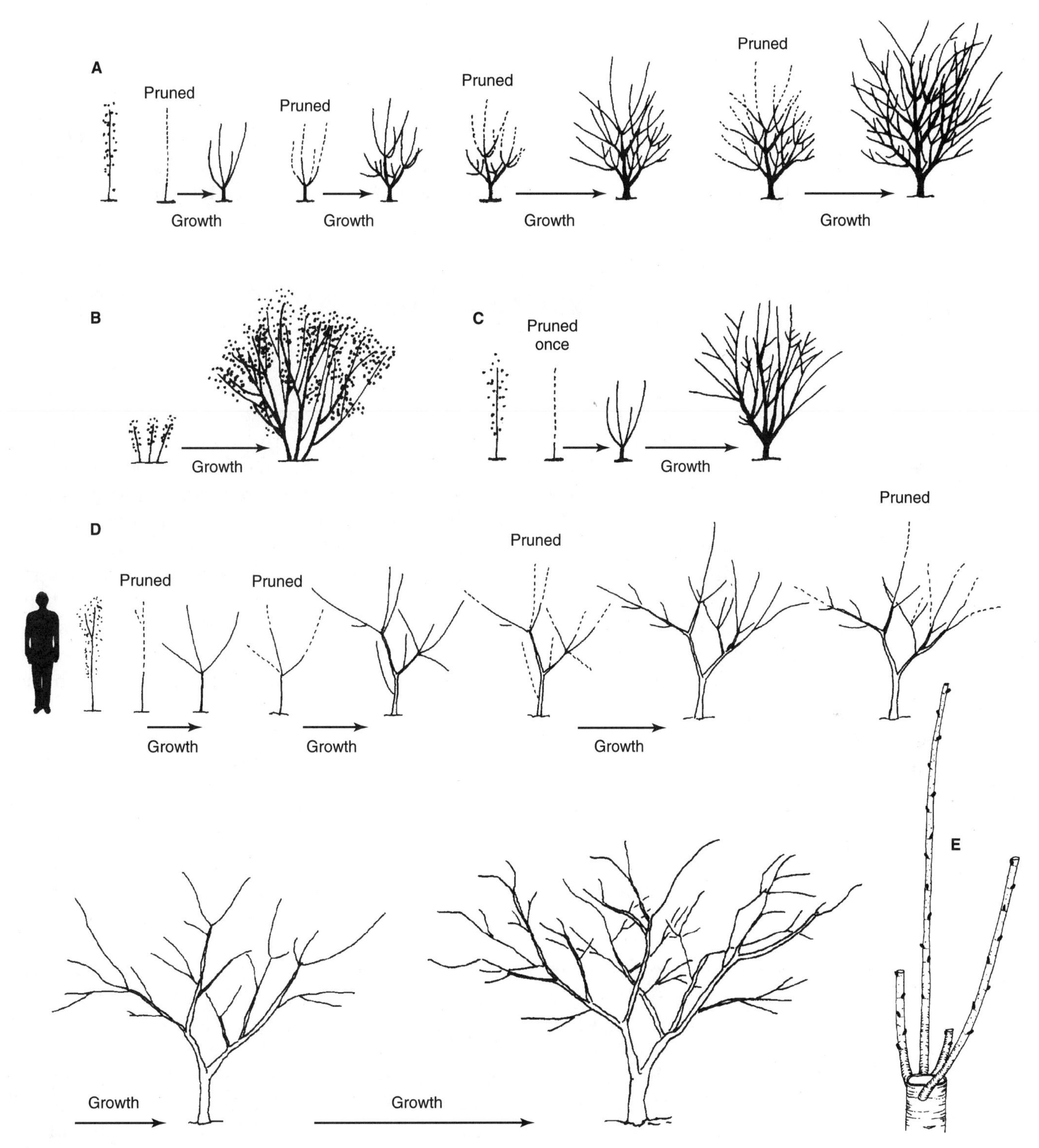

FIGURE 10-3. A small-maturing, multi-trunked tree can be developed in at least four ways. The technique on the top (A) represents the strongest structure, and the one at center-right (C), the weakest structure. On small-maturing trees, several seedlings can be planted close together (B). Many fruit trees are pruned with an open center or vase shape, represented in the bottom two rows (D). Sprouts from heading cuts are headed to different lengths to develop the strongest structure in techniques (A and D) with the fullest, best looking canopy (E).

A third method is to head the leader once and let the tree grow with no further training (Figure 10-3, option C). This creates a nicely formed tree, but it will have a weak structure because included bark will often develop in the unions where stems are clustered together. This could cause the tree to split apart as it grows bigger.

A fourth method simply allows the plant to grow with little pruning the first couple years (Figure 10-2, top). Then remove lower branches from main stems to produce a multi-trunked effect. As long as the tree does not develop included bark nor grow too large, this structure is usually strong enough to hold the tree together for a period of years.

DEVELOPING FRUIT TREES

Deciduous fruit trees are typically trained to one of two forms. Apple trees are often trained to a **modified central leader** whereas peaches, nectarines, plums, and apricots are commonly trained with an open center. There are some orchards that train peaches to a single central leader with no reduction in fruit yield compared to the open-center system (this book does not examine the details of commercial fruit production). Others train plants as espalier or to other forms. Trees trained to the modified central leader (Figure 10-3, option A) system usually have from five to eight **scaffold branches** spaced 6 to 10 inches apart along a central trunk that can be straight or may zig-zag up through the canopy. Scaffold branches are those that are the largest diameter on the tree. Limbs with wider crotch angles support more fruit than upright limbs, often because included bark forms in the crotches of vigorous, upright limbs. Branches with included bark can break easily. If major branches are forming at desirable positions on the tree, but they are at a narrow angle with the trunk, consider increasing the crotch angle by inserting a wooden dowel sharpened to a point at both ends in the crotch. A nail can be driven into each end of the dowel and sharpened to hold it in place. (The sharpened ends help keep the dowel from slipping off the branch and trunk.) This could slow the growth rate of the branch by directing it to grow more horizontally and helps secure it to the trunk. Subordination cuts (Figure 8-3) also will help reduce the splitting due to included bark by slowing the growth rate on the cut branch, and they can be used in conjunction with **crotch spreading.** Branches that are subordinated and spread may generate new sprouts along the tops of the branches.

Trees trained to an **open-center** or **open-vase system** (Figure 10-3, option D) have three to four scaffold branches originating from a short (2- to 4-foot-long) central leader. The best structure places scaffold branches at least 4 to 6 inches apart (Figure 10-1, top center, center left) so the collar can develop properly around the base of the branches. Scaffold development should begin within the first year or two after planting. Remove root suckers and vigorous sprouts as they emerge early in the growing season in order to direct growth into the developing scaffold branches. Scaffold branches should be selected and should dominate the tree structure by the second or third year after planting. Once the main scaffold branches have been established, in subsequent years you may have to shorten or remove vigorous sprouts in the interior of the canopy in order to keep the tree open in the center. The open-vase system can be more profitable than the central leader system for certain orchards.

DEVELOPING AND MAINTAINING A STANDARD FORM

A tree or large shrub trained to one short, straight trunk with a dense, round canopy is commonly referred to as a **standard.** A number of small trees are suited for this type of training (Appendix 3). The canopy of a standard has numerous branches. Frequently pinched or headed, the canopy develops into a dense mass of foliage originating from many small-diameter branches. Numerous branches often originate from the same point on the trunk, which would be undesirable if the tree were to be grown to a large mature specimen; however, standards are meant to be maintained with regular clipping and are not designed to become large plants. Therefore, this branch attachment is fine. *Never purchase standards if they will not be maintained as such because their branch structure is weak and not suitable for growth into large trees.* In many ways, the 'Bradford' Callery pear is grown as a large standard that becomes weak as it grows older.

When necessary, the trunk of a young standard is held straight with a stake or by some other means. This ensures that the tree is secured in an upright position to create a straight trunk. The support is removed as soon as the tree is able to stand under its own weight. Until the tree can stand erect on its own, low-vigor, temporary lateral branches should be left on the lower trunk to increase trunk diameter and strength. Leave as many as possible. Keep them clipped to about 12 inches long to prevent a large-diameter branch from forming that would leave a large wound when removed. Immediately remove any that grow to more than about pinky width. These temporary low branches will be pruned back to the trunk prior to sale in the nursery. Allowing the wind to move the trunk will also help make it stronger.

Standards can take on different forms depending on how they are trained and pruned in the nursery. The easiest method of training is to make a heading cut on the trunk of a young sapling and let the sprouts grow (Figure 10-4, option B). Each sprout can be headed again and again to create a nicely shaped, compact plant (option C). Unfortunately both methods result in a weak plant that is susceptible to splitting at the point where the four main branches meet the trunk, especially if the tree will grow to be taller than about 15 feet.

A better method is to head the leader and then head each of the sprouts to a different length (option A). The sprout cut back the least becomes the leader, the one cut back slightly more becomes the major branch at that point on the trunk, and the other two become much less dominant because they were cut back severely (Figure 10-3, E). Head each of the sprouts that originates from these cuts in the same manner (i.e., cut them back to different lengths). Continue heading in this fashion about every 6 inches up the trunk until the plant develops the form and size you want. Notice in Figure 10-4, option A, that the two branches indicated with arrows are larger and have more secondary branches than the others. These are spaced several inches apart on the trunk, not clustered together as in options B and C. This should help the tree stay together in a wind storm and make it more sturdy overall.

If a standard is to be planted near a street or in a parking lot or other location where pedestrians or vehicles will walk or operate beneath it, the canopy or head should begin at least 6 to 7 feet from the ground. Provided that the tree can stand up without support, the trunk below this point should be clear of branches.

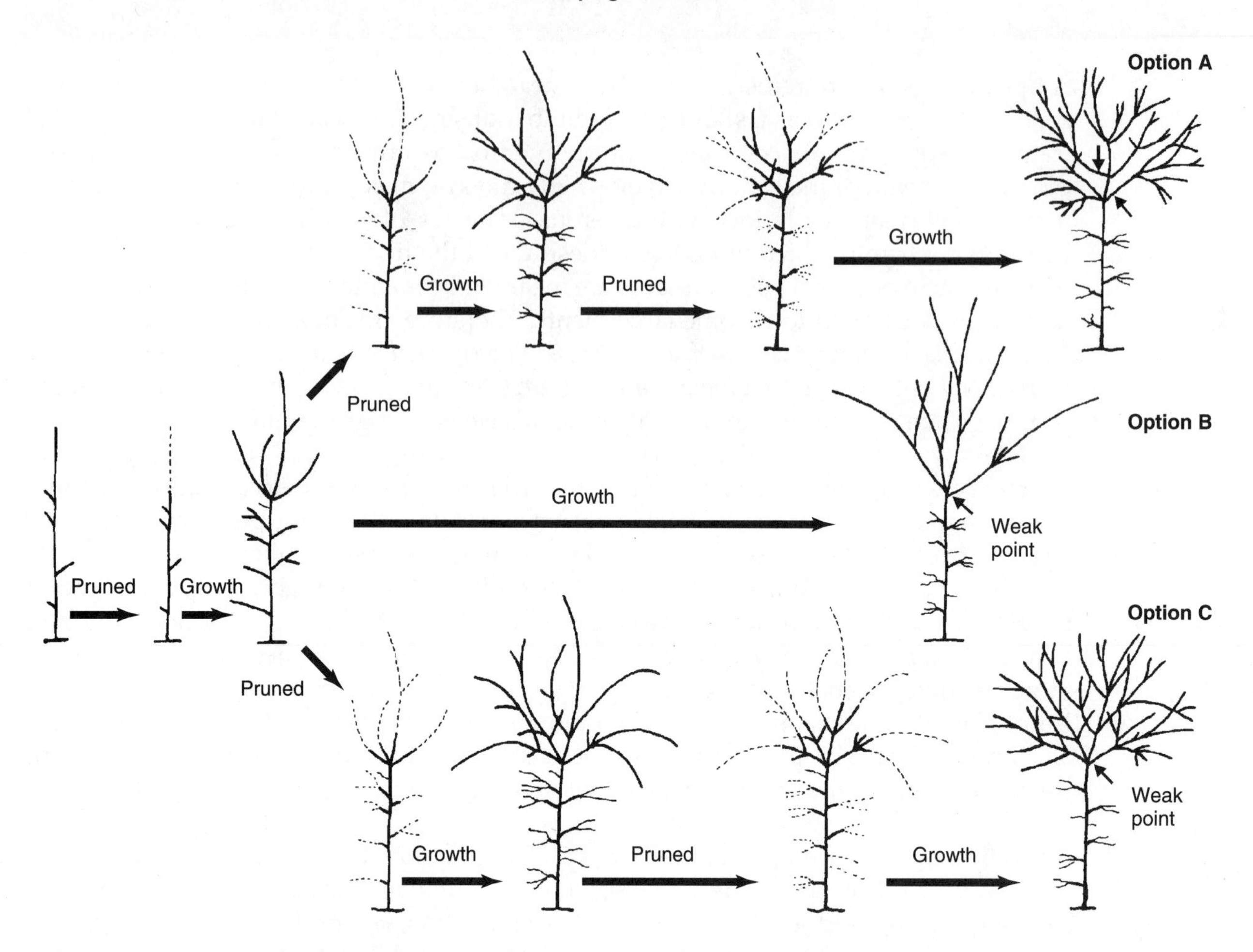

FIGURE 10-4. Standards can be trained in a variety of ways. Option A provides for the strongest tree. Notice that some branches are larger than others (see arrows). Be sure to leave lower branches on the trunk until the tree develops the strength to hold itself erect. These also help increase overall growth rate.

CREATING SPECIAL EFFECTS

So-called **architectural pruning** shapes and maintains trees unnaturally to a specific form and size with regular pruning. For example, **pleaching** interlaces or twines young branches and trunks on saplings together to form a hedge, archway, or tunnel. **Curtain** creates a flat wall surface with annual heading that is typically accomplished with power shears. **Topiary** creates an animal, column, ball, or other shape with regular (at least annual) heading cuts. Heading cuts are usually made with power shears forming a smooth surface of foliage on the outside of the canopy. Please note that rounding over (topping) trees every three to five years beneath power lines or in other situations is not topiary because shoots more than a year or two old are headed. Many trees are suited for topiary, including those in the following genera: *Acer, Cupressus, Eugenia, Ilex, Podocarpus, Prunus, Quercus* (small-leaf types such as live oak), and *Taxus.* **Bonsai** is the art of maintaining small trees that involves root pruning and directing shoot growth with pruning and wires. Espalier and pollarding are explained next in detail.

DEVELOPING AN ESPALIER

Plants can be located close to a structure and trained into an espalier. **Espalier** is a specialized, high maintenance pruning technique requiring patience and regular pruning. A shrub or tree (see Appendix 4) is trained to grow more-or-less flat against a wooden or metal trellis, garden, garage or house wall, fence, or other support. It is a great way to shade the west or east side of a home or office building from the summer sun. Consider painting or mending the wall before planting.

Start the process by installing a young plant 6 to 36 inches away from the support. In deciding how close to plant, be sure to account for trunk and root flare growth over time. Branches growing toward and away from the structure can be bent and tied parallel to the structure. Those that are too big to bend or are in the wrong position may have to be removed back to the trunk; or you could leave a stub with a few buds and train the new sprouts to the desired position. Leave most or all branches growing parallel to the structure.

There are several ways to train an espalier. One is to attempt to space branches several inches or more apart along a single, straight trunk. No branches are opposite each other. Another system locates branches directly opposite each other. And yet another system develops several trunks equal in diameter with branches along each one. There are many more ways to do this.

Develop the basic framework early, removing appropriate branches and pruning with heading cuts regularly and moderately. Light to moderate pruning prevents excessive sprouting. On the other hand, there may be other cases where you want sprouting to occur and can use this to your advantage. If necessary, attach branches using string ties to a decay resistant support to encourage a flat, two-dimensional structure. Replace or adjust ties regularly to prevent them from girdling secured branches.

Any number of shapes and sizes can be developed. Once the main framework and desired size is in place, some horticulturists use pollarding cuts to maintain the espaliered plant at a certain size; others use heading and reduction cuts. Like any pruning technique, this one is best mastered with practice.

DEVELOPING AND MAINTAINING A POLLARD

For centuries in Europe, trees were maintained at a designated height with regular pruning. This practice, called **pollarding** maintains a tree at any specified height, sometimes for centuries, and provides a formal look to the landscape. Not everyone cares for the look of a pollarded tree, especially in winter. Traditionally, trees have been maintained at 20 to 30 feet tall. Pollarding can be used to keep a large-maturing tree small if it was improperly located in a restricted soil space such as a planter, narrow soil strip, parking lot island, or sidewalk cutout. It is also useful to control size if improperly planted too close to structures such as a building, street light, or electric wire. Once begun, it is essential that pollarding continue.

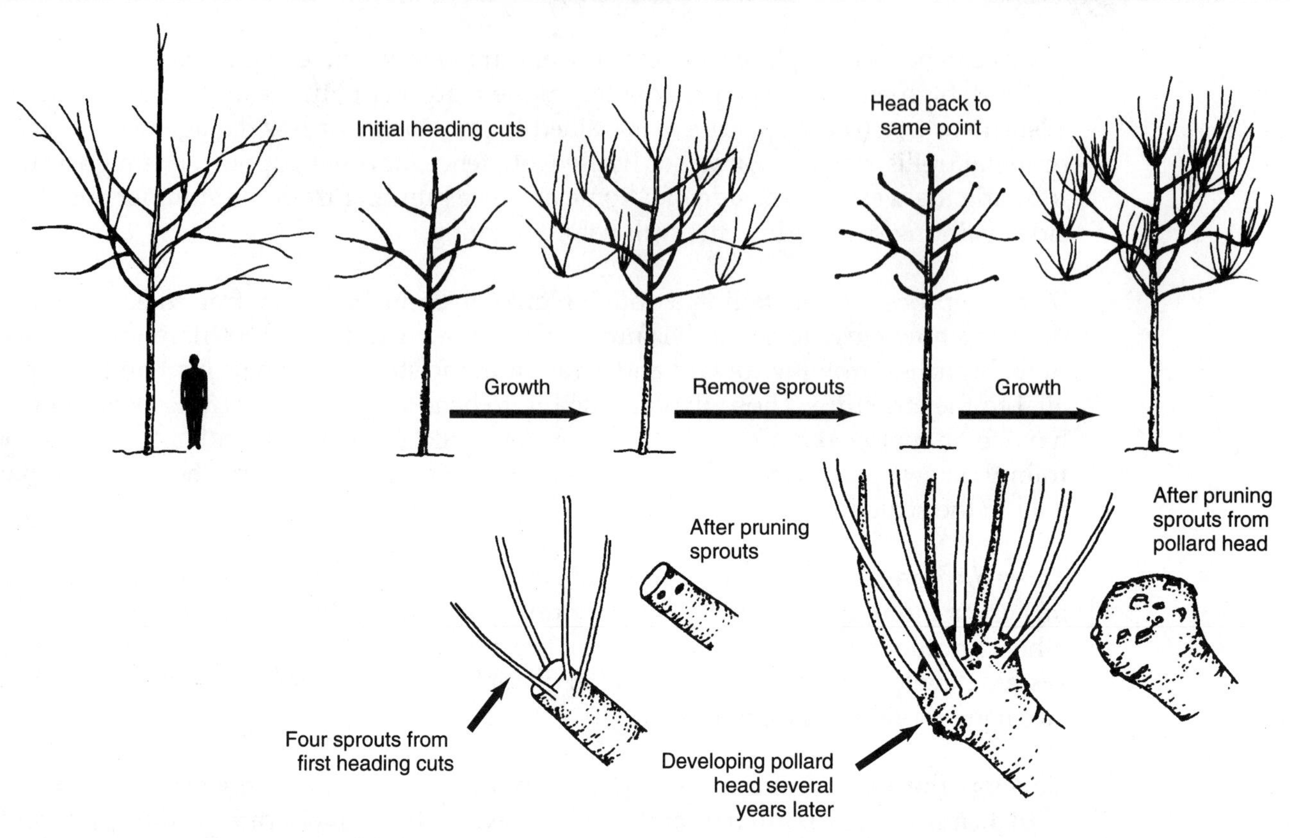

FIGURE 10-5. After you establish the main structure of the tree (top left), head one- or two-year-old wood at the position you would like pollarding to start (top center-left and bottom left). In each subsequent year or two, cut back to this same point (top center-right). Eventually, a knob of tissue called a pollard head develops at the location of the cuts (bottom right). See Figure 10-6 for pollarding detail.

Preferably, the pollarding process begins when the tree is very young. Over a six- to ten-year period, and before pollarding begins, the main tree architecture can be developed. In the typical form, upright or horizontal branches with wide angles of attachment spaced along a dominant trunk are chosen for the main scaffold limbs. Many other forms can be created. After this, young stems are headed with a slanted cut at the exact position that the tree will be cut to at each subsequent pruning. *Choose these points carefully because they cannot be repositioned at a later time* and be sure each is located so sprouts will receive adequate sunlight. If shaded, that part of the plant could decline. In the dormant season, all sprouts but one or two from the original heading cuts are removed back to the point where the original heading cut was made (Figures 10-5 and 10-6). Always remove the most aggressive sprouts. Some people remove all sprouts back to the original heading cut. For the first or second prunings, leave 1 to 2 inches of stub if necessary to ensure that there will be buds to initiate next year's sprouts.

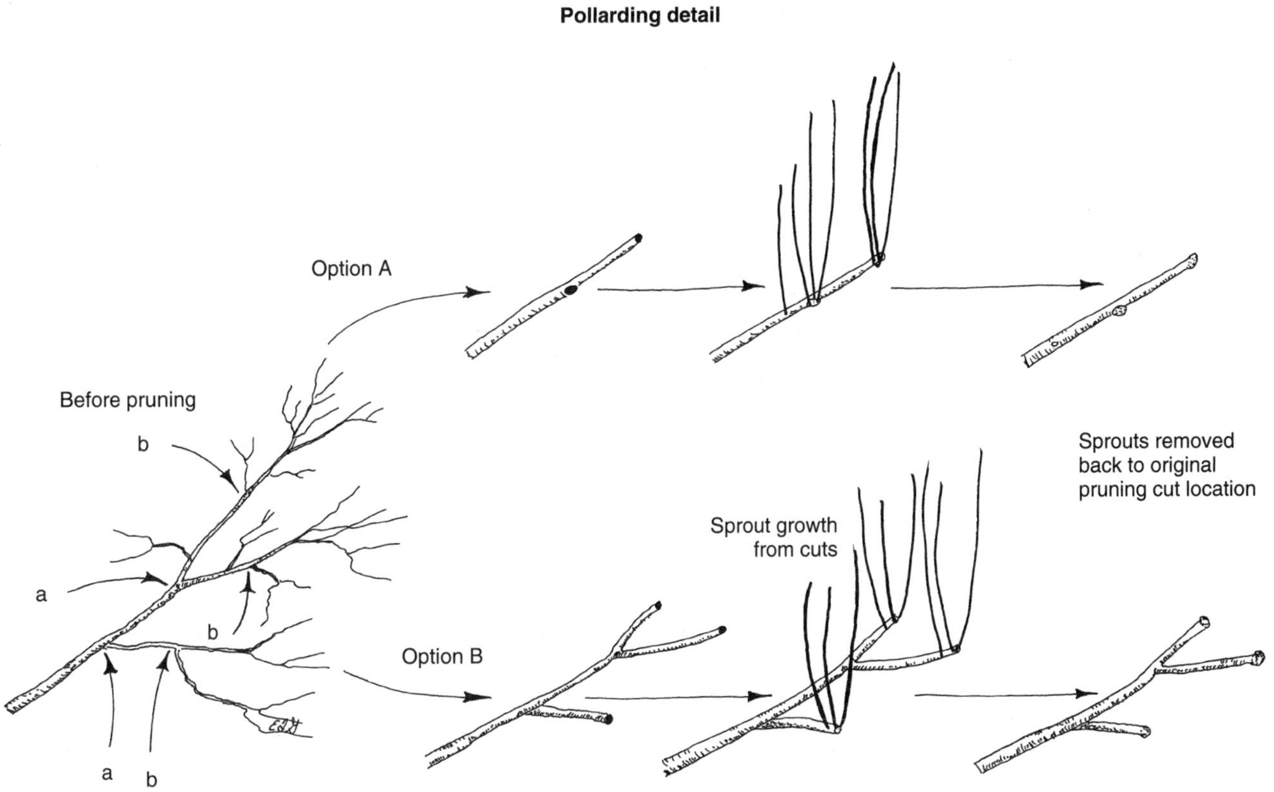

Pollarding detail

Option A

Before pruning

b

a

b

a b

Option B

Sprout growth
from cuts

Sprouts removed
back to original
pruning cut location

FIGURE 10-6. There are at least two different methods of making initial cuts to begin the pollarding process. In option A (top), two cuts were made. One was a removal cut and the other a heading cut. In option B, heading cuts were made farther out on the branches. Cuts were made just below a branch union. There is a collection of buds at this point on certain trees that will result in the generation of many sprouts.

A knob of tissue called a **pollard head** resembling a ball develops several years after the first cut was made. A starch-rich pollard head is a collection of buds, callus, and collars. *The head is never to be removed.* Most shoots grow from this tissue, which enlarges slightly each year. Most are oriented upright, do not branch, and grow at a rapid rate. They are cut back to this knob at each pruning. Shoots originating below the pollard head should be removed each time the tree is pruned. Do not discard removed shoots; they make great kindling for fires and can be used to tan leather, for animal fodder, and weaving.

A clear distinction must be made between pollarding and topping (Table 10-1). Topping heads branches and stems regardless of their age and diameter and initiates cracks and decay inside the tree, whereas pollarding cuts branches less than 2 years old and minimizes decay. Pollarding is a high-maintenance practice requiring annual or biannual pruning, but it can create unique trees that live for a very long time. Because of the regular pruning requirement, many European communities are abandoning the practice and planting small-maturing trees instead. Others are topping the trees periodically

TABLE 10-1. Comparison of topping and pollarding.

Topping
• Commonly practiced, yet inappropriate, technique • Make heading cuts through regardless of diameter or age • Practiced on trees regardless of age • Tree may or may not be pruned again • Practiced on many types of trees • Initiates decay in cut branches and trunk • Leads to weak structure and could place landscape at risk • Can lead to short life
Pollarding
• Rarely practiced, yet appropriate, technique • Make heading cuts through one to a few years old, no more than about 1 inch in diameter • Training begins when tree is young • All shoots removed every one to three years back to the same position on the tree, never below the original cut • Appropriate species include sycamore, lindens, horsechestnuts, and others • Does not initiate trunk decay • Maintains good structure • Pollarded trees can live for centuries

(e.g., every five years). Subsequent topping occurs a foot or more out from the last topping (Figure 10-7). Although decay results from this treatment, trees remain small and some can live for many decades before declining or failing. Pollarding has a place in landscapes and should be tried more often than it is currently. Trees that respond well to pollarding and have lived a long time under this treatment are included in the following genera: *Acer, Aesculus, Alnus, Crataegus, Fraxinus, Liriodendron, Morus, Platanus, Quercus, Tilia,* and *Ulmus.* Small-maturing trees such as crape-myrtle also respond well to pollarding (Figure 10-8).

DEVELOPING WEEPING PLANTS

Many trees are grown in ornamental weeping forms (Table 10-2). Some growers may refer to this as a mop-top form because the canopy resembles a mop. Some of these are budded or grafted on top of a straight stem (Figure 10-9). Weeping trees left un-pruned can develop long, stringy branches and a thin canopy that might not be appealing to customers (Figure 10-9, bottom). The canopy density can be increased and the form improved by reducing the length of branches using heading cuts (Figure 10-9, top center). This is best done in the dormant season. Several shoots typically form just behind the pruning cuts. These new shoots are shorter than the removed portion of the cut branch resulting in a more compact, attractive plant.

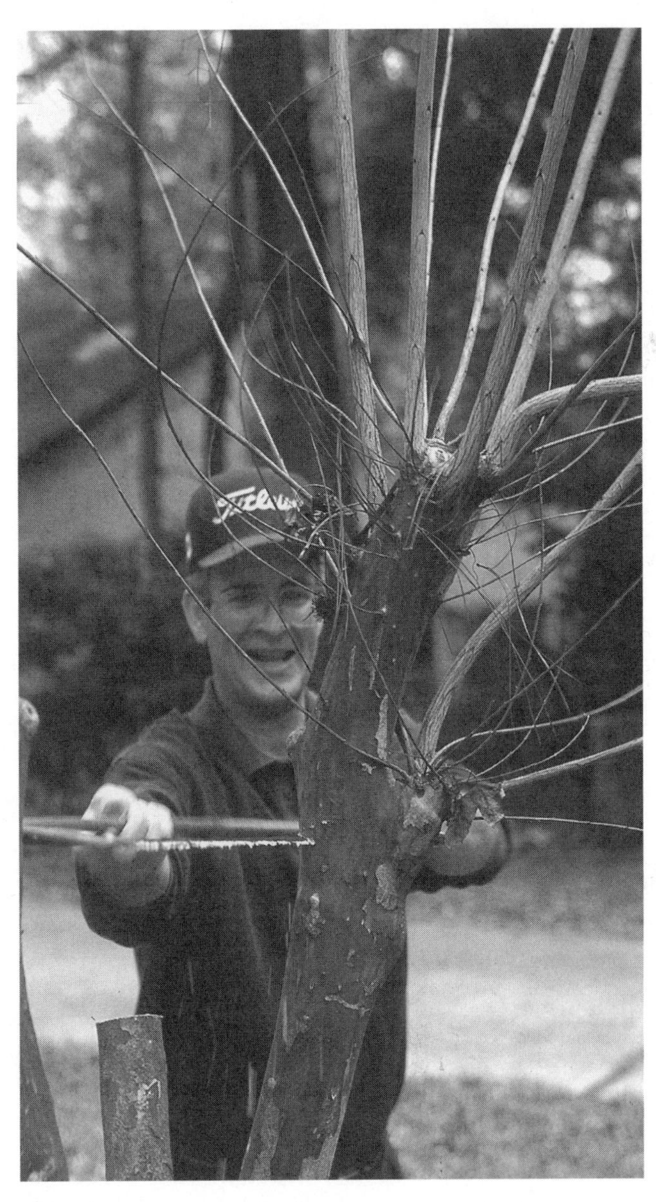

FIGURE 10-7. Progressively topping a tree at higher and higher positions in the canopy results in an unsightly mess (left). Choose pollarding instead for a cleaner, more professional job. This tree was topped 3 or 4 times, each at a slightly higher position. Eventually, many of these trees are cut below the original topping cut (right). This can lead to a downward spiral in tree health. Some people enjoy this annual ritual.

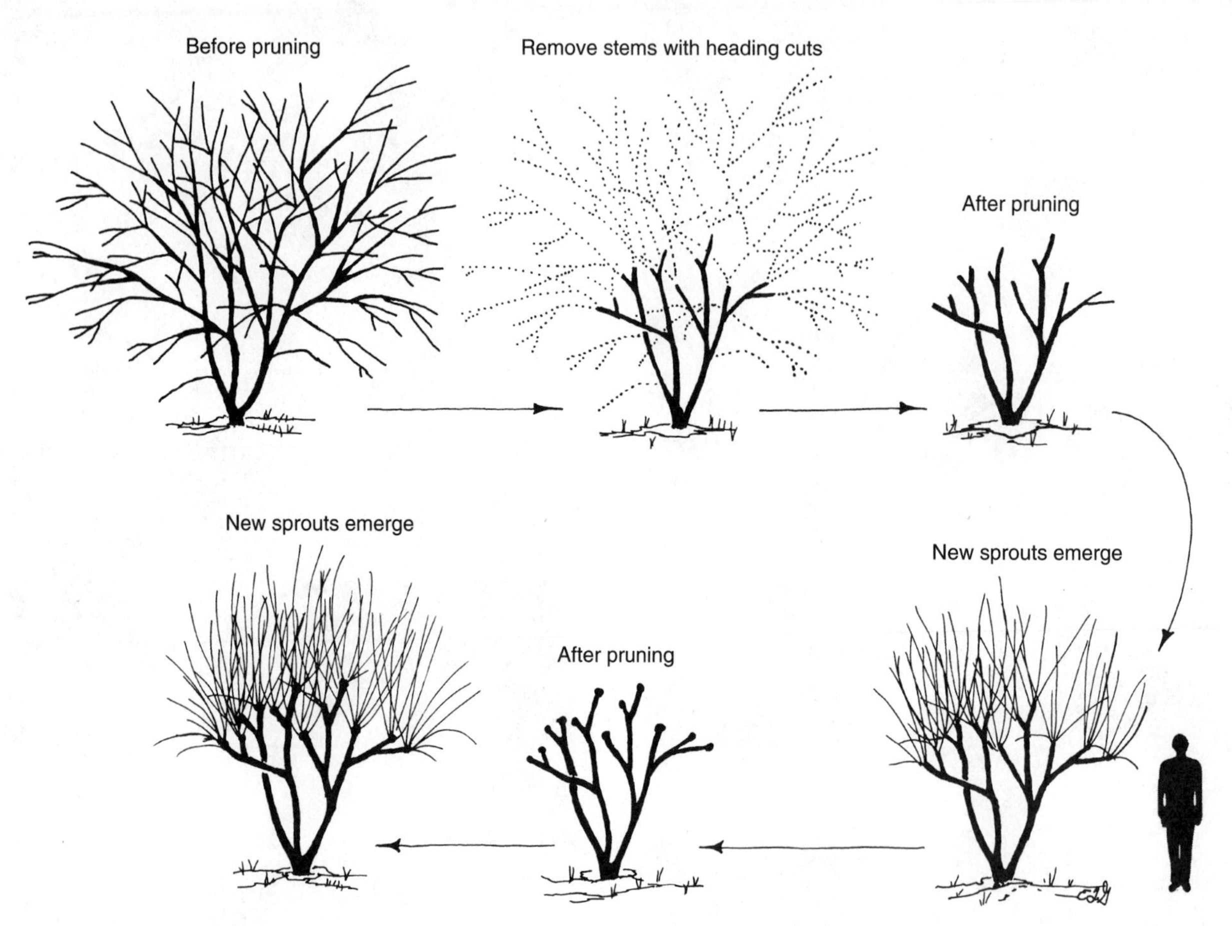

FIGURE 10-8. Small ornamental trees can be pollarded to keep them small and compact. Trees are initially headed at the position that they will be pruned to annually (top). New growth from heading cuts is vigorous and upright with few branches. Each year, new sprouts are removed back to the original heading cut (bottom).

TABLE 10-2. Some weeping trees suitable for pruning into a mop-top.

Acacia pendula	*Larix* x *marschlinsii* 'Varied Direction'
Acer palmatum 'Dissectum'	*Malus* sp
Callistemon viminalis	*Morus alba* 'Chaparral'
Caragana arborescens 'Pendula'	*Pinus densiflora* 'Pendula'
Caragana pygmaea 'Pendula'	*Pinus strobus* 'Pendula'
Catalpa bignonioides 'Nana'	*Prunus subhirtella* 'Pendula'
Cedrus deodara 'Pendula'	*Robinia pseudoacacia*
Cercidiphyllum japonicum 'Pendula'	*Salix caprea* 'Pendula'
Duranta erecta	*Sophora japonica* 'Pendula'
Gleditsia triacanthos 'Emerald Kascade'	*Tsuga canadensis* 'Pendula'
Ilex vomitoria 'Pendula'	*Ulmus alata* 'Lace Parasol'
Juniperus scopulorum 'Tolleson's Blue Weeping'	*Ulmus glabra* 'Camperdownii'

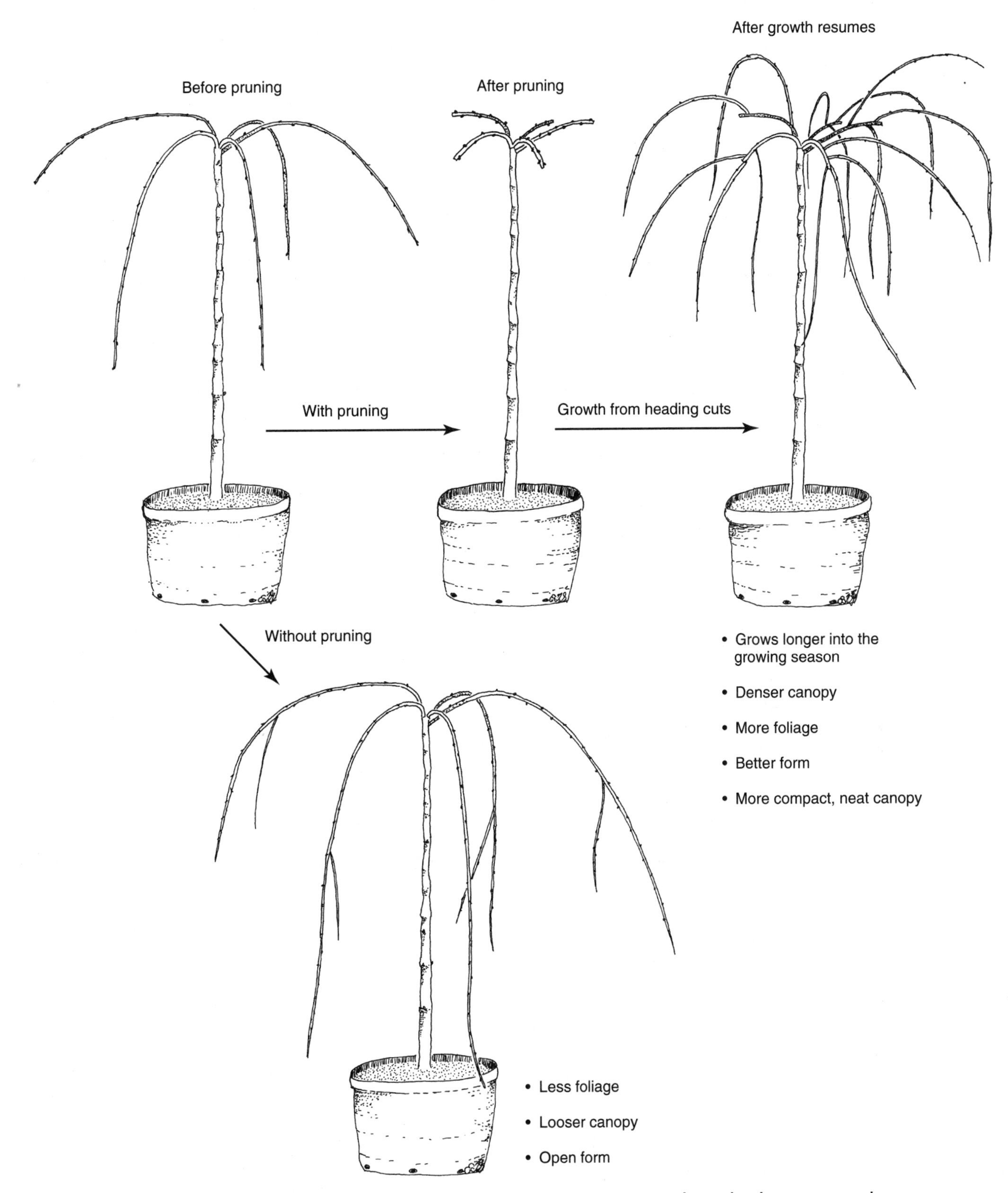

FIGURE 10-9. Weeping trees can grow to the ground becoming sprawling shrubs or ground covers (bottom). Heading the main branches when they are one or two years old can generate a more upright weeping habit. This makes the plant more presentable in the nursery or landscape (top).

CHECK YOUR KNOWLEDGE

1) The practice of removing all shoots every one or two years back to the same position on the tree is called:
   a. heading.
   b. rounding over.
   c. stag horning.
   d. pollarding.
2) For the strongest multi-trunked tree:
   a. develop at least 4 to 6 inches of vertical space between multi-trunks arising from a short stem.
   b. trees should not be grown with multi-trunks.
   c. all trunks should arise from the same point on top of a short stem.
   d. use reduction cuts to subordinate all stems.
3) Most "standards" are not suited for growing into large specimens because:
   a. species chosen for this treatment are usually weak wooded.
   b. main branches usually originate from one point on the trunk.
   c. they were headed in the nursery.
   d. they were staked for too long in the nursery.
4) A tree or large shrub trained to one short, straight trunk with a dense head of foliage is called a:
   a. pollard.
   b. topiary.
   c. curtain.
   d. standard.
5) Pollarding:
   a. is another word for topping.
   b. can be used to keep a tree or shrub at a specified size.
   c. should begin in the nursery for best results.
   d. is often practiced under power lines and other utilities.
6) Which of the following may respond poorly to pollarding?
   a. *Tilia*
   b. *Quercus*
   c. *Alnus*
   d. *Pinus*

Answers: d, a, b, d, b, d

CHALLENGE QUESTIONS

1) Small-maturing trees can be trained to grow to a multi-trunked habit. What would be the disadvantages of allowing a large-maturing tree such as an oak to grow with multiple trunks?
2) Describe several urban situations where pollarding might be a good option for managing trees.

SUGGESTED EXERCISES

1) Find a young, large-maturing tree such as a sycamore, oak, or similar plant and make appropriate pollarding cuts to begin the pollarding process.
2) Take a young nursery tree that normally is grown with several stems arising from one location on the trunk and train it to grow with the main stems spaced 4 to 6 inches along a main trunk.
3) Stand next to a standard and discuss why it is not suited for growing into a large specimen.

CHAPTER 11

STRUCTURAL PRUNING OF SHADE TREES IN THE LANDSCAPE: OBJECTIVES

OBJECTIVES

1) Describe the problems that result from no pruning program.
2) Show where single-trunked and multi-trunked trees are most appropriate.
3) Discuss the benefits of a long-term management approach to tree care.
4) Describe the benefits of keeping low branches small in diameter in urban sites.
5) Determine why trees need structural pruning.
6) Describe the objectives of structural pruning.

KEY WORDS

Good architecture
Live crown ratio
Lowest permanent limb
Maximum critical diameter
Medium-aged shade trees
Natural tree form
Permanent canopy
Preventive arboriculture
Structural pruning
Scaffold limbs

INTRODUCTION

There are three basic problems with not placing planted shade trees on a management program. The problems are (1) codominant stems develop and they could split from the tree; (2) aggressive branches and stems develop low on the tree, they droop and require removal resulting in large pruning wounds; and (3) aggressive branches develop low on the tree, they grow too long and fail. Many trees can develop these problems (Appendix 8). All three problems result in tree stress, could shorten tree life, and could place people and property at risk (Figure 11-1). *Pruning programs on young and medium-aged shade trees should be devoted to preventing and solving these three problems.* In this sense, **structural pruning** in the first 25 years after planting in the landscape is an extension of nursery production pruning. The structure on medium-aged and mature trees is more difficult to influence because the tree is significantly committed to its form. Mature tree pruning is covered in Chapters 13 and 14.

The objectives of structural pruning to help solve these problems include (1) develop or retain a dominant leader; (2) identify the lowest branch in the permanent canopy; (3) space main branches along the dominant trunk; (4) prevent or suppress formation of included bark; and (5) prevent branches below the permanent canopy from getting too large. In short, encourage growth in the portion of the tree you want to become dominant by reducing the length of or removing other live portions of the tree. Unlike cabling and bracing limbs together which is designed to *treat* problems, structural pruning attempts to *prevent* problems from occurring. This is **preventive arboriculture.**

Medium-aged shade trees can be defined as trees that are not yet mature. In practical terms they can be thought of as trees less than about 30 to 40 years old, depending on species. Many of these trees are young enough to meaningfully divert and direct growth to desirable portions of the tree by removing other living parts of the tree. The art of structural pruning is to determine how much to remove (the pruning dose) and where to remove it from. Dose is the most difficult aspect of tree care to teach.

One way to conceive a plan for pruning is to decide what size removal cut and what size reduction cut you are willing to make on certain species. More problems develop following larger pruning cuts than small cuts. This decision would be based on the reaction of trees to pruning in your region and may vary by species. Once this decision has been made from your observation, research, and discussions with others, you can develop strategies to keep cuts under this **maximum critical diameter** while accomplishing the five objectives listed previously. Maximum critical diameter is the largest diameter cut you are willing to make on a certain species or size tree. This diameter should be smaller on trees that compartmentalize decay poorly (see Chapter 4).

For example, if you were to place trees on a ten-year pruning cycle (each tree is pruned every ten years), larger cuts than your critical diameter (e.g., 3 inches) might have to be made in order to accomplish your objectives. You would conclude that the trees should be placed on a shorter pruning cycle so cuts will be less than 3 inches. In other words, they need to be pruned more often. Pruning cycles might have to be shorter in the warmer climates than in cooler climates.

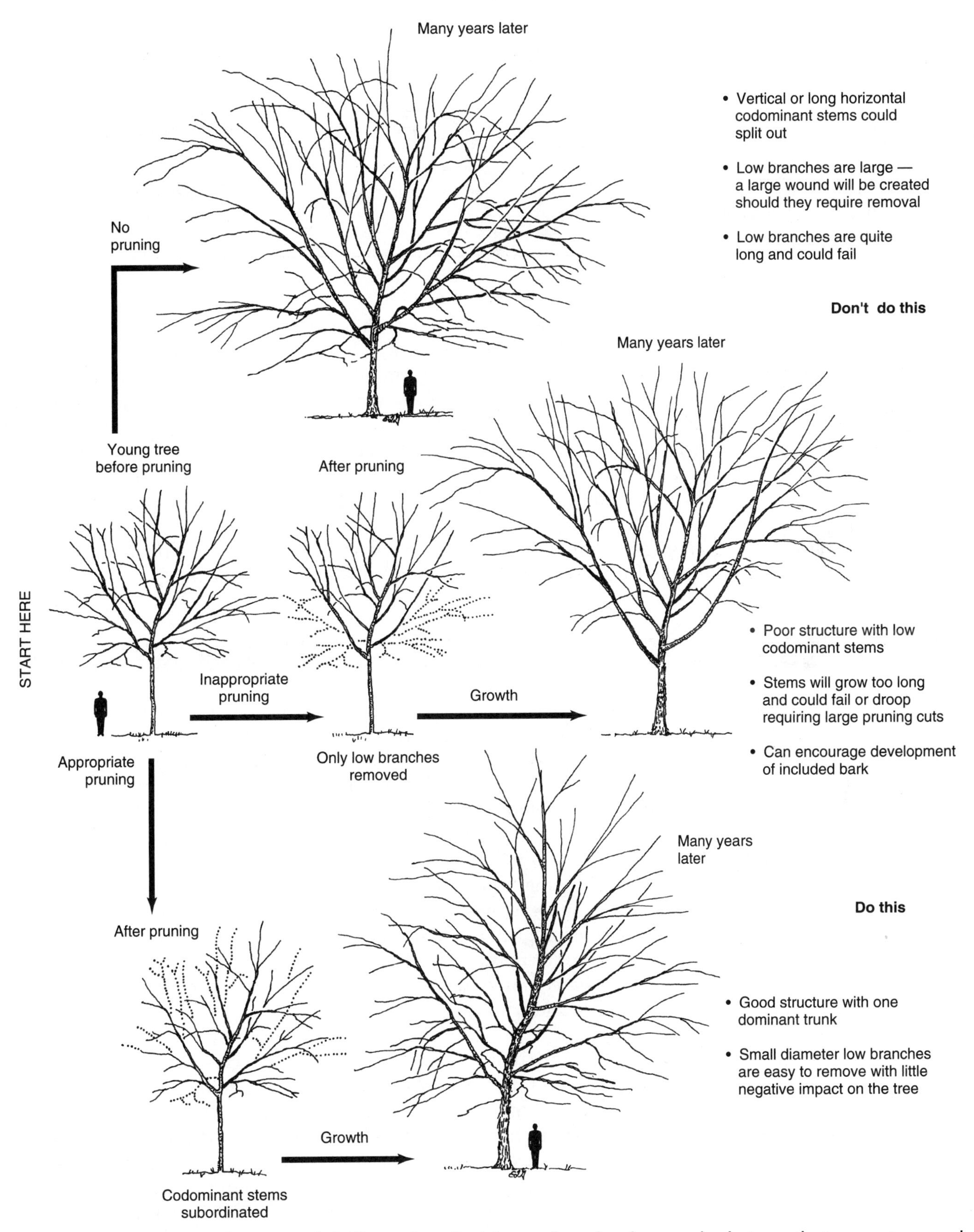

FIGURE 11-1. Start at the center left illustration. Problems often develop on shade trees that are not pruned or are pruned inappropriately (top and center). Appropriate pruning leads to trees with strong structure that is sustainable. Subordinate codominant stems and aggressive low branches to keep them smaller than half the trunk diameter (bottom).

TABLE 11-1. Removing branches of different sizes back to the trunk, or removing lateral branches back to main branches.

Branch Size	Consequences of Removal	Recommended Action
Less than ⅓ trunk diameter	Few consequences	Remove if needed
⅓ to ½ trunk diameter	Some trunk defects could result	Consider shortening instead
More than half trunk diameter	Defects likely	Shorten instead of removing
Large enough to have heartwood	Defects likely	Shorten instead of removing

There are two components to branch size: diameter relative to the trunk diameter and actual diameter of the removed branch. Branches that are smaller than about half the trunk diameter generally can form a branch protection zone in the base of the branch where it meets the trunk. Removing small branches usually causes little harm (Table 11-1). Since old, large-diameter branches could have heartwood or wood that is unable to react to injury, there is little to retard the spread of decay-causing organisms into the trunk following large branch removal. If you suspect that heartwood is present, consider shortening the branch instead of removing it.

Tip: *Encourage growth in the portion of the tree you want to become dominant by reducing the length of or removing other living portions of the tree.*

WHY TREES NEED STRUCTURAL PRUNING

Many shade tree species used in urban and suburban landscapes evolved close to each other over millions of years in a forest. Trunks on trees in the forest typically grow upright reaching for sunlight at the top of the forest canopy. Large-maturing trees spaced closer than about 15 feet in landscapes or along streets often grow like this as well provided they are planted with a single dominant leader (Figure 3-2). In the forest and on very closely spaced planted trees lateral branches close to the ground are shaded, die, and are shed from the tree. This natural process often forms a tree that has a dominant trunk with small-diameter branches. Most of the major branches and codominant stems in the forest emerge in the top half to two-thirds of the maturing tree. Trees evolved physiological mechanisms for developing and maintaining strong structure with this typical or **natural tree form.** These mechanisms are: (1) branch protection zones at the base of relatively small branches and (2) strong connections between small branches and trunk.

Many trees we care for in urban landscapes were not planted close to each other so they develop an unnatural open form. Few large-maturing trees in moist climates evolved in open landscape settings without other trees nearby. Therefore, it is not surprising that these trees lack good mechanisms for handling the spreading low-branching habit that results from growing in the open urban or suburban landscape (Figure 11-2). Without a planned pruning program three problems develop: (1) protection zones are rare, (2) low branches grow to become quite large, and (3) included bark often forms in branch unions.

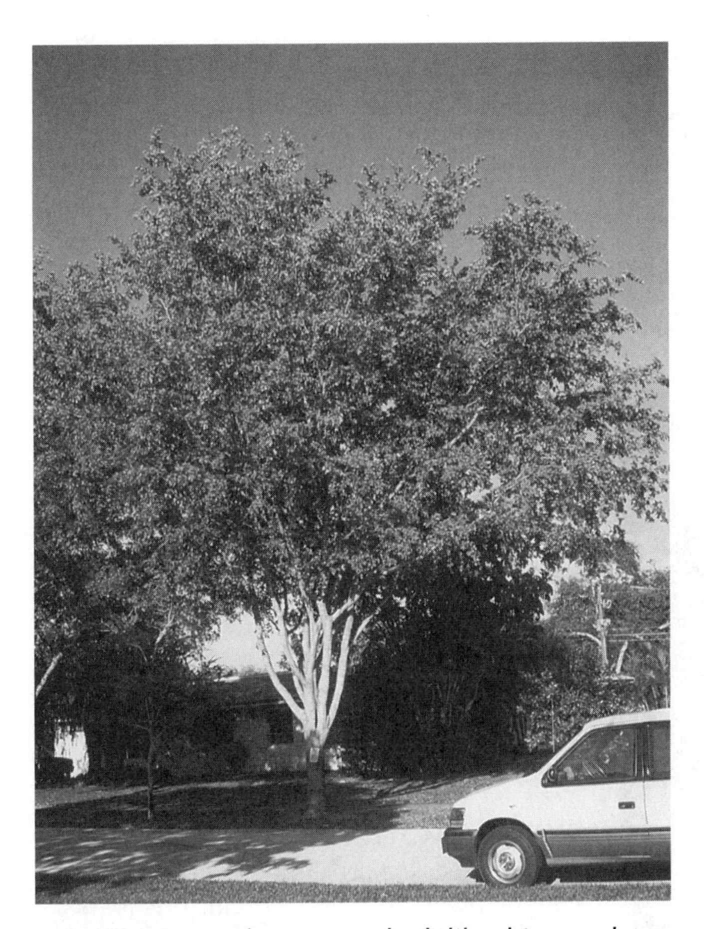

FIGURE 11-2. If your trees look like this, you have a problem. The codominant stems arising from the same spot on the trunk are likely to split from the trunk as it grows larger. Deciding which cuts to make on trees with this form can challenge even the most skilled arborist.

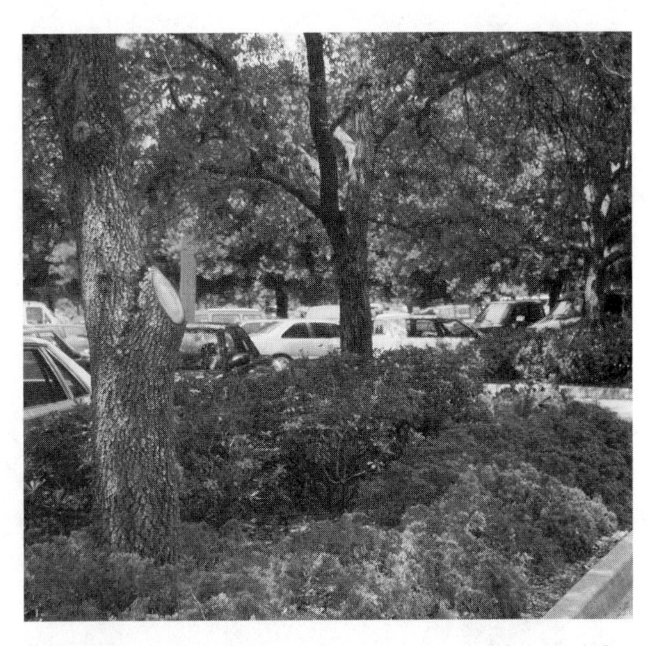

FIGURE 11-3. Notice the large wound created several feet off the ground after removal of the low limb on the tree at left. Large, low limbs on other trees in this photograph will have to be removed in a year or two as they continue to droop toward the ground. Removing large, low limbs can initiate trunk cracks and decay. Avoid this by keeping low limbs small in diameter with subordination. This will keep branches small in relation to the trunk when they are eventually removed.

These three problems are not good for long-term tree health in urban landscapes. As low branches and stems grow up into the **permanent canopy** (Figure 11-2) they can crack or break as they grow larger and taller. (The permanent canopy is that portion of the tree that will remain for a long time.) This results from the stems and branches growing too long and/or formation of included bark. Other branches may droop toward the ground and get in the way requiring their removal after they have grown to be quite large (Figure 11-3). Both conditions can lead to cracks and decay in the trunk and broken trees (Figure 11-4). Properly trained trees with **good architecture** are likely to live longer and provide more benefits to us than untrained trees (Figure 11-5).

Tip: *Reduce risk by instituting a structural pruning program in your community.*

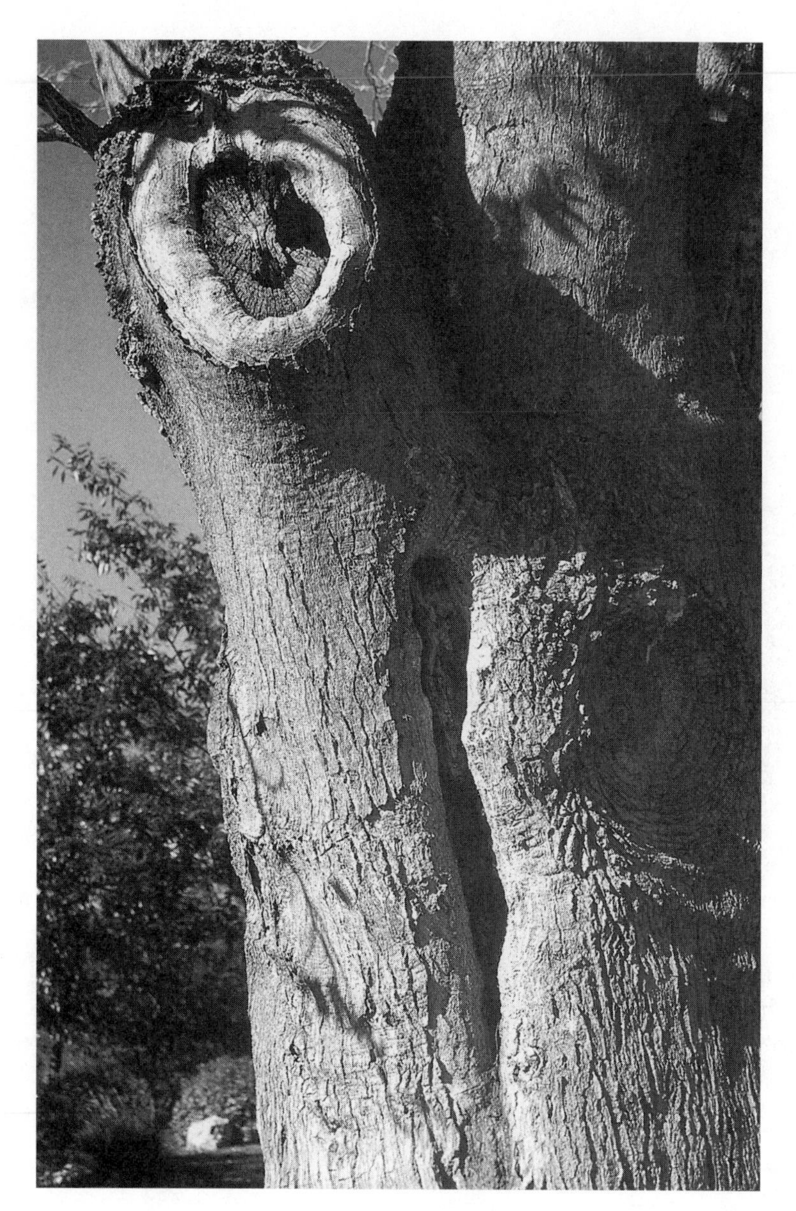

FIGURE 11-4. Removing the large branch on the upper left resulted in a large dead spot below the pruning wound. This is common when large branches are removed from certain trees. Cracks and decay occur at this location because there is no branch protection zone at the base of branches that are larger than about half the trunk diameter. Removing large branches can also result in these problems, even if they are small relative to the trunk.

Trees native to more open settings such as a prairie, savannah, and desert environments evolved without trees nearby. Spacing between trees is much greater than in a moist forest. The result is a low-growing tree with a spreading canopy. Low branches often become quite large on these trees. In their native habitat, these trees, such as the Paloverde tree (*Cercidium floridum*), hawthorns (*Crataegus*), and others, often develop included bark in branch unions. This is typically of little concern for the plant in its native habitat because low branches can touch the ground to support themselves. This helps prevent the low branches and stems from breaking from the tree.

When we plant these trees in an urban or suburban landscape, low branches usually cannot be left on the tree because they get in the way of cars, people, lawn mowers, and other activities. Therefore, we should keep these branches small so that when they are removed, a protection zone forms at the base of the branch. Since large branches with included bark cannot rest on the ground to support themselves in most landscapes, we should take measures to prevent them from splitting from the tree. Measures include

FIGURE 11-5. Trees trained to grow with one dominant trunk and branches smaller than the trunk have a strong structure. The ease of training trees to this ideal structure depends on the species or cultivar. Low branches are kept small so when they droop and get in the way only a small wound is left following removal. The pruning wound closes over with no ill effects on the tree.

keeping these branches small by shortening them regularly in the first twenty to thirty years after planting.

Some tree cultivars have been developed for urban landscapes. Many of these have an upright habit. Unless a cultivar has been developed and maintained for its dominant leader habit most cultivars need structural pruning as any other tree would. This includes shortening upright stems, especially those with included bark so they remain less than half the trunk diameter. These shortened, upright stems often remain on the tree for a long time.

Structural pruning may be less crucial on trees that, due to adverse site conditions, are expected to live for only about twenty years. For example, trees in 4 × 4 foot planting pits along a street are more likely to die from adverse site conditions than from the effects of poor structure. These trees may not benefit greatly from structural pruning. Pollarding, topiary, and other techniques may be most appropriate for these short-lived trees. Trees along greenways, nature preserves, and other locations where relatively few people visit are usually not structurally pruned. The risk of injury from tree failure is considered less than in urban locations.

The dominant leader structure is far less important for trees that mature at less than about 30 feet tall. Their parts do not grow as big and as heavy as large-maturing shade trees. If a branch or stem with poor structure or a poor connection should fail, injury or property damage may be less than for a large tree. Although training to a dominant leader structure is probably best, few people bother doing it.

OBJECTIVES OF STRUCTURAL PRUNING

The first priority in caring for young trees is to develop a plan (Appendix 6). The goal of the plan should be to ensure that trees are strong enough to stand against most wind, snow, and ice storms as they grow older. The plan will be designed to prevent poor tree form resulting in only poor pruning options later (Table 11-2).

A dominant leader on medium- and large-maturing shade trees such as oaks, honeylocust, linden, elms, and many of the large maples should extend up into the canopy as far as possible, preferably at least 25 to 40 feet from the ground if not more. The leader does not have to be straight, but it should be dominant. That means it should be considerably larger in diameter than all the branches (Figure 11-6). As a rule of thumb, *branches should be no more than about half the diameter of the trunk* when measured directly above the branch union. Arborists report that trees maintained in this manner often sustain less storm damage than trees allowed to develop other forms.

Trees growing to more than about 35 feet tall should have five to ten major scaffold limbs spaced along one dominant leader as far up into the canopy as possible. **Scaffold limbs** are the largest diameter, permanent branches that comprise the main structure of trees without needles. They should be spaced apart a distance equal to at least 5 percent of ultimate tree height in two rotations around the trunk (Figure 11-6, Table 11-3). Many old trees have scaffold branches spaced much farther apart than 5 percent of ultimate tree height. This might be one of the reasons they have managed to live so long. Of course there are exceptions, where old trees have branches closer than this.

If there are too many branches close together, the trunk and leader above may not be able to develop correctly. The trunk tissue above the branch cluster may have a difficult time growing around the base of the branches. As a result, resources may be shuttled to the branches and may not move around them to reach the leader above. The branches essentially steal the water, elements, and growth regulators from the leader causing it to grow slowly and become less dominant. In some cases, the leader declines and dies while the branches flourish. This results in poor structure and unhealthy trees.

The selection and development of the scaffold limbs and hence the permanent structure on the tree may take fifteen to forty years depending on the species or cultivar, climate, mature tree size, growth rate, location in the landscape, and desired tree form.

TABLE 11-2. Objectives of structural pruning of shade trees in the landscape.

1. Develop or retain a dominant leader
2. Identify the lowest branch in the permanent canopy
3. Prevent branches below the permanent canopy from getting too large
4. Space main branches along the dominant trunk
5. Keep all branches less than half the trunk diameter
6. Suppress formation of included bark
7. Maintain a live crown ratio of 0.6 or greater
8. Increase the quality of wood in the trunk should trees be harvested for lumber

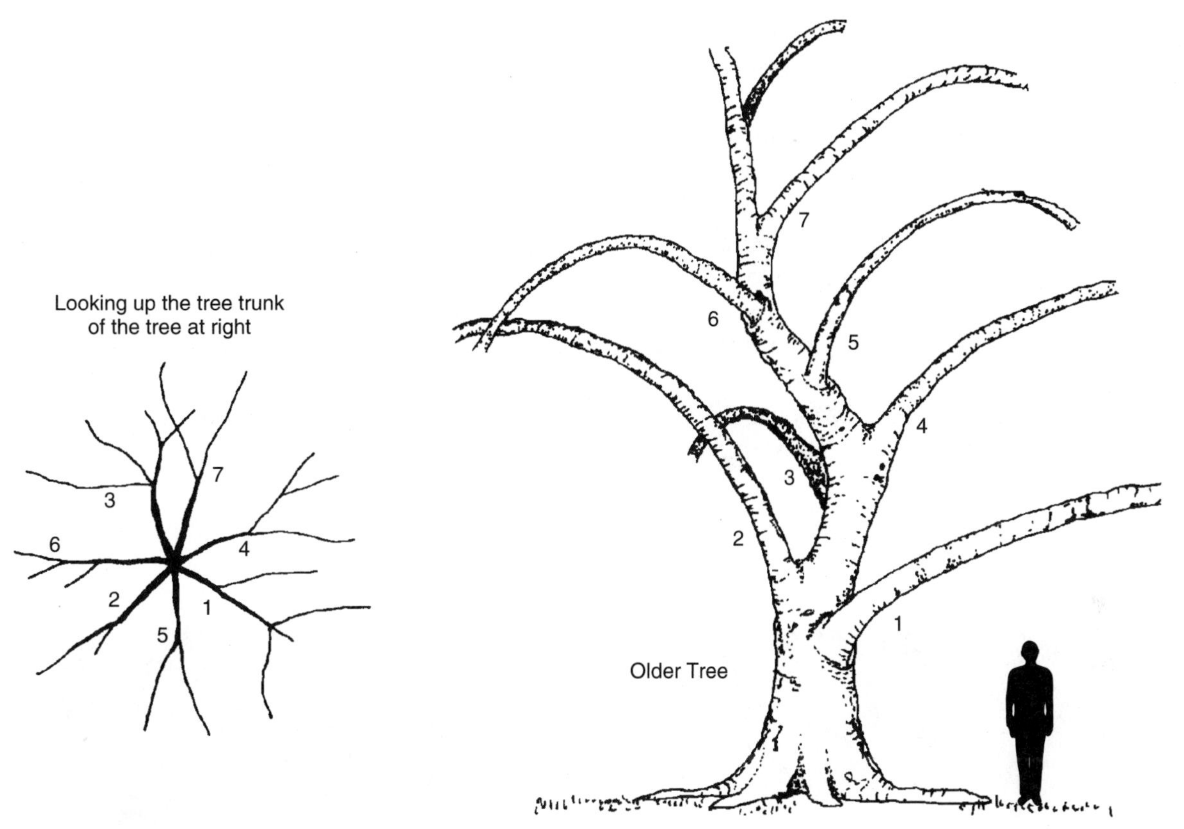

FIGURE 11-6. Scaffold limbs on medium- and large-maturing trees should be spaced along a dominant trunk. A minimum of 2 to 3 feet of spacing is best for older trees (right). Notice that scaffolds are distributed radially in two rotations all around the canopy (left). None are located directly above another. Smaller branches between the scaffolds are not shown in this drawing.

TABLE 11-3. Spacing scaffold branches along a dominant trunk.

Ultimate Tree Height	Recommended Minimum Distance Between Scaffold Branches
20′	1′
30′	1.5′
40′	2′
50′	2.5′
60′	3′
70′	3.5′
80′	4′
90′	4.5′
100′	5′

Select and position them carefully since they will remain on the tree for a long time. Although an even radial distribution of scaffold limbs around the trunk is ideal, with none directly above the other, this is rarely achieved (Figure 11-6). On most needle-leaf trees, allow development of no more than about five large branches at each node.

If large stems or upright branches form in what will become the lower third of the canopy when it nears maturity, (Figure 11-7), storms could break these trees easier than those pruned to a stronger structure (Figure 11-8). Prune to shorten these stems when the tree is young to prevent formation of codominant leaders and low aggressive branches. Do not clear the lower trunk of branches too high up into the canopy too soon. Strive to maintain a **live crown ratio** of at least 0.6 which means there is live foliage in the lower half of the tree at all times (Figure 11-9).

Tip: *Prevent branches originating in the lower 15 to 20 feet of shade trees from sending stems up into the permanent canopy.*

(a)

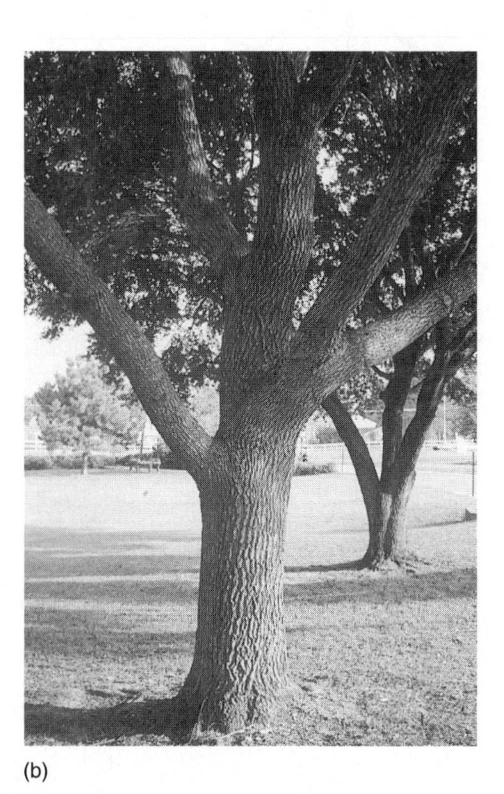

(b)

FIGURE 11-7. Trees, such as Chinese elm, that do not readily form a straight trunk can be trained with scaffold branches spaced along one dominant trunk (left). Aggressive low stems should be subordinated so that they grow more horizontally and develop into lateral branches rather than upright codominant stems. Early training to this optimum form will pay off because the trees will develop a strong structure with age (right). The lowest 4 or 5 branches on the elm (left) should not be allowed to grow to a large size nor up into the permanent canopy. They are too low to be part of the permanent canopy. Remove the upright and end portion of these branches to ensure they remain temporary branches and are removed latter as the tree grows. Weak form can be seen on the tree in the background on the extreme right.

There is one more reason to keep low branches small in diameter. Some communities are beginning a program to harvest urban trees for their timber value because some of the planted species have valuable wood characteristics for hobby and professional woodworkers. Some woodworkers look for wood that is free of knots, cracks, and decay. One of the best methods of producing this "product" is to keep low branches small and prevent formation of codominant stems.

Trees that are too tall for their trunk diameter are subject to blow over in storms. The ratio of tree height (in feet) to trunk diameter (in feet) can be used as a guideline. More than one recent study has shown that when this ratio is greater than 50 trees are prone to failure. A tree height: trunk diameter ratio of 20 to 30 is considered optimum (Claus Mattack). Consider reduction or removal on exposed trees that exceed height: trunk diameter of 50.

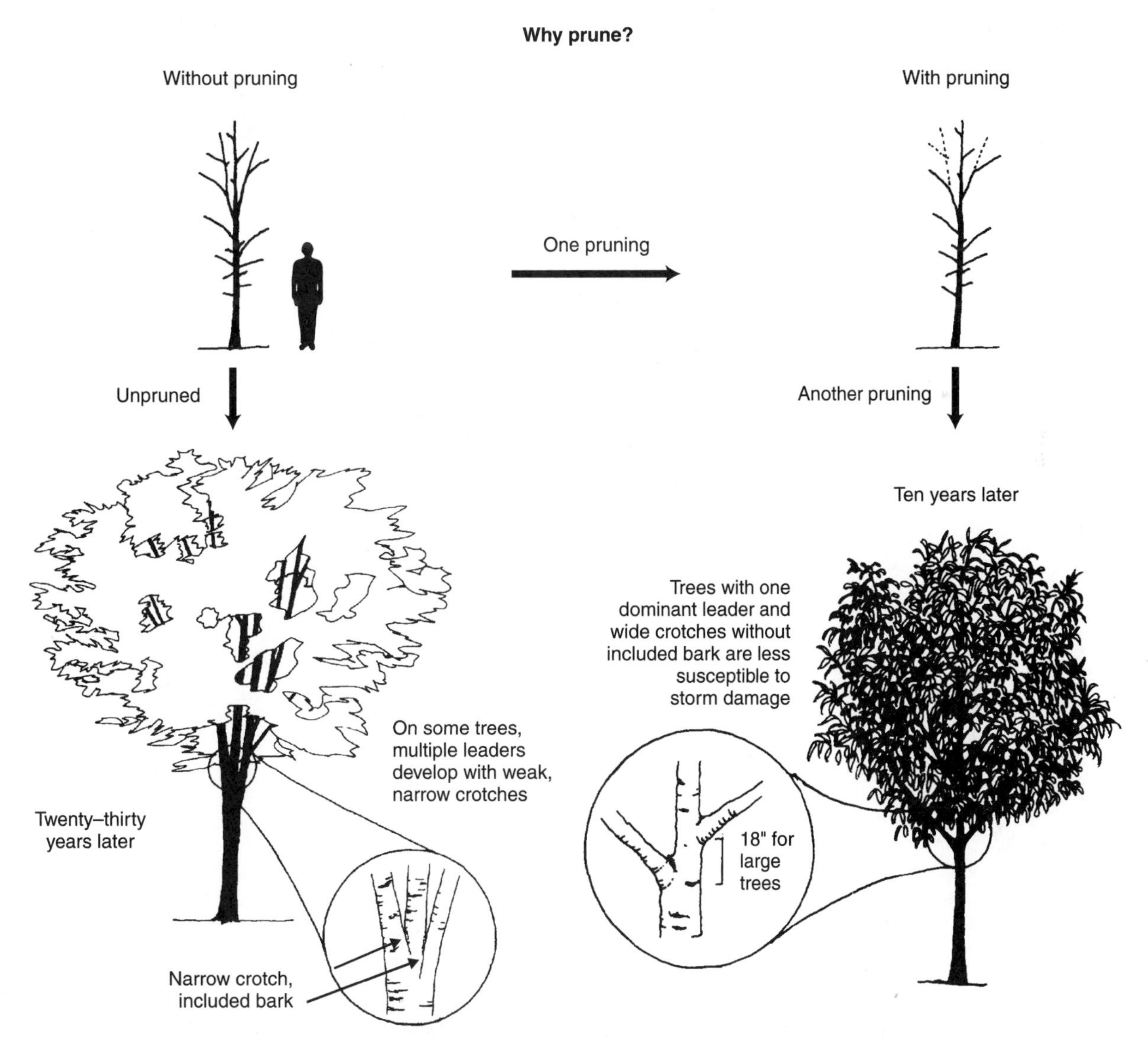

FIGURE 11-8. Training trees that mature at a large size to grow with one dominant leader (right) can minimize the possibility of branch failures (left). *(Illustration continued on next page)*

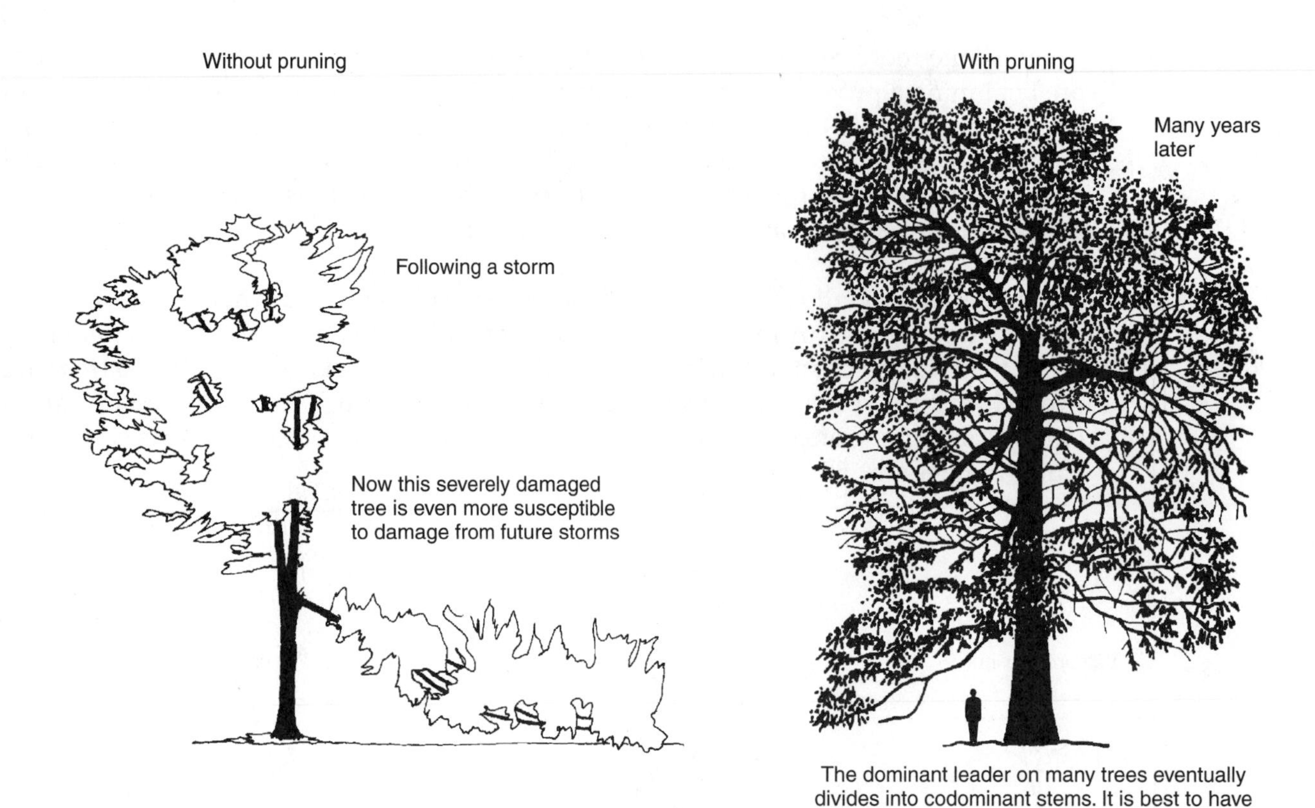

FIGURE 11-8. (continued) Training trees that mature at a large size to grow with one dominant leader (right) can minimize the possibility of branch failures (left).

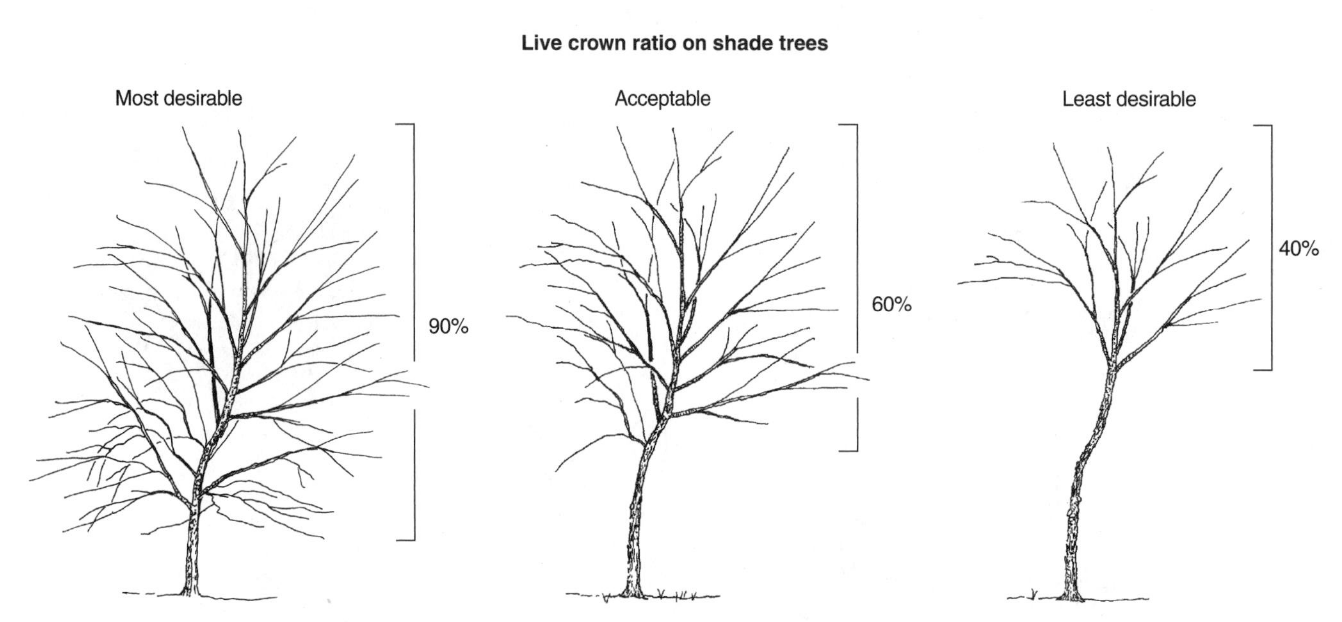

FIGURE 11-9. Live crown ratio is the percentage of the tree height with live foliage. A tree with a live crown ratio of 90% has foliage nearly to the ground (left). Ideally, trees should be maintained with at least a 60% live crown ratio. When too many branches are removed, a small ratio results (right).

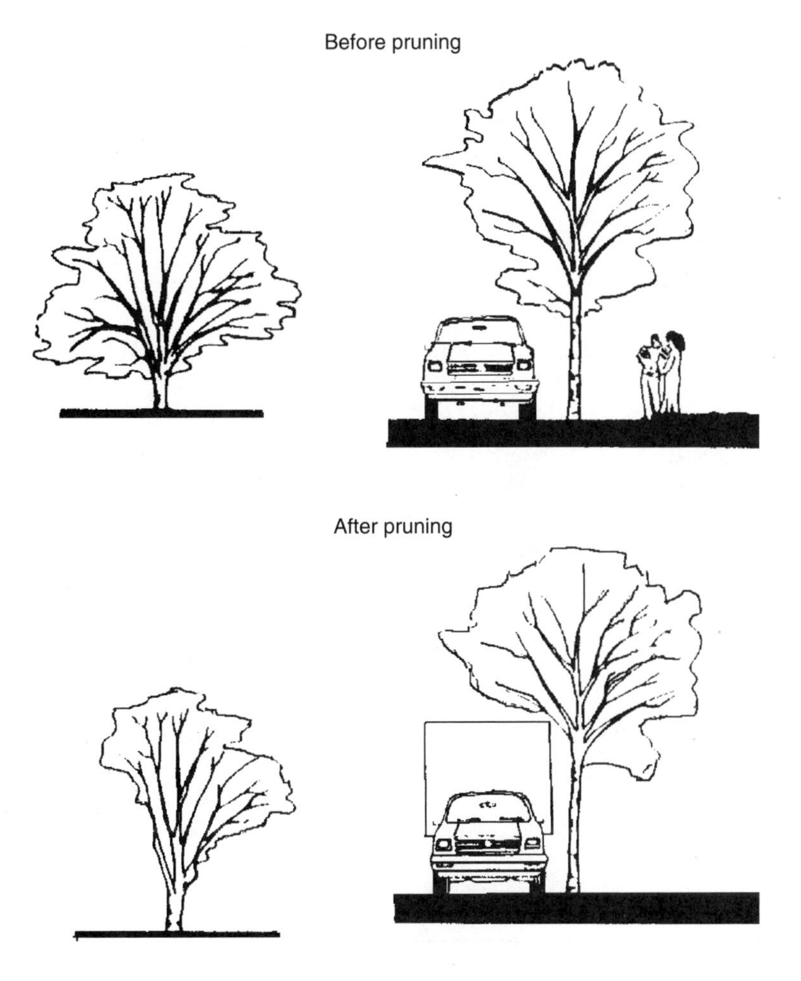

FIGURE 11-10. Single-trunked trees (right) are preferable to trees that are multi-trunked (left) for planting along streets and sidewalks and in many landscape situations. Branches and trunks on multi-trunked trees droop into traffic and must be removed. This disfigures the canopy (lower left) and encourages growth in upright stems which can result in poor form and included bark. These upright stems may grow too tall and split from the tree. They may also droop in the way of traffic on certain species and have to be removed leaving a huge wound. This can lead to trunk cracks, decay, and poor health.

DECIDE WHAT FORM IS MOST SUITABLE

Trees display a variety of shapes depending on their genetic makeup, spacing, growing conditions, access to sunlight, previous pruning, and cultural practices. However, their form can be modified greatly by training and pruning. Many trees grown in open landscapes with unlimited access to sunlight develop low branches that sweep the ground unless shortened or removed. Their form is much like that of very large shrubs. This low-branched form is suited for open lawn areas and other spots that have plenty of room for lateral spread of the canopy close to the ground. However, most landscapes where people live, play, and work including parks, yards, parking lots, and streets cannot support trees with this form.

When they are young, these same trees can be trained in the nursery and landscape to grow one trunk with no aggressive lower branches. This produces a more upright growth habit and a slightly taller tree. These trees are easier to walk, sit, or drive a vehicle under and they mimic the tree's more natural, forest grown form (Figure 11-10). Executing this pruning program may take twenty-five years, or more. In other words, it takes about as long to develop good structure in shade trees as it does to put good structure in a child (Appendix 6). Trees with this form are easier to manage than trees with large, low branches and codominant stems. This makes them the best choices for sustainable landscapes. Table 11-4 lists some of the suitable locations for trees with different forms.

TABLE 11-4. Suitable locations for shade trees with one dominant trunk, those with low aggressive drooping branches, and those with multiple trunks.

Trees with one dominant trunk are best:
Street trees
Parking lot islands and buffer strips
Residential properties
Near areas where people and vehicles frequent
Theme parks
Plazas
Golf courses
Near buildings
Low aggressive branches are OK for:
Arboreta and botanical gardens
Commercial building grounds away from buildings
Play grounds, outdoor sports facilities, and recreation areas
Residential properties where under clearance is not needed
Theme parks
Open lawn areas
Codominant stems are OK for:
Portions of parks where people do not frequent
Woods and forested areas
Reclamation areas
Retention ponds
Locations that will lead to short tree life
Pollarded trees
Topiaries

The tree shape and habit best suited for a particular site depends on the location of the tree, the desired effect, and the clearance needed beneath the canopy. Identify which branch will become the **lowest permanent limb** as soon as possible in the life of the tree. The lowest permanent limb is the lowest large branch or scaffold limb that will remain on the tree for a long time. The lowest limb can be chosen in the first ten years on some trees. More time is needed on other trees and in cold climates. Large-growing trees that have not attained 20 feet in height may have no permanent limbs on the tree. All may be temporary and will eventually be pruned off or shaded out.

All branches below the lowest permanent limb are temporary and they should not remain on the tree for more than a few years (Figure 11-11). Do not allow branches below the lowest permanent limb to grow up into the permanent canopy. The presence of included bark in temporary branch unions is of no consequence since the branches will eventually be removed. *However, permanent limbs should not have included bark!*

Single-trunked trees

Training trees near a single-story building to a single dominant trunk with no large branches on the lower 20 feet of the trunk is recommended if the branches will be trained to grow over the building. Aggressive branches on the lower 20 feet of the

FIGURE 11-11. Many trees along streets and in other locations have no permanent branches on the tree at planting. All branches are temporary in the sense that they will be or should be removed as the tree grows taller. On all but the most upright tree types, do not allow temporary branches to grow upright because they are too low on the trunk to become a permanent part of the canopy. Permanent branches originating low on the trunk can cause problems (see Figure 11-1).

trunk are difficult to train over the building. They will often crowd the side of the building or droop onto the roof as they increase in spread and diameter. Trees next to taller buildings should be trained to one dominant trunk as well to prevent branches from interfering.

Trees with no large lower branches and one dominant trunk are also the best choice near a street or sidewalk or in parking lots. This will allow easy pruning to create clearance for people and vehicles to pass under the canopy (Figure 11-10). If there are branches on the lower 20 feet of trees with a spreading form planted along a street, then they are likely to droop toward the ground and require removal later when they are large. Large branch removal can cause serious tree injury. Keep them shortened to prevent this.

It is easy to remove small-diameter, drooping low branches as needed on trees with a single dominant trunk. Removal of small branches mimics the natural branch-shedding process that occurs in the forest, so trees are well prepared for and capable of handling this treatment. Not much of the canopy is removed so the tree looks good after pruning. Small branches also form a branch protection zone so cracks and decay are less likely to enter the tree compared to removing large branches.

Low-branched, multi-trunked trees

It is not advisable to allow aggressive, upright low branches and stems to develop on trees that mature at greater than about 35 feet tall (Figure 11-2). This form may not be strong enough to hold together. Weak included bark could form in the branch unions of the low aggressive branches. This can result in broken trees. Aggressive horizontal branches should not be allowed to develop either if they will require removal later.

Tip: *When branches from the lower half of the canopy grow up into the upper third of the canopy, cut them back to live lateral branches.*

Recognizing the appropriate measures to take requires some imagination, foresight, and practice. When low branches on young trees grow upright, they appear as though they will not droop to get in the way later. But the low branches on many spreading type trees will grow large and get in the way because they usually have access to abundant sunlight from the side of the canopy. To determine if the low branches are likely to droop later, look at older trees of the same species or cultivar. If they will be in the way later, prune to shorten them now.

CHECK YOUR KNOWLEDGE

1) Which of the following is NOT a main objective of structural pruning in the landscape?
   a. develop or retain a dominant leader
   b. identify the lowest branch in the permanent canopy
   c. space main branches along the dominant trunk
   d. thin the canopy

2) For street tree planting, single-trunked trees are preferable over multi-trunked trees because:
   a. single-trunked trees provide more shade.
   b. branches and trunks on multi-trunked trees droop into traffic and must be removed.
   c. many multi-trunked trees do not have a strong structure.
   d. single-trunked trees are more vigorous than multi-trunked trees.

3) The dominant leader structure important for large-maturing trees is less important on small-maturing trees because small-maturing trees:
   a. are less likely to cause injury or property damage if they fail.
   b. usually do not have included bark.
   c. cannot be grown with a dominant leader.
   d. grow slower than large-maturing trees.

4) It is less important to prune trees to a good structure in a small cutout in a sidewalk because:
    a. these trees are under stress and grow slowly.
    b. these trees are more likely to die from adverse site conditions than from structural problems.
    c. trees chosen for these locations are usually small-maturing trees.
    d. these street trees usually have good structure to begin with.

5) If low codominant stems are allowed to develop on the trunk:
    a. trees usually grow fast.
    b. trees usually develop a spreading habit.
    c. included bark always develops in the unions of these codominant stems.
    d. trunks can decay if the stems need to be removed later.

6) Scaffold limbs on large-maturing shade trees should be:
    a. chosen as soon as possible but certainly within the first ten years in the life of the tree.
    b. spaced apart along the trunk a distance equal to about 5 percent of the ultimate height of the tree.
    c. at least 12 inches apart.
    d. oriented upright so they do not droop to the ground.

7) The permanent branches on the tree are called:
    a. lateral branches.
    b. secondary branches.
    c. scaffold limbs.
    d. codominant stems.

8) Decurrent trees with a spreading form that are not placed on a regular pruning program:
    a. have a short life.
    b. develop included bark.
    c. are likely to require large branch or stem removal on the lower trunk.
    d. develop codominant stems.

9) One of the objectives of structural pruning is to keep branches less than what diameter compared to the trunk?
    a. two-thirds
    b. three-quarters
    c. one-third
    d. one-half

Answers: d, b or c, a, b, d, b, c, c or d, d

CHALLENGE QUESTIONS

1) What are the major objectives when pruning young and medium-aged trees in urban and suburban landscapes?
2) Describe several situations where trees with low branches would be suitable.
3) Make a sketch (without looking at this book) of the ideal form on a medium-aged shade tree.
4) Describe an appropriate program for pruning 4–inch caliper street trees that are 15 feet tall with many aggressive upright branches.

SUGGESTED EXERCISES

1) Take a group out to a young street tree or parking lot planting. Have the group visualize and discuss the form on the trees twenty years from now if no structural pruning was performed. Now discuss the general strategies for pruning these trees taking into account potential drooping branches, included bark, and codominant stems.
2) Do the same as in the previous exercise but for medium-aged trees.
3) Take a group to a park where trees are planted in open areas and discuss the strategies for creating trees with strong structure.

CHAPTER 12

STRUCTURAL PRUNING OF SHADE TREES IN THE LANDSCAPE: EXECUTION

OBJECTIVES

1) Determine how to develop and maintain a dominant leader.
2) Determine how to space main limbs apart.
3) Determine how to improve structure in young and medium-aged trees.
4) Contrast limb management on landscape trees with street trees.
5) Determine where to position the lowest permanent limb.
6) Contrast pruning excurrent with decurrent canopy forms.

KEY WORDS

Lateral branch
Moment of stress
Pruning dose
Suppression

DEVELOPING AND MAINTAINING A DOMINANT LEADER

There are three basic steps to developing and maintaining a dominant leader. First identify the stem that will make the best dominant trunk (Table 12-1). It should not have cracks, openings, large pruning wounds, cankers or other defects that could compromise its strength. Next identify the stems and branches that are competing with this stem. Lastly, subordinate competing stems and branches by shortening or removing them. *This will have to be repeated at regular intervals for the first twenty-five years or so after planting in order to develop sustainable structure in shade trees.* If done properly, the tree may greatly benefit from five or six prunings more or less evenly spaced during this period (Miller, 1981).

The choice between removal and shortening of competing stems, in other words the **pruning dose,** depends on customer expectations, the size of the stems, and the pruning cycle. The dose is the amount of live tissue removed at one pruning. Administering a large dose removes a great deal of live tissue from a stem, or might remove the stem altogether. A small dose removes a lesser amount of live tissue (Figure 12-1). Following subordination on a stem or branch, a weak union may still be weak but failure is less likely because the stem was shortened. This reduces the moment of stress on the union. Failure potential may diminish in future years as new wood forms around the base of the subordinated stem.

Like many aspects of tree care there is a trade-off. Removing a large amount of tissue on competing stems with a reduction cut or a combination of reduction and removal cuts (i.e., applying a large pruning dose) encourages more growth in the leader, but this can leave a large void in the canopy. A smaller void is left in the canopy by applying a smaller dose, but this sends less of a growth-encouraging signal to the leader. The tree appears less pruned with the small dose. Horticulturists and professional arborists balance these to make appropriate prescriptions (Table 12-2).

You can apply a larger pruning dose if customers are willing to live with a large void in the canopy or if the pruning cycle is more than about three to five years. In other words, make a reduction cut farther down into the canopy, or remove the competing stem entirely (Figure 12-2). However, keep the cut smaller than your maximum critical diameter. Apply a smaller dose by pruning higher in the canopy if only one or two years will lapse between prunings. This will result in smaller pruning cuts than a larger dose. With a smaller dose, you will return at regular intervals to make small-diameter reduction cuts in the top half of the codominant stems. The tree may look thinner for a while on the side of the canopy that was subordinated. If this is a concern, the other side of the canopy can be thinned to visually balance the canopy, if desired (Figures 12-3 and 12-4g-h).

TABLE 12-1. Steps to pruning trees to a dominant leader.

Step	Action
Step one:	Identify a stem that will make a strong dominant leader
Step two:	Identify stems and branches that are competing with this stem
Step three:	Decide on the pruning dose and either remove or shorten competing stems and branches

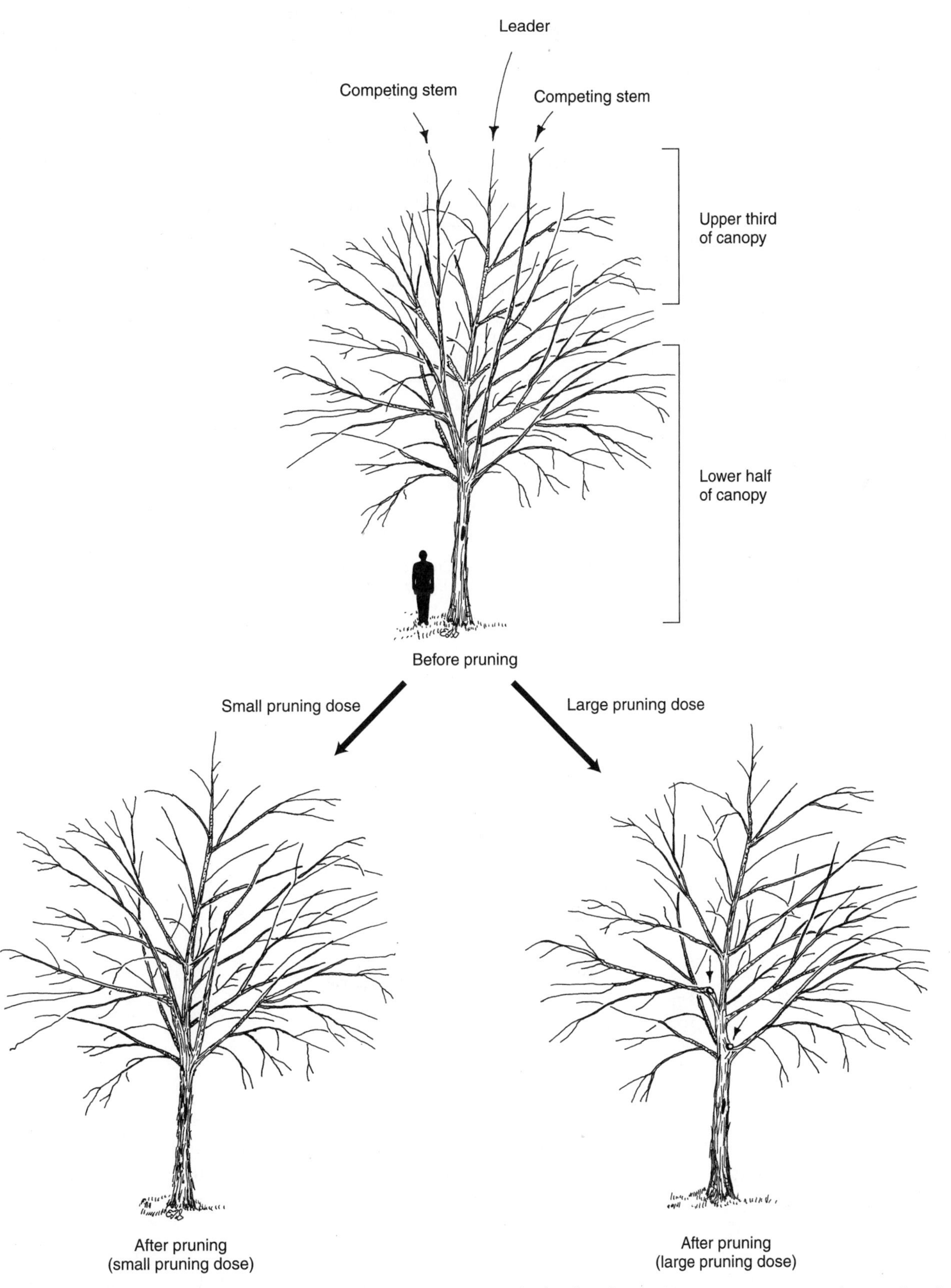

FIGURE 12-1. Shorten stems and branches competing with the leader (see two arrows—top) growing into the upper portion of the canopy which originate in the lower half of the tree. A small pruning dose (lower left) shortens competing stems less than a large pruning dose (lower right). A large dose removes more live tissue (see two large arrows) leaving a larger void in the canopy.

TABLE 12-2. Comparing pruning doses (see Figure 12-1).

Appropriate Pruning Dose for Specific Applications	
Large Pruning Dose	**Small Pruning Dose**
Municipality	Residences, commercial properties
Long pruning cycle	Short pruning cycle
Places where aesthetics is less of a concern	Places where aesthetics are a concern
Effects on the Tree from Applying Pruning Doses	
Large Pruning Dose	**Small Pruning Dose**
Results in a larger pruning wound	Results in a smaller pruning wound
Larger void in the canopy	Smaller void in the canopy
Greatly encourages growth in unpruned portions of tree	Encourages some growth in unpruned portions of the tree

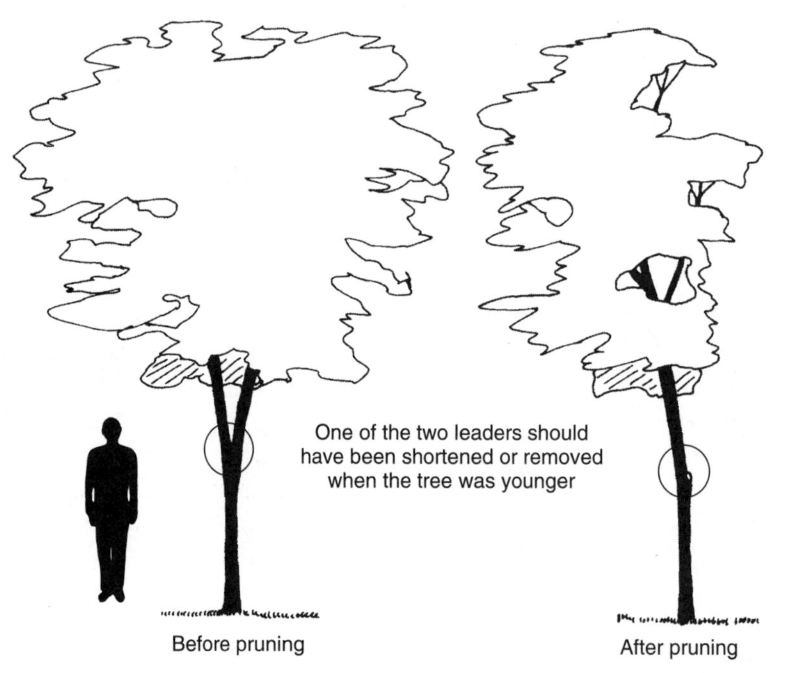

FIGURE 12-2. One side of a double leader can be removed from trees when this condition is recognized, especially on trees that grow to a large size. This improves structure because now there is only one dominant leader but it leaves a large void in the canopy. Subordination is usually more appropriate than removal.

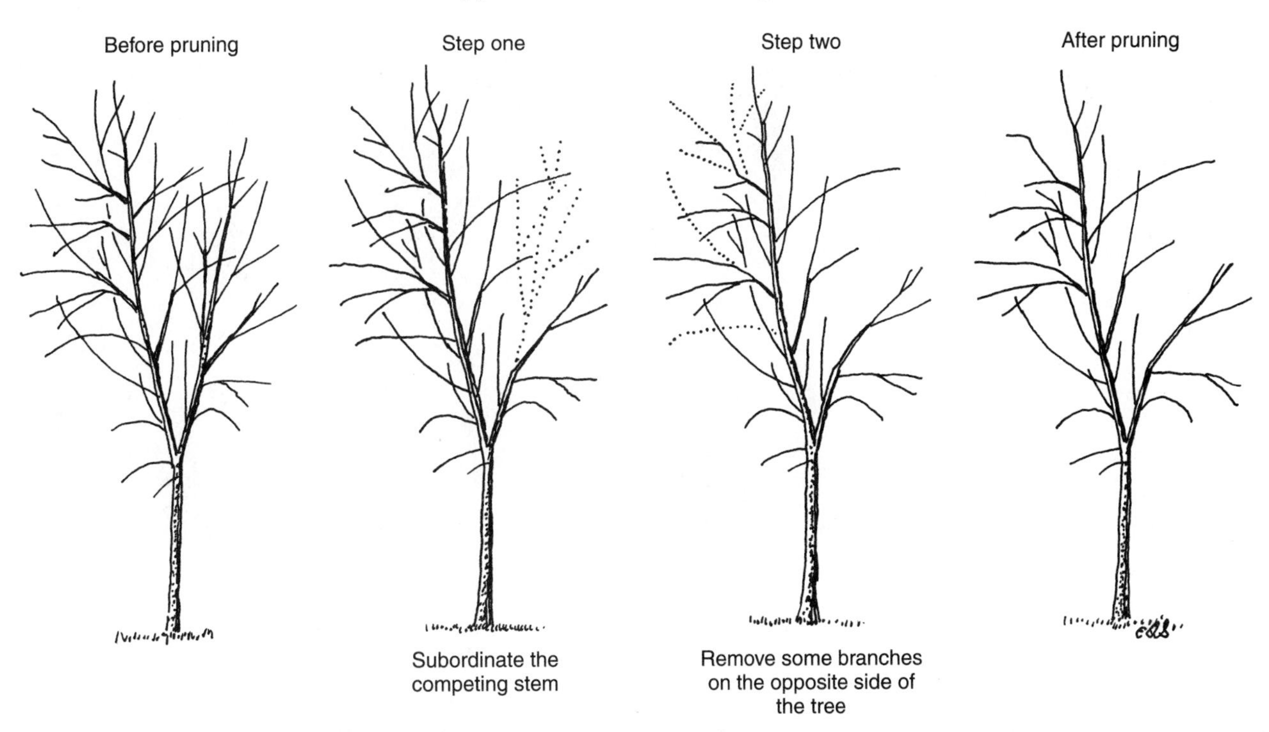

FIGURE 12-3. Thinning the side of the canopy opposite of the subordinating reduction cut improves the appearance of the tree by balancing the canopy. This could help a customer accept the subordination technique.

Tip: *A dominant leader is produced by removing or shortening stems and branches that compete with the stem you think will make the best trunk.*

Properly executed, a tree can be structurally pruned usually without disfiguring the canopy (Figures 12-4 and 12-5). In many cases, the key is to remove moderate amounts of tissue with reduction cuts and some removal cuts instead of removing entire stems back to the trunk. The subordinated stems may or may not be removed at a later time, but *this decision does not have to be made now.*

Before pruning the 10-inch diameter elm in Figure 12-4a, there was an aggressive stem growing into the upper third of the canopy that originated from the lower left side of the trunk. There was another aggressive stem on the lower right originating several feet from the ground. Both stems were shortened with reduction cuts to encourage the stem in the center of the tree to become dominant (Figure 12-4b). Two years later, both shortened stems were much smaller in diameter compared to the trunk diameter (Figure 12-4c). They were not growing in the upper third of the canopy indicating that they were sufficiently subordinated. This encouraged the stem in the center to become dominant. Now there was another aggressive stem originating about halfway up the tree on the left side. The arborist shortened this stem with a reduction cut (Figure 12-4d). This process should continue at regular intervals until a sound structure develops. This is a structural pruning.

FIGURE 12-4. a) Before pruning an elm. b) After subordinating one aggressive stem on each side of the tree. c) Two years later on the same elm tree another stem on the upper left needs to be shortened. d) After shortening this stem, the leader appears to dominant the structure. e) Before pruning a tree that was rounded over and shaped in the nursery and planted in this landscape five years ago and never pruned. f) After pruning, a leader appears in the center of the canopy to the top of the tree. Many branches were shortened or removed back to the trunk to allow those remaining to grow lateral branches. g) Before pruning an oak in a parking lot. h) The right-hand stem was subordinated with a reduction cut and the left side of the canopy thinned to balance the void left by the reduction cut.

FIGURE 12-5. Problem: There are three equally dominant stems originating about 8′ from the ground on this 12 inch diameter large-maturing shade tree (left). They are way too low to allow them all to grow up into the permanent canopy. **Solution:** Established trees benefit from the subordination of codominant stems. Making single reduction cuts on the left and right stems will cause the stem (in the center) that was not cut to dominate the tree structure (center). Both subordinated stems can be removed at a later time, if needed. The tree would look bare if both stems were entirely removed now; this would be overpruning. The photograph at right shows a close-up of the reduction cut. Sprouts that originate from the cut can be shortened if they become too aggressive.

Trees that were inappropriately topped and shaped into an oval or ball in the nursery often generate hundreds of sprouts when planted out into the landscape (Figure 12-4e). These sprouts are in a race for the edge of the canopy because that is where the sunlight is. They grow with no taper because few lateral branches are produced along their length. No leader will dominate in this type of tree without appropriate pruning at regular intervals for about a decade following planting. Many branches should be removed (with removal cuts) and shortened (with reduction cuts) to encourage a leader to dominate the structure (Figure 12-4f) and to space branches apart (see Chapter 13). Figure 12-6 provides several examples of using reduction and some removal cuts to improve structure on established landscape trees.

Tip: *When finished with structural pruning on young trees, you should be able to see the tip of the dominant leader while walking around the tree about as far away from the trunk as the tree is tall. Shorten stems that block your view of the leader.*

Older trees can benefit from subordination as well but they respond much slower than younger trees (Figure 12-7). Reduce the length of stems and branches that compete with the leader using reduction cuts. Remove some branches with removal cuts for

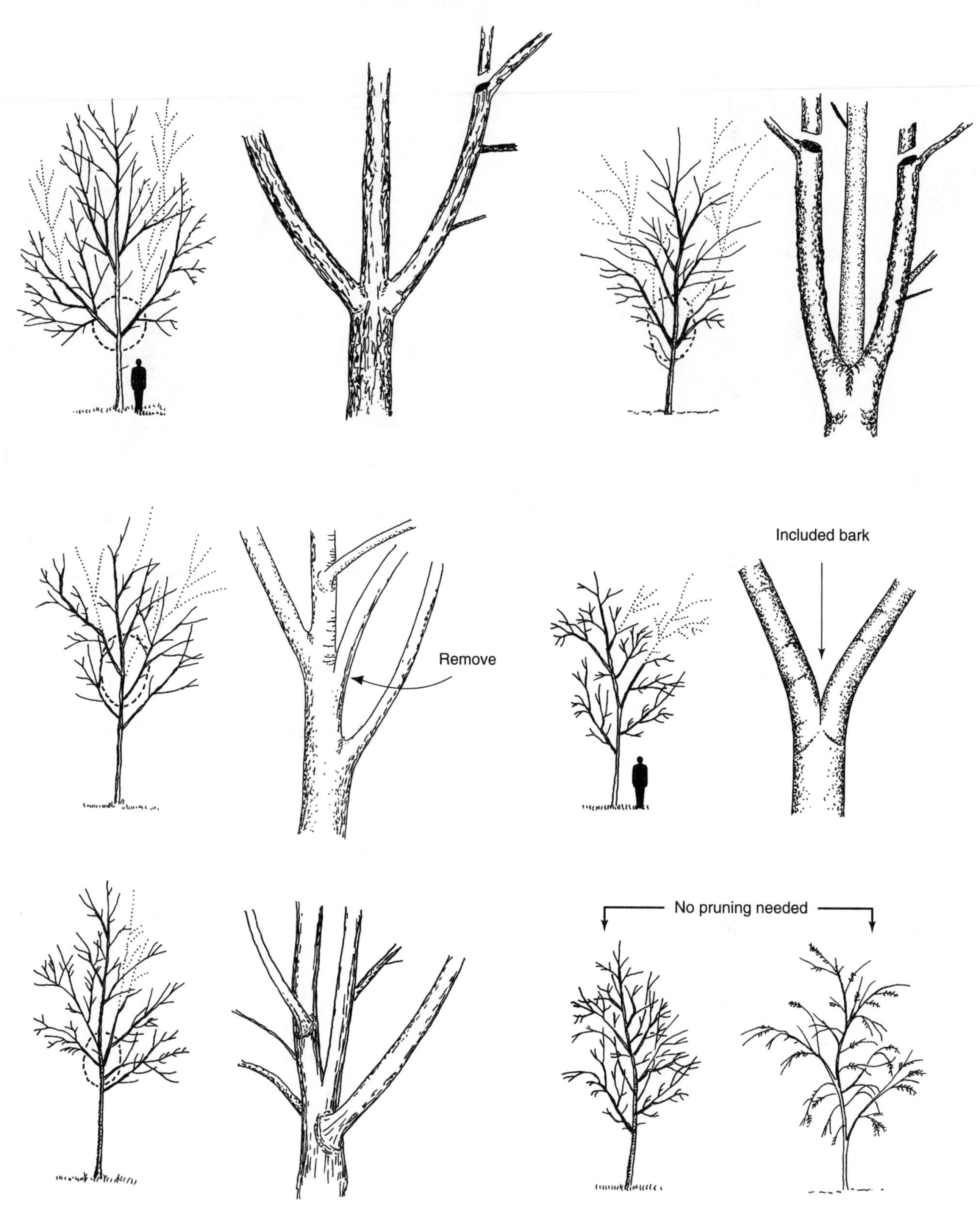

FIGURE 12-6. Remove stems and branches indicated with dotted lines. Circled area on each full tree drawing is shown in the detailed view to the right of each tree. There are two large diameter branches opposite each other low in the canopy. A reduction cut on each competing stem will slow their growth with the intention of keeping them smaller than half the trunk diameter (upper left). The second example is similar except than an additional reduction cut was made at the top of the canopy (upper right). In addition to a reduction cut to slow an aggressive stem half way up the left side of the tree, and entire stem was removed back to the trunk on the right hand side of the tree on the middle left. Stems with included bark should be reduced to lessen the chances for breakage (middle right and bottom left). If there is already a dominant trunk on the tree, little or no structural pruning may be necessary (bottom right).

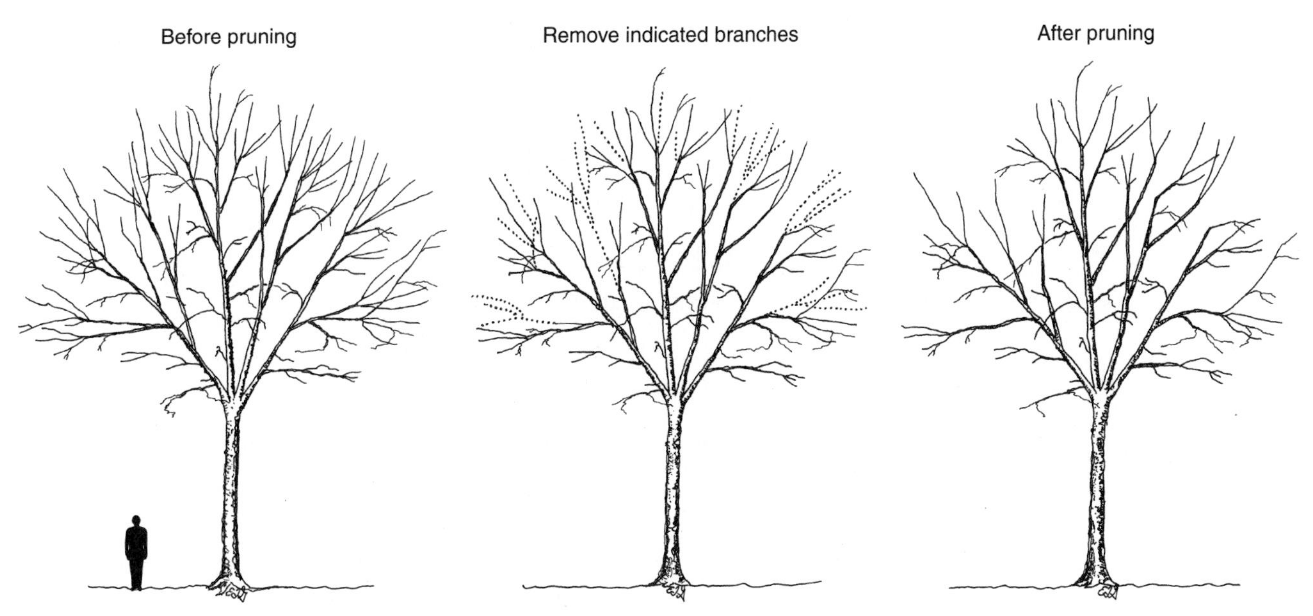

FIGURE 12-7. Older trees can be subordinated to influence structure. Remove some branches from the outer portion of the canopy on stems that are competing with the main trunk (center). After pruning, the canopy will be more open and there will be some unevenness at the edge of the canopy (right). If there is no main trunk, shorten as many stems as possible so that one can dominate. Many prunings over a period of years may be required to significantly influence structure.

further suppression of growth. *Reduce the length of those with included bark or other major defects.* Keep reduction cuts less than a maximum critical diameter chosen for that species. Instead of making reduction cuts on mature trees, consider removing lateral branches from stems and branches that compete with the leader to help slow their growth. Removal cuts usually cause less decay and minimize other potential problems in mature trees compared to reduction cuts. Remove lateral branches from the outer (distal) half of the competing stems and branches leaving those closest to the trunk. The problem with removing laterals to subordinate is that you do not get as much immediate reduction in the **moment of stress** on a poor branch union, and you may force the branch to grow too long. Moment of stress is the force applied to the branch union.

DEVELOPING MAIN LATERAL LIMBS

General strategy

Before pruning a young to medium-aged tree, determine the position of the lowest scaffold limb in the permanent canopy. Do not allow any branches or stems below this point to grow up into the canopy because you do not want them to get large. If they grow up into the permanent canopy, they are likely to get large. Table 12-3 lists the strategies for branch management over the life of a tree.

TABLE 12-3. Branch management strategies for shade trees (see Figure 12-8).

First 5 Years
Shorten all aggressive branches to keep them less than half trunk diameter
5–20 Years
Shorten all aggressive branches originating from the lower 10–20 feet of the tree Keep all branches less than half trunk diameter Identify lowest couple of scaffold limbs Prevent lower branches from growing up into the permanent canopy Shorten branches with included bark
20–40 Years
Identify and develop most of the scaffold branches Space scaffold branches appropriately Remove many of the branches below the lowest scaffold Shorten branches with included bark
50 + Years
All low temporary branches removed All scaffold branches are developing and are spaced apart Shorten branches with included bark

In most instances on young and medium-aged vigorous trees less than about 25 years old, consider shortening or removing vigorous upright branches, even those growing from scaffold limbs. Upright branches are the ones growing more or less vertically into the upper portion of the canopy. Pay particular attention to shortening those below what will become the permanent canopy (Figure 12-8). When shortening upright branches reduce back to a more horizontal or smaller **lateral branch.** If there are no branches on the tree that will become part of the permanent canopy because the tree is very young, shorten all aggressive branches on the tree to slow their growth (Figure 12-8, left and right). This forces growth into the leader and speeds development of the permanent canopy because less growth goes into lower branches that will eventually be removed. Some trees, such as Jacaranda (*Jacaranda mimosifolia*) crabapples (*Malus sp*) and others, produce many vertical branches from the main branches as part of their natural habit. It would be best to shorten these instead of removing them because removing them could prune off too much living tissue. This could cause stress and more sprouting.

Tip: *Shorten all branches below the lowest scaffold limb often enough so they remain smaller than half the trunk diameter.*

Identify and encourage growth in scaffold limbs as the tree grows taller. Since scaffolds will remain on the tree for a long time they should be free of serious defects such as included bark, openings, cankers, large pruning wounds and cracks, and they should be kept less than about half the trunk diameter. There can be smaller branches on the trunk between scaffold limbs. Keep in mind that all branches below the permanent canopy, in other words those below the lowest scaffold limb, will eventually be removed from the tree—so *keep them small through pruning.*

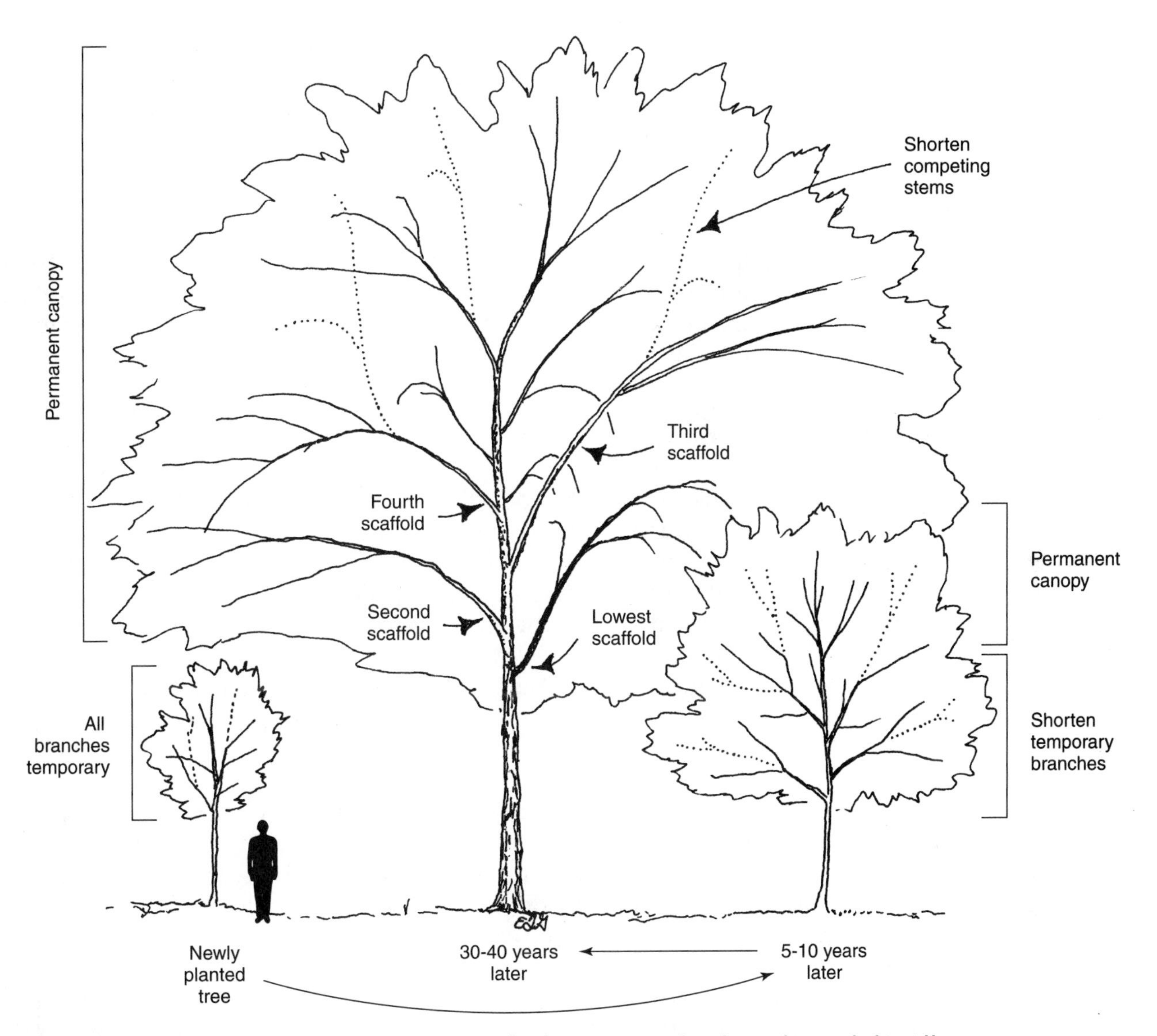

FIGURE 12-8. Often, there are no permanent limbs on a newly planted tree (left). All are temporary and will (or should) be removed as the tree grows taller. Once the tree is established for a few years, the tree may grow tall enough for one or two scaffold limbs to appear (right). Each time the tree is pruned, temporary branches should be shortened to prevent them from growing up into the canopy. By 30 to 40 years after planting, all temporary branches have been removed and scaffold limbs have been spaced apart with appropriate pruning (center). Continue to shorten the upright portion of branches, especially those in the top portion of the tree.

It is the responsibility of the urban forester, arborist, municipality, landscape management firm, and others to choose and develop the scaffold limbs. This cannot be done in the nursery because the trees are too small to have more than one or two scaffold limbs. Many nursery trees are too small to have *any* scaffold limbs. This lack of attention in the landscape is where the systematic pruning cycle often breaks down and is the source of many of our urban tree problems.

A good pruning program in the nursery that creates a well-structured finished nursery tree will do little if anything to ensure good form develops in the landscape. A well-structured nursery tree only provides a good basis to continue to build on—much like elementary school provides a child with a good basis to build on. A good elementary school education does not ensure success in high school without continued training in high school. The twenty-five year training program required for creation of good, strong tree architecture must be carried out by the landscape manager (Appendix 6).

Any large branch touching another large branch should be pruned so they no longer touch. Rubbing branches can injure one another causing deformities, internal decay, and cracks. Prune the one that is not positioned well on the tree or the one that will be removed after a while, anyway.

Branch management on young trees

Pruning strategy in the tree's first two to five years after planting into the landscape should focus almost exclusively on developing a dominant leader. By about the fifth year, some attention should be paid to forming branches properly. Vigorous branches should be pruned, but only enough to prevent them from overtaking the leader, growing up into the permanent canopy, or growing too large. The early shortening then removal of low, vigorous branches allows the trunk to quickly close over the small wounds created by pruning. It also encourages rapid growth in the permanent portion of the tree canopy above. Keeping less-dominant branches along the trunk may help close pruning wounds nearby.

The lowest scaffold branch or two and hence the beginnings of the permanent structure can usually be chosen by the fifth to tenth year after planting to an open landscape where there is no need for vehicular traffic passage beneath the tree canopy. The lowest scaffold limb may take longer to identify for a street or parking lot situation because higher clearance is required under the canopy.

If there are many branches close together on the trunk of a young tree, remove some so the remaining ones are spaced farther apart. This will allow the leader and branches to develop appropriately (Figure 12-9, left) and can help increase height by stimulating leader and branch growth in the permanent canopy. As the tree grows to 20 and 30 years old, continue to remove some main branches back to the trunk so the remaining ones can develop into scaffold branches (Figure 12-9, right). In a sense, some branches temporarily "rent" space on the trunk for a period of years. We eventually remove these. Those that "purchase" space remain on the tree for a long time and we call these scaffold limbs.

Tip: *Temporary branches rent space on the trunk; scaffold branches purchase space.*

On many trees, some branches grow faster than others and eventually dominate others. Choose these for scaffolds if they are in the correct position as described earlier. If a branch grows in the wrong direction, make a reduction cut back to a lateral pointed in the appropriate direction. If one is growing too close to another, reduce its length or remove it. Figure 12-10 shows reducing the length of main branches competing with

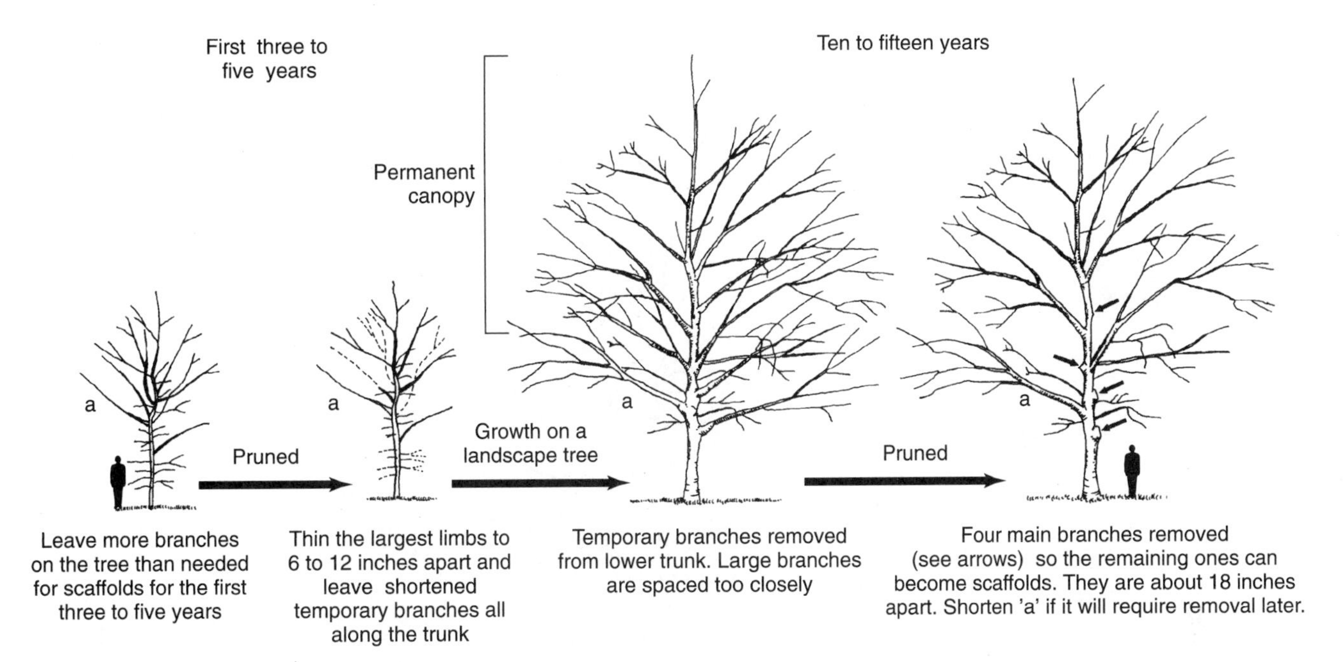

FIGURE 12-9. Space scaffold limbs apart along a dominant trunk. On trees that do not require under clearance, the lowest permanent scaffold can be close to the ground as shown above. Shorten 'a' and other low scaffolds if they will require removal later.

branches you want to develop into scaffold limbs. Reducing branches instead of removing them (shown in Figure 12-9) keeps the canopy full of foliage. In most instances the reduced branches will eventually be shaded out and their growth will be slowed. This will allow the scaffold limbs to dominate that position on the trunk.

Branch management on sheared trees

In many regions it is popular for nursery managers to shear all branches back in a hedge-like manner without regard to developing scaffold branches. This is done to meet market demand for this type of candyland, picture-book form. No effort is made to encourage scaffold development, and the trunk is crowded with branches. A look inside the canopy of these trees reveals that many branches are the same diameter. This may be acceptable in some cases for young trees since all these branches may be removed. It is a questionable practice on a decurrent tree with branches that could develop into scaffold limbs.

Some trees hedged in this manner look quite odd when they begin to grow out in the landscape (Figure 12-11). The outline of the clipped canopy can be seen through the thinner canopy that develops after frequent clipping stops. Other than looking odd for a period of years, this may have little lasting effect on trees with an excurrent growth habit such as the holly shown in Figure 12-11. However, the leader may not dominate, so subordination practices may need to be employed for five or more years to build durable structure. Trees that have been hedged in this manner may require more time for several years after planting to develop good structure.

Reducing the length of branches competing with scaffold limbs

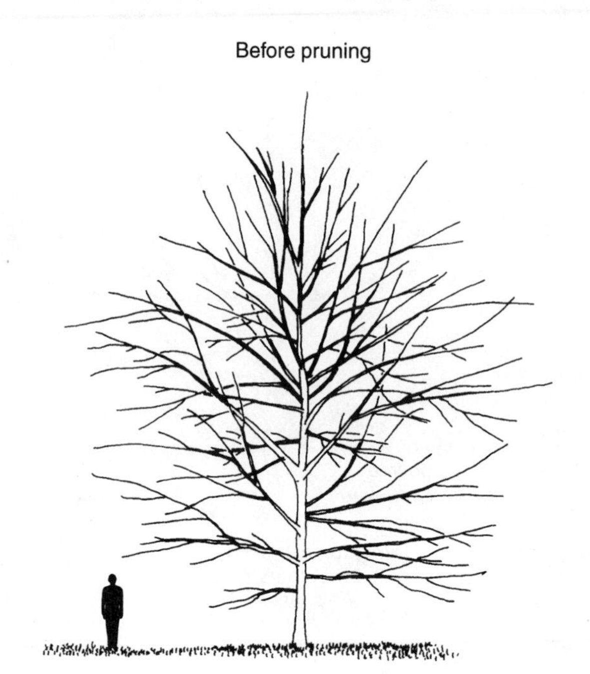

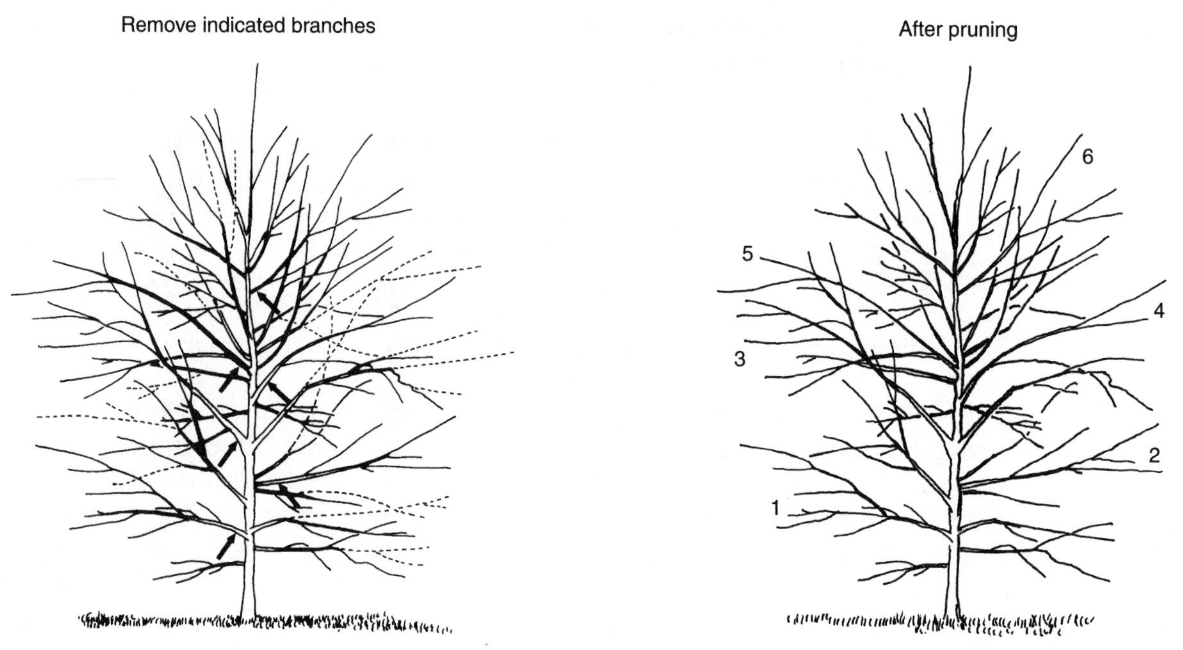

FIGURE 12-10. Decide which branches will become the scaffold limbs (arrows). Then shorten the other large-diameter branches to slow their growth rate (lower left). Six scaffold limbs are now spaced along a dominant trunk (lower right). The distance between scaffold limbs should be at least 5% of ultimate tree height.

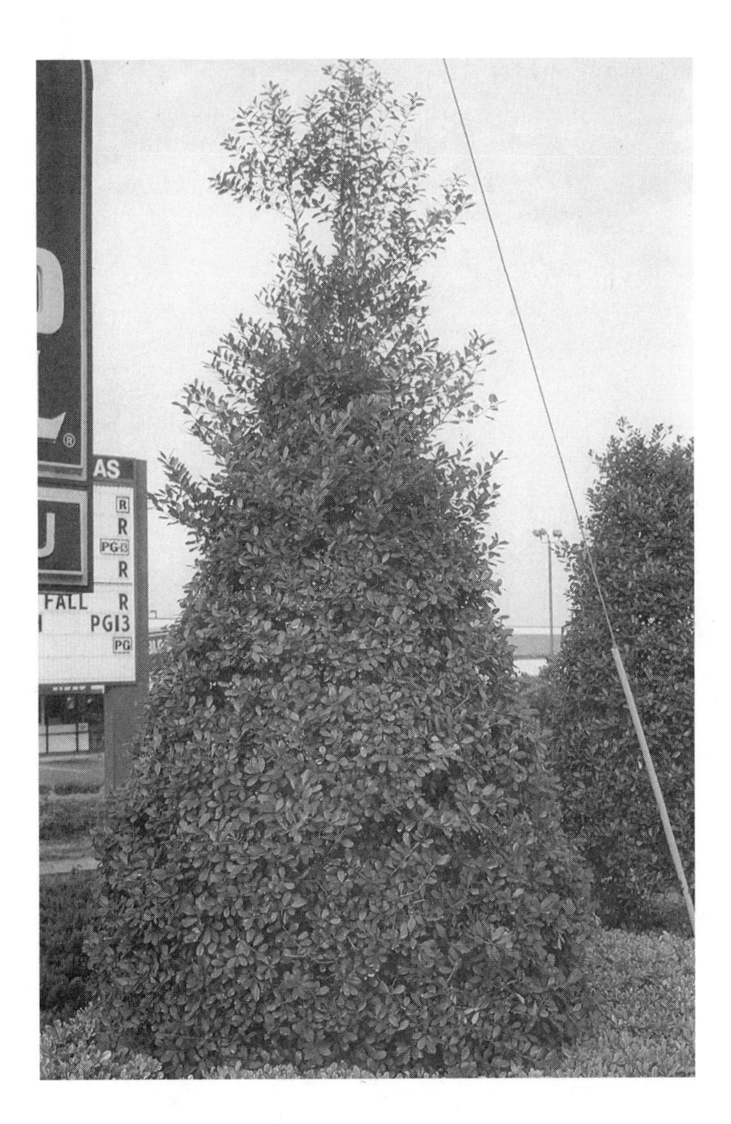

FIGURE 12-11. Trees that are sheared in the nursery will look odd for several years, until they have reestablished their natural form. Notice that the smooth outline of the sheared form is visible through the canopy (top half of photo).

To encourage scaffold formation in sheared trees with a decurrent habit, look inside the canopy and identify which branches would make the best scaffolds. Remove or shorten other branches (leave the small ones intact) located within 6 to 12 inches of the desirable scaffold limbs so they do not compete with the scaffolds (Figure 12-10). Remove or shorten branches that touch or cross over a scaffold or those with included bark. Once you have done this allow the tree to grow a year or two and then continue spacing scaffolds apart.

Branch management on street trees

Low branches on trees with an upright habit can remain on the trunk for a very long time, perhaps indefinitely provided they remain less than half the trunk diameter. Some upright cultivars used as street trees were developed for this purpose. Low branches usually do not droop to get in the way of pedestrians and vehicles. Use reduction cuts to shorten any branches that develop included bark.

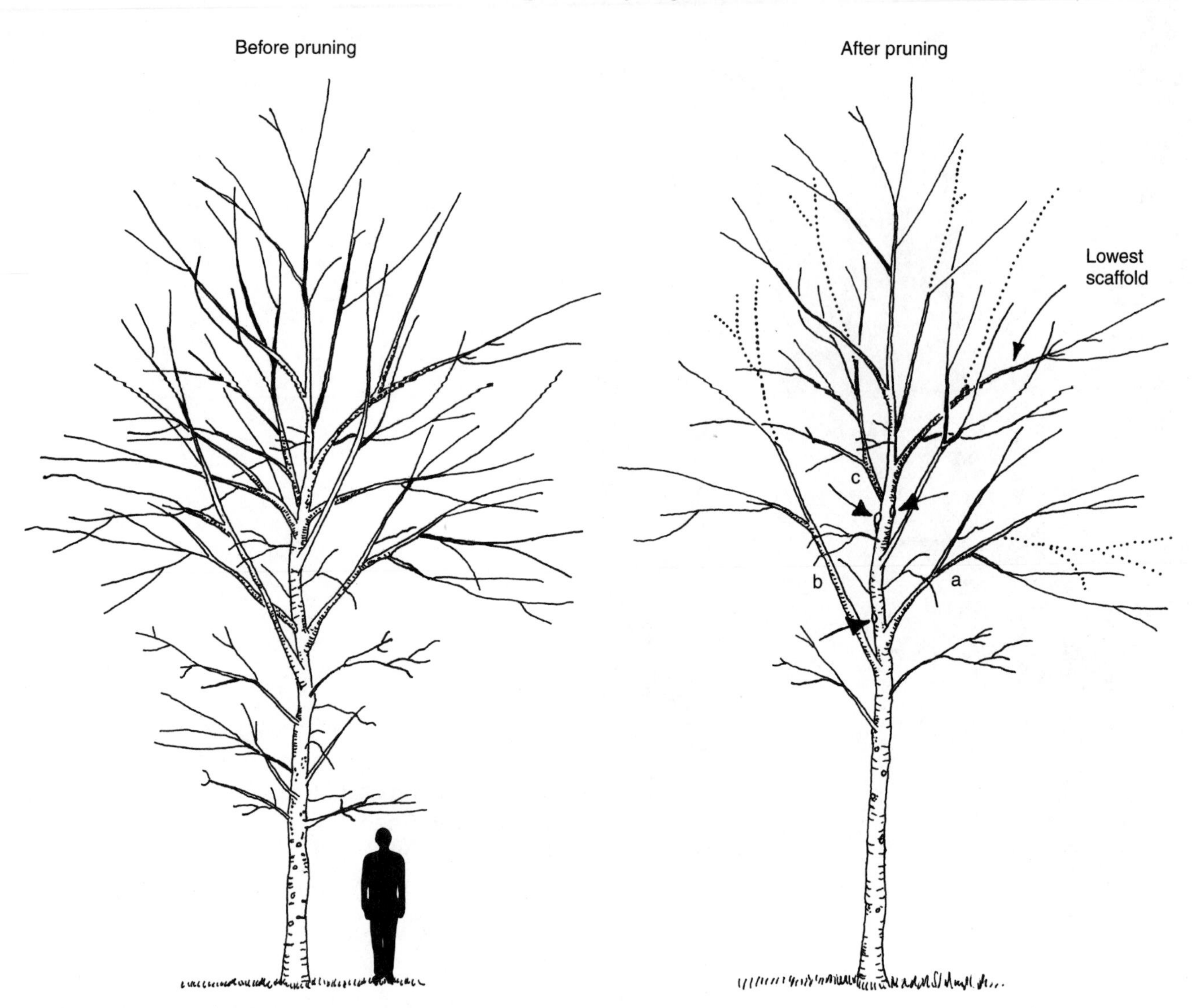

FIGURE 12-12. Shorten aggressive branches below the lowest scaffold limb to prepare the tree for their eventual removal later. That's 'a', 'b' and 'c' in this example (right). Remove some branches back to the trunk to encourage growth in the permanent portion of the canopy. Three were removed in this example (see arrows). The permanent portion of the canopy begins at the lowest scaffold limb. Two reduction cuts were made in the permanent canopy to help the leader dominant the structure.

Trees with a round, oval, or spreading habit planted along streets, in parking lots, and in other areas that will require eventual removal of all branches in the lower 15 to 20 feet of the trunk should be trained in a special manner (Figure 12-12). In short, branches below 20 feet should be kept shortened so they develop slowly and do not grow into the permanent canopy (Figure 12-13). The size of the reduction cut, and hence any resulting decay, is of little concern since the pruned low branches will be removed later anyway. Keeping low branches shortened should result in smaller pruning wounds later and craft healthier trees.

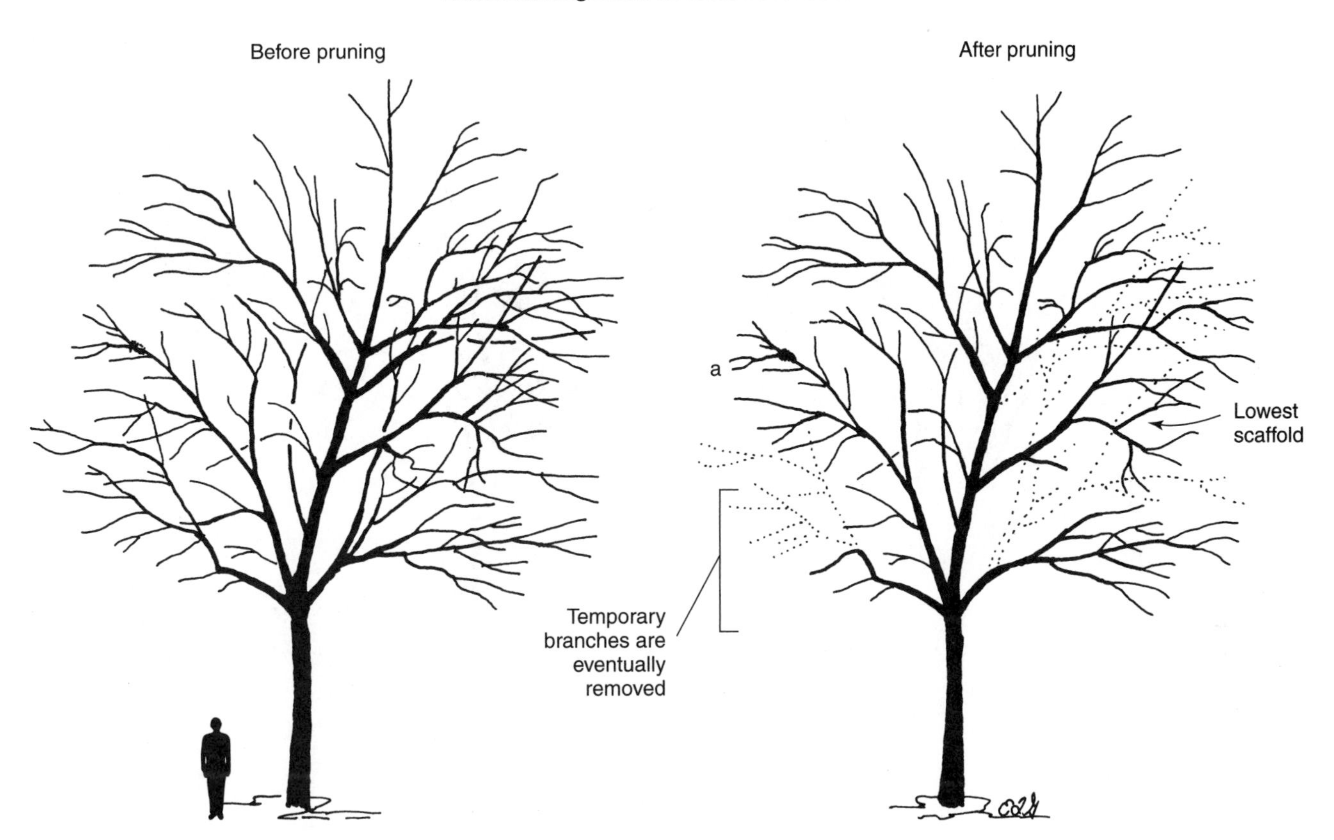

FIGURE 12-13. **Problem:** 1) Branches are too low on this street tree; 2) a low branch is sending an aggressive branch up into the canopy; 3) two large branches are opposite one another half way up the tree. **Solution:** 1) Shorten (as shown) or remove the low branch on the left; 2) remove the upright portion of the low branch on the right; 3) remove one of the two opposite branches half way up the tree. Now there is only one scaffold limb at each position on the dominant trunk. Branch 'a' could be shortened now to prepare for its eventual removal. This will prevent it from growing too large.

Branch management on medium-aged trees

Large branches opposite or close to other large branches may need to be shortened (Figure 12-14). Shortening is preferable to removal if the branch is half the trunk diameter or greater, or if branches are larger than about 3 inches diameter regardless of trunk size. Large branches may have heartwood that might not have good ability to retard the spread of decay when exposed to oxygen. Large branches with bark inclusions and other serious defects should also be shortened. Small branches with bark inclusions can be removed if they pose a danger, but this may not happen due to their small size.

Figure 12-13 shows three large branches at the same position in the lower canopy. The lower left and lower right branches were shortened so that the remaining one could dominate at that position on the trunk. The upper portion of the branch on the right was removed because it was growing up into the permanent canopy. It is usually best to remove the upright and leave the lower portion when shortening branches because this

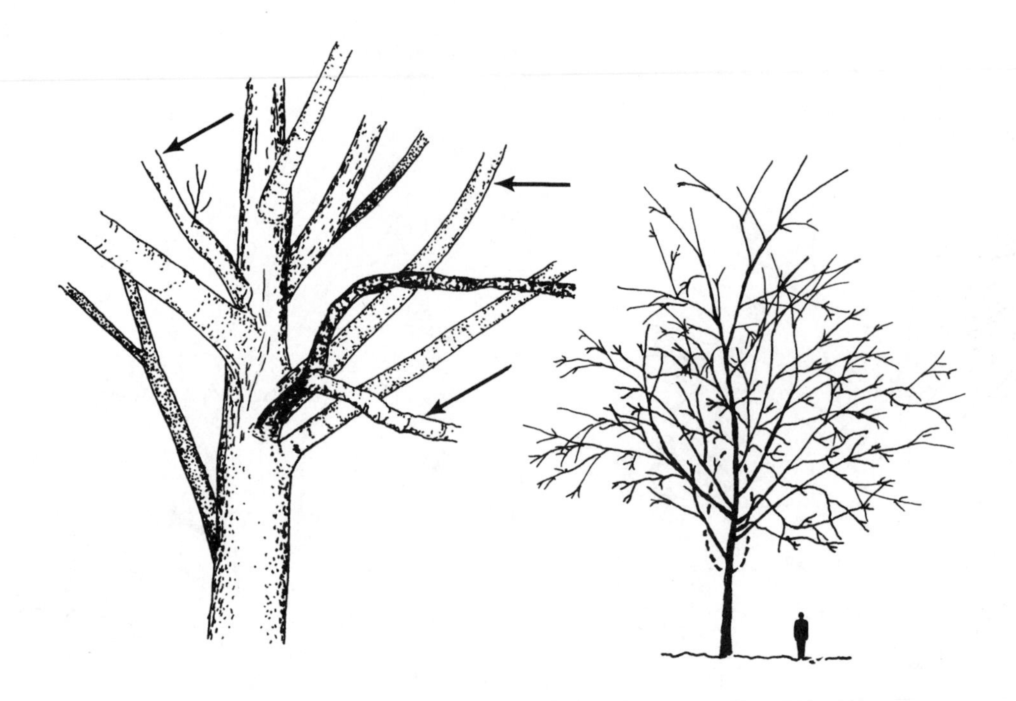

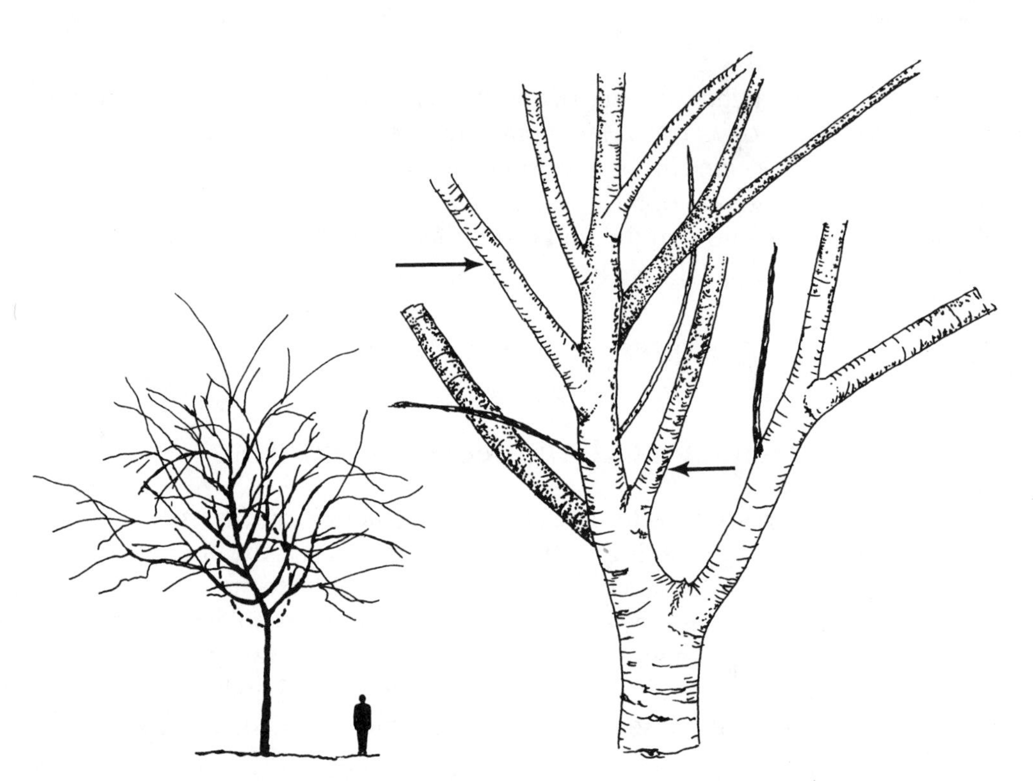

FIGURE 12-14. Developing scaffold limbs on trees that have not been pruned for a decade or more can be challenging. Remove the branches indicated with arrows back to the trunk (they are too close to other large branches). This allows uncut limbs to develop. (The close-up drawings provide a detailed view of the circled area of the full tree drawing.)

results in the most suppression of growth. Some smaller branches can be removed below the reduction cut for further suppression. Removing both branches back to the trunk would have created a large void in the lower canopy, removed too much foliage, and could have caused cracks in the trunk. This technique should be used throughout the tree on young and medium-aged trees. Notice that one branch was removed entirely back to the trunk halfway up the canopy. There were enough nearby branches on the trunk to fill the void created by removal and the branch was small.

Decurrent trees occasionally grow without generating dominant branches from which to select scaffold limbs. All branches may grow at about the same rate. When this occurs, well-spaced, existing branches can be encouraged to dominate as scaffold limbs by removing (Figure 12-9) or shortening (Figure 12-10) nearby branches. Equal sized branches on conifers and some excurrent deciduous trees, such as larch and baldcypress, are normal and this condition usually does not need correction.

Clustered branches

On decurrent trees, the worst branch arrangement resulting in the weakest tree is one where two or more large equal sized branches or stems arise from the same position or level on the trunk. This arrangement can result from heading in the nursery or landscape, or simply from neglect. Remove or reduce branches so there is only one aggressive main branch at any one level or position on the trunk (Figure 12-15). Reducing is generally more appropriate. All necessary pruning may not be possible at one time. You may have to return to the tree several times to correct the problems over a period of years (Figure 12-16).

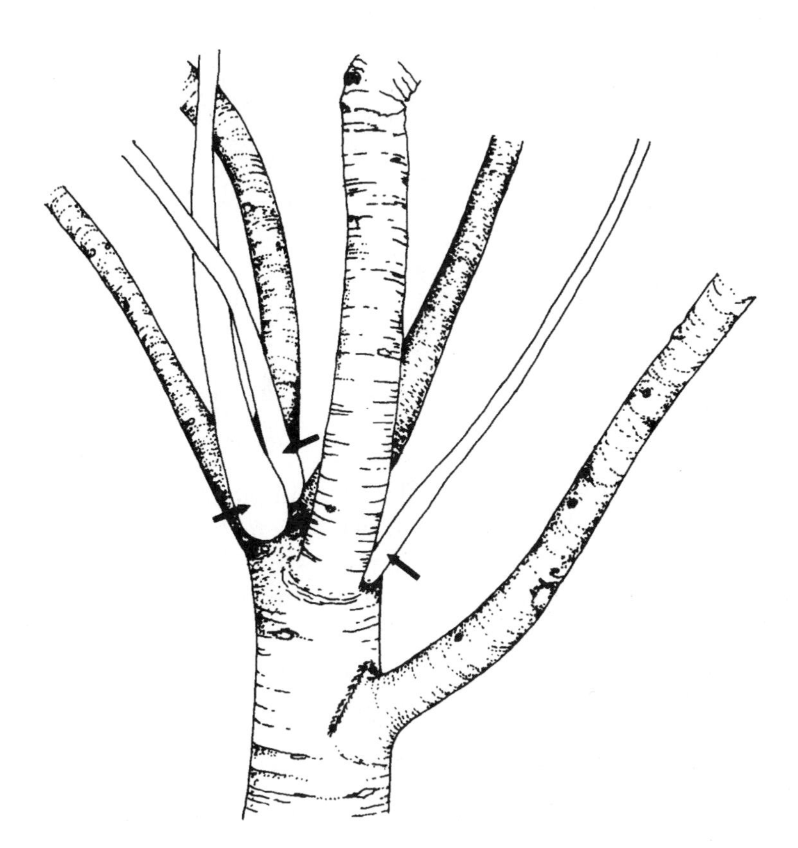

FIGURE 12-15. Clustered branches are poorly attached to the trunk. On young trees, you should immediately remove branches like those marked with an arrow (there are 3 shown). A couple of years later, reduce two of the remaining three branches so that only the main trunk and one large branch grow large at this point on the trunk. If the tree has grown with this form for about 5 years or more, correction is very difficult.

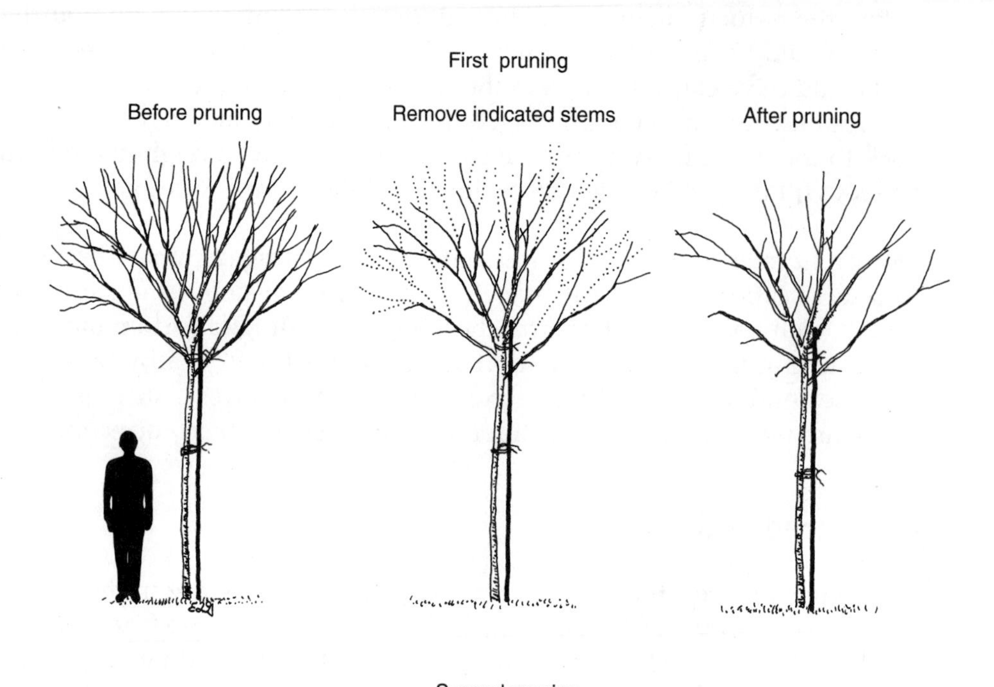

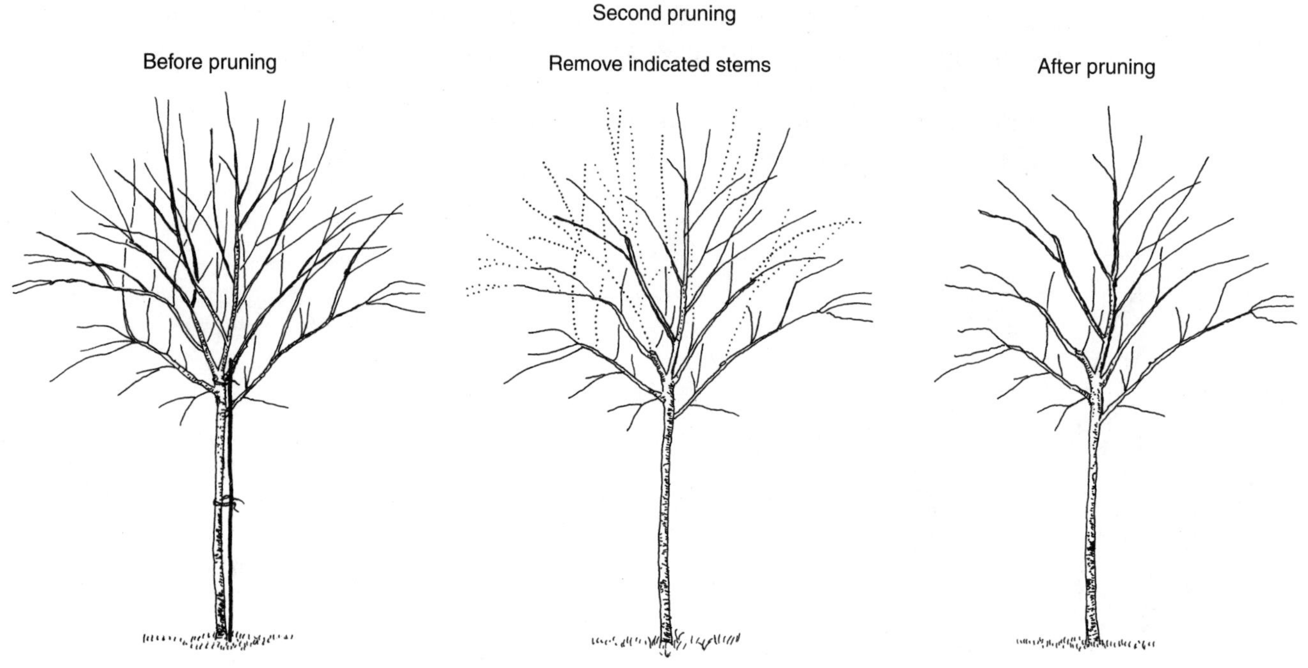

FIGURE 12-16. Trees such as oaks, ficus, jacaranda, elms, tristania, goldenrain tree and others are routinely topped into globes in the nursery in certain regions of the country (top left). To create structure in this poor-quality nursery stock, reduce some of the stems back to a lateral branch if possible; head them if there are no lateral branches to cut back to (top center and right). Except for the one stem you pick as the leader, most or all of the other stems and branches will eventually be removed from the tree. Continue this procedure each year or two until a good structure develops (bottom). It is much easier to start with good-quality nursery stock than to perform this pruning procedure on poor-quality trees.

DEVELOPING AND MANAGING LOWER BRANCHES

Developing low branches

Low-branching trees can be trained so that the lowest branches grow slightly upright by removing the *end* portion of the main branches and lateral growing toward the ground. This will help prevent branches from drooping to the ground as they increase in weight. It also allows for easier passage beneath the canopy. Leave most or all of the small lateral branches along the main branches. This will help the main branches increase in diameter toward their base which will help prevent them from drooping. This technique can be used on amur maple (*Acer ginnala*), silver buttonwood (*Conocarpus erectus*), crabapples (*Malus*), crape-myrtles (*Lagerstroemia*), lilacs (*Syringa*), ligustrum (*Ligustrum*), many fruit trees, and other small-maturing trees. It can also be used for large trees.

If passage beneath the canopy is not needed, the lower branches can be trained and encouraged to sweep the ground. At each pruning, remove some of the upright growth along the main branches. Always leave some secondary branches spaced along the main branches. These help the branches store energy for various jobs including fighting pests and compartmentalizing decay. It also helps them increase strength because they will have greater taper. Low branches sweeping the ground may not be appropriate on trees susceptible to diseases harbored on fallen foliage.

Managing low branches

Lower branches on many trees that are spaced apart from each other eventually droop toward the ground as they increase in length and weight. They often have to be removed to provide clearance along a street, walk, or other passageway. If that branch was quite large, a large wound on the trunk is created with all its associated problems. To prevent this, develop and execute a plan with regular pruning beginning when the tree is young and continuing for a couple of decades (Figure 12-17). Suppressing lower branches by reducing their length forces growth into the upper portion of the tree, where it belongs. Much like a water balloon, when you squeeze the bottom of the balloon the top pushes out. Similarly, when you "squeeze" the bottom of the canopy by reducing lower branch length, the top pushes up and out because it grows more. In many ways, this is the essence of pruning trees in urban and suburban landscapes!

Tip: *Suppressing low branches and stems is the essence of pruning shade trees in many situations.*

The details of lower branch **suppression** are quite easy to master. One of the best methods is to remove the upper and lower portion of the branch while leaving the middle intact (Figure 12-18). This accomplishes two objectives. Removal of the upper portion ensures that the low branch will not become part of the permanent canopy. Removal of the lower portion provides some clearance beneath the canopy. This is a good way to prepare that branch for eventual removal from the trunk.

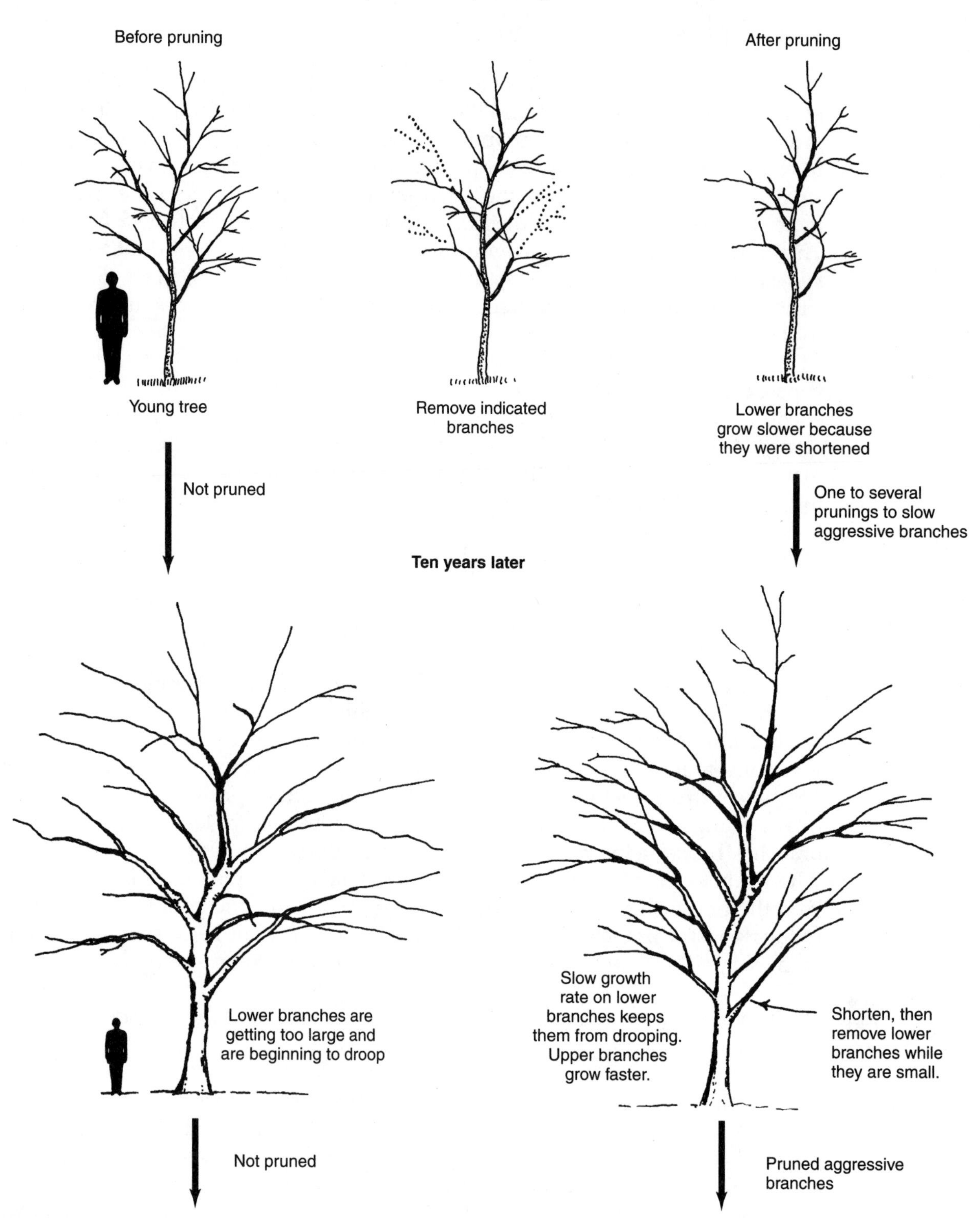

FIGURE 12-17. *(Illustration continues on next page)*

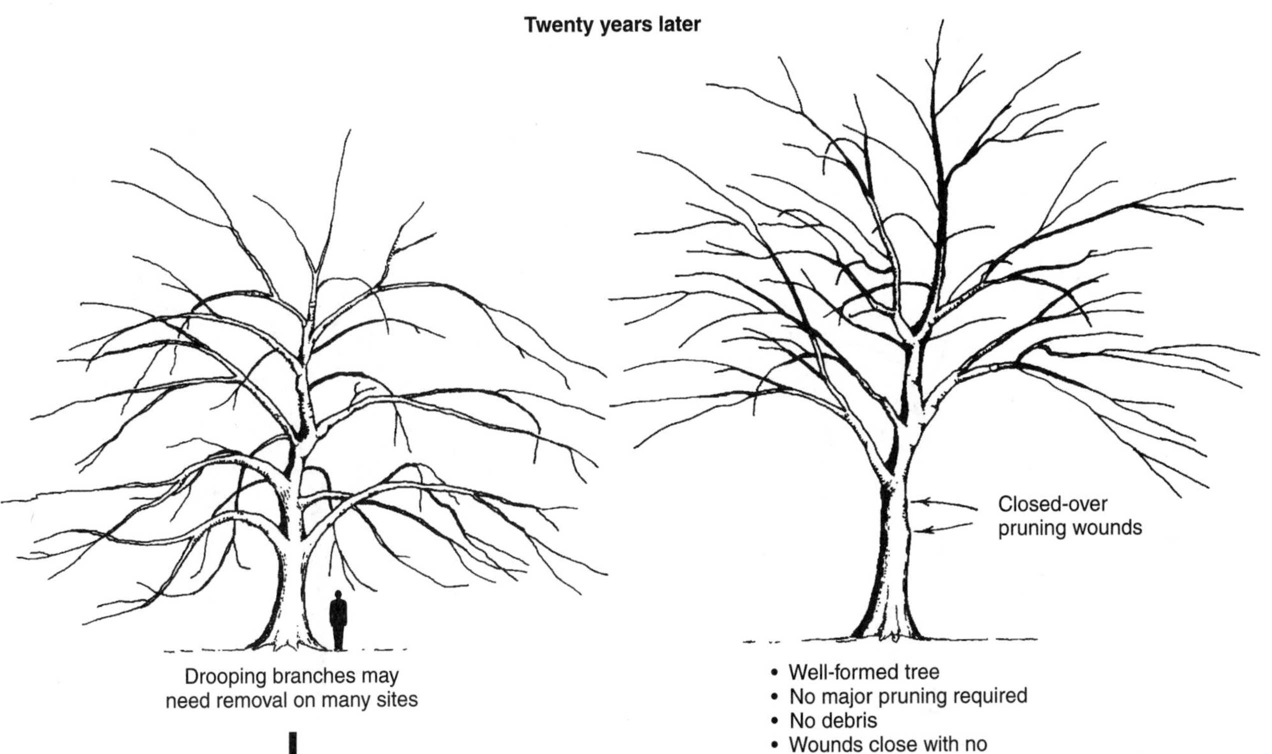

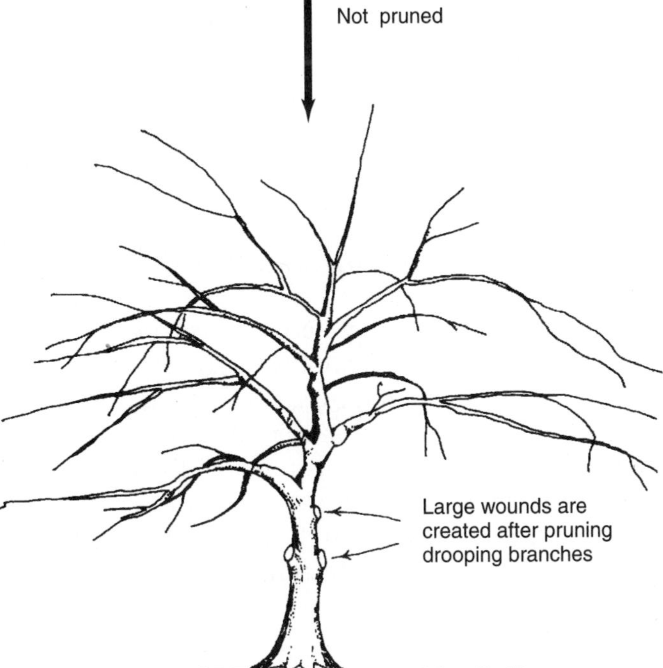

FIGURE 12-17. (continued) Branches on many young trees emerge at an upright angle and appear to be growing out of the way (top, previous page). However, they begin to droop as the branch becomes older and heavier (bottom left, previous page). About twenty years after planting, the lower branches may need to be removed to make room beneath the canopy (top left). Trunk decay and cracks may start at the large pruning wounds created on the trunk after the large branches are removed (bottom). Lower branches could have been shortened earlier to keep them small and encourage growth in the upper canopy. When small lower branches are removed, the size of the wound is also small, which reduces the incidence of trunk decay (right).

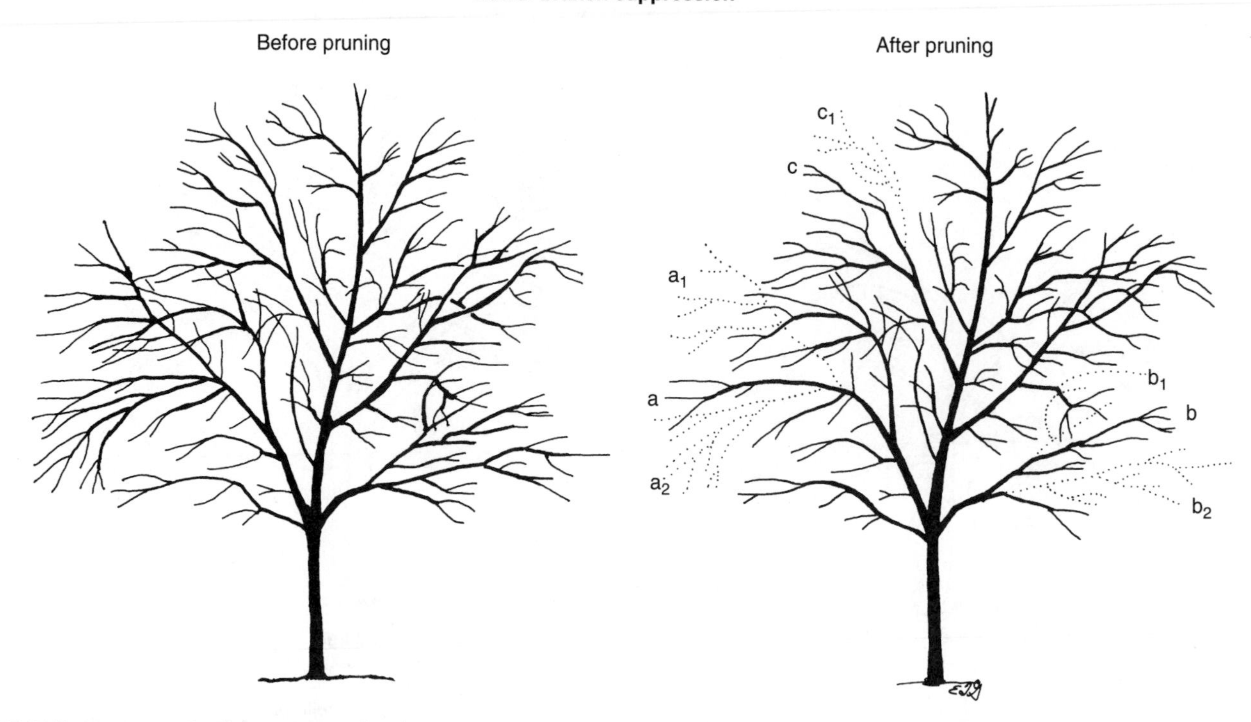

FIGURE 12-18. Problem: Low limbs are too aggressive and require suppression. A codominant stem is developing in the top portion of the canopy. **Solution:** One of the best methods of suppressing low limbs is to remove the upper (a_1 and b_1) and lower (a_2 and b_2) branches from the ends of limbs (right). Suppress the codominant stem by removing c_1. That will encourage 'c' to become a branch, not a codominant stem.

AERIAL ROOTS ON TREES

Some tropical trees, such as figs (*Ficus*) naturally develop aerial roots from branches. As these touch the ground they root into the soil and eventually become trunks. These help hold the branches on the trunk. If aerial roots are removed, the branch from which they originate may not have the support it needs to hold itself on the tree. It could split from the tree due to excessive end weight. Allow aerial roots to touch the ground and root in to help ensure that these trees remain intact. If roots are regularly removed, keep horizontal branches shortened with reduction cuts and thin the branches by removing some from the edge of the canopy. This may reduce the likelihood of these branches failing.

CONIFERS AND OTHER EXCURRENT TREES

Branch management on conifers and many other excurrent trees is quite simple. Reduce (on older stems) or remove (on younger stems) any upright stems or branches, especially those with bark inclusions. Double leaders and codominant stems with included bark are very prone to splitting out in storms.

CHECK YOUR KNOWLEDGE

1) What is the best way to prevent formation of codominant stems?

   a. shorten or remove live branches that compete with the potentially dominant leader
   b. remove live lateral branches from main branches so main branches remain less than three-quarters the diameter of the trunk
   c. prune the canopy to encourage more light penetration to interior branches
   d. root-prune on the side of the tree with a large branch

2) The best way to treat a 12-year-old large-maturing shade tree with three codominant stems located on the front lawn of a residence is to:

   a. remove two codominant stems so only one leader remains.
   b. root-prune the tree moderately to slow its growth.
   c. subordinate two stems so one will eventually dominate.
   d. thin the canopy and remove dead branches.

3) When subordinating a stem on a young landscape tree, what type of pruning cut is most effective and appropriate?

   a. heading cut
   b. thinning cut
   c. jump cut
   d. reduction cut

4) What is the best permanent solution to the problem of codominant stems on a 20-year-old tree?

   a. removal cuts over a period of time
   b. inserting a threaded screw rod through the crotch to hold it together
   c. cabling the stems to hold them together
   d. subordinating stems with reduction cuts over a period of time

5) About how many prunings are typically required to develop a good structure in the first twenty-five years after planting a large-maturing shade tree with a decurrent growth habit into the landscape?

   a. none
   b. two
   c. five
   d. ten

6) The lowest scaffold limb:

   a. is usually located about 10 feet from the ground.
   b. is located from the ground a distance that depends on the location of the tree in the landscape.
   c. is typically on the tree when the tree leaves the nursery.
   d. is removed after several years due to shading from scaffold limbs above.

7) By five to ten years after planting in the landscape:
   a. most scaffold limbs are well developed.
   b. one or two scaffold limbs may be present.
   c. the tree is too young to have scaffold limbs.
   d. it may be too late to develop scaffold limbs now.

8) To prevent large codominant stems from forming low on the trunk:
   a. prune all but one of them regularly to slow their growth.
   b. remove branches and stems as they get too large.
   c. prune only in the dormant season.
   d. thin the canopy.

Answers: a, c, d, d, c, b, b, a

CHALLENGE QUESTIONS

1) Describe the concept of pruning dose. Show a colleague by pruning a tree using a small dose, then prune the tree with a larger dose.
2) Why should low branches be kept shortened on young trees planted along a street?

SUGGESTED EXERCISES

1) Take a group to a row of large-maturing street trees planted about twenty years ago. Discuss the strategies needed to bring good, strong structure into these trees.
2) Do the same as in the previous exercise but for a row planted about five years ago. Compare the strategies.
3) Take a group to a park with recently planted trees. Discuss how to manage the structure and form on these trees. What actions should be taken now to improve structure? Contrast this with management of street trees.
4) Locate several trees in a park or along a quiet street. Divide into groups of about five to eight people and give each group a role of a uniquely colored ribbon. One arborist should be positioned in each tree. Position one group by each tree. Each group will have the arborist wrap the stems and branches they would remove assuming a pruning cycle of five years (or use another number). When the groups are finished with their first tree, have them switch trees. Continue until all groups have marked all trees. Then gather together around one tree and critique and discuss the decisions each group made. Make the appropriate pruning cuts and discuss why you made them.

CHAPTER 13

PRUNING TYPES ON ESTABLISHED TREES

OBJECTIVES

1) Determine why the tree needs to be pruned.
2) Establish priorities.
3) Contrast methods for reducing tree size.
4) Increase light penetration and reduce storm damage.
5) Show how to preserve trees with pruning.
6) Show the quickest method to improve tree appearance.
7) Show how to reduce hazardous conditions in the tree.
8) Restore structure in damaged and neglected trees.
9) Direct growth away from structures.

KEY WORDS

Cleaning
Directional pruning
Hazardous conditions
Lateral pruning
Lions-tailing
Pruning types
Raising
Reducing
Restoration
Thinning
Tipping
Vista pruning

INTRODUCTION

The primary goals when managing established trees in urban landscapes are public safety, tree health and integrity, amenity values, and aesthetics, in that order of importance. These priorities should guide management plans for trees of all ages and size (Table 13-1). **Pruning types** on established trees include cleaning, thinning, reducing, raising, and balancing the canopy. Trees can also be pruned to channel and focus a view that allows an important object to be seen (called **vista pruning**); to reduce risk of catastrophic failures; or to restore the canopy following a storm, abusive or incorrect practices, or vandalism (Table 13-2). Appropriate pruning type depends on the species or cultivar; age, size and condition of the tree; use and location in the landscape; and other factors.

Medium-aged and mature trees are most often the recipients of the pruning types described in this chapter. With the possible exception of directional pruning, root pruning, and raising, young newly established trees typically require little of the pruning types discussed in this chapter. Structural pruning (Chapters 11 and 12) is the most appropriate pruning for these young shade trees.

Here is a strategy for approaching trees you have not pruned previously. *First consider any structural pruning the tree might require to treat defects.* This could include subordination of codominant stems or shortening of low aggressive (fast-growing) branches to

TABLE 13-1. Top priorities when pruning established trees.

Remove dead branches greater than a specified diameter (e.g., 1.5 inches diameter)
Perform structural pruning where appropriate
Shorten or remove branches with included bark
Shorten or remove branches with defects
Remove parasites
Minimize risk
Maximize biological efficiency

TABLE 13-2. Pruning types on established trees.

Clean—removes dead, broken, rubbing, or diseased branches, and foreign objects
Thin—increases light and air penetration, or reduces weight
Reduce—decreases height (extent) or spread (reach) on entire tree or on one section only
Raise—provides clearance under canopy by reducing live crown ratio
Balance—removes live branches to redistribute wind and gravity loads in the canopy
Risk reduction—reduces conditions that could lead to catastrophic branch or tree loss
Restoration—improves structure and biological effectiveness on damaged trees
Directional pruning—directs future growth by removing live branches on trees too close to other objects
Vista–allows for a specific view
Root pruning—removes circling and girdling roots around trunk base or roots negatively impacting infrastructure; prepares tree for transplanting
Eradication—removes tissue with pests
Shaping—pollarding, espalier, pleaching, curtain, and other techniques that create special effects

slow their growth rate. Then perform a canopy cleaning, removing dead branches greater than a specified diameter. Next, shorten branches with cracks, those with bark inclusions, and those that are too long, to reduce the likelihood of them splitting from the tree. Then choose from the other pruning types in Table 13-2 to complete the job. Consider some of the basic issues before pruning (Table 13-3).

CLEANING

Cleaning removes dead, dying, diseased, damaged, rubbing, broken, and out-of-place branches, and perhaps some water sprouts (Figure 13-1). This is the most commonly practiced type of pruning on mature trees because it improves their value, appearance, and health and can reduce risk. If two branches are rubbing, remove the one that is most injured by the rubbing or the one that is most out of place. If they are both deformed, both may need to be removed. Suckers from the base of the tree, below grafts, and from the roots, as well as water sprouts, can become very aggressive and, if left unpruned, can spoil the habit and beauty of the tree and disrupt internal allocation of resources.

Before removing water sprouts, you must judge whether sprout removal will benefit the tree. Removing all sprouts and other branches only from the interior of the canopy is considered improper pruning and is not recommended (Figure 13-2). Removing all sprouts from along a limb causes more sprouting and can cause harm by unnecessarily reducing photosynthesis. This usually violates the American National Standards Institute A300 pruning standard (see Chapter 15). Sprouts should be left on branches in a tree that has been overpruned, injured by root removal, or stressed in other ways. This helps rebuild energy reserves of the tree (see "Restoration" later in this chapter).

Water sprouts on certain trees such as goldenraintree, some oaks and maples, jacaranda, and many others are regularly produced as limbs bend and droop under their own weight. Sprouting results from growth regulators unevenly moving to the underside of the limbs stimulating buds on top of the branch to produce new sprouts. These sprouts often grow to become an important part of the canopy and they should not all be removed. If sprouts are young and numerous, remove some so the remaining ones are spaced apart. This will allow each to develop appropriatel (see "Restoration" later in this chapter).

This is a wonderful time to remove foreign objects from the canopy, trunk, and roots such as nails, wire, and string. If left on the tree, foreign objects could cause health problems later. An arborist should check for defects (see Chapter 1) in the tree that are

TABLE 13-3. Basic pruning issues on established trees.

1. Remove live tissue only for a good reason. What is to be accomplished? Will tree structure and public safety be improved with the proposed treatment?
2. Do not make flush cuts—always leave the collar intact.
3. Avoid removing branches that are large in relation to the stem and those with heartwood.
4. Avoid removing many branches that are close to each other, unless they are small.
5. Consider shortening branches instead of removing them—perhaps remove them in a subsequent year.
6. Prune live branches from drought stressed, flood-damaged, or pest-damaged trees sparingly, if at all.

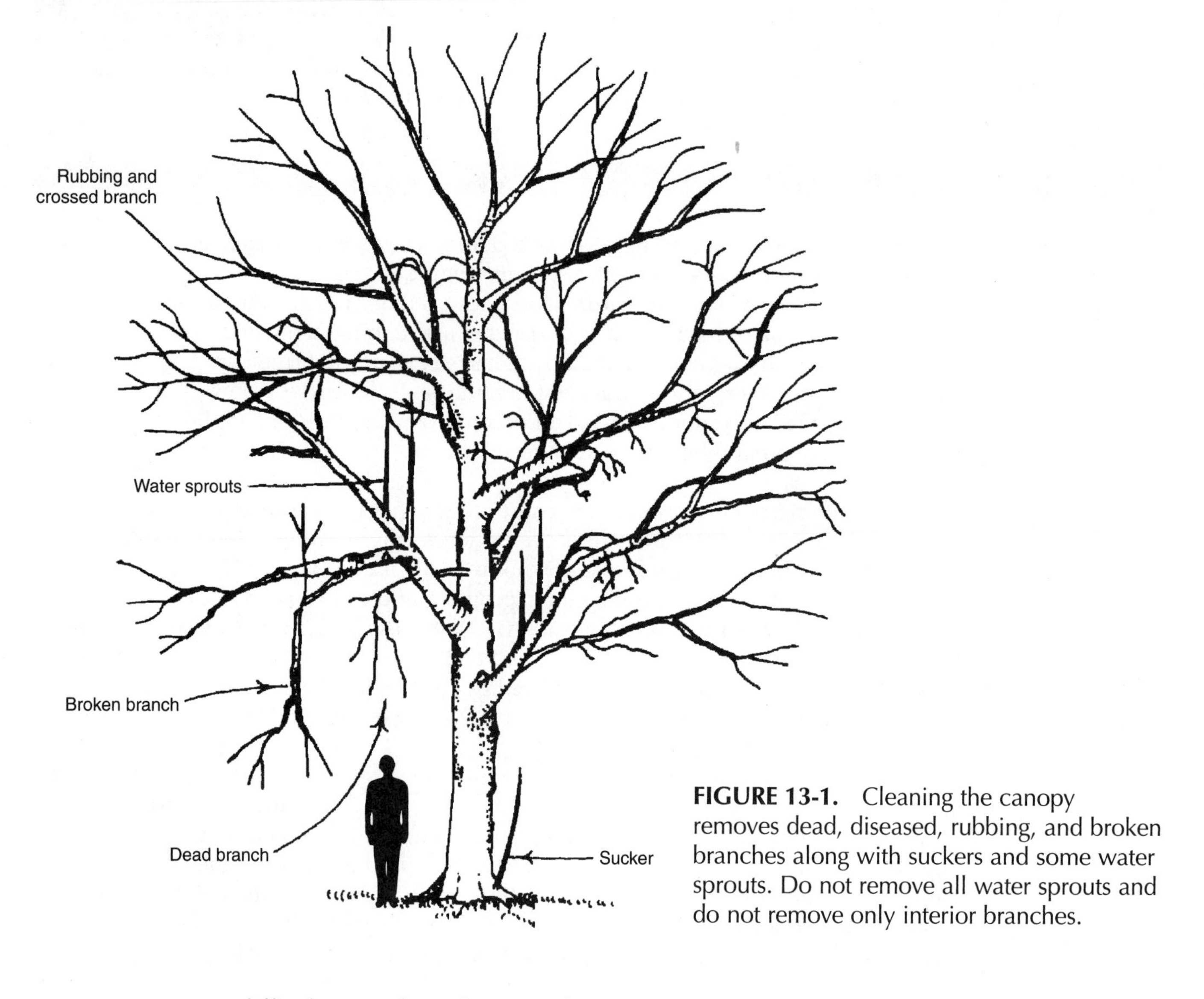

FIGURE 13-1. Cleaning the canopy removes dead, diseased, rubbing, and broken branches along with suckers and some water sprouts. Do not remove all water sprouts and do not remove only interior branches.

difficult to see from the ground. Parasites such as mistletoe, vines, other higher plants, and mosses that reduce photosynthetic capacity and increase mechanical loads (like Spanish moss) can also be reduced or removed.

THINNING

Trees in the forest provide protection for each other from the damaging effects of wind. Trees in an open setting, such as along a street or in a park or yard, and those at the edges of forests are open to the effects of wind, especially when they are planted far apart from each other. The canopies on planted trees often thicken because light reaches all parts of the plant. A thick canopy catches wind which, under certain circumstances, can damage the tree by breaking branches or blow it over in a storm.

Thinning has been the conventional method to minimize damage caused by storm winds (Figure 13-2). **Thinning** removes lateral and parallel branches, especially from

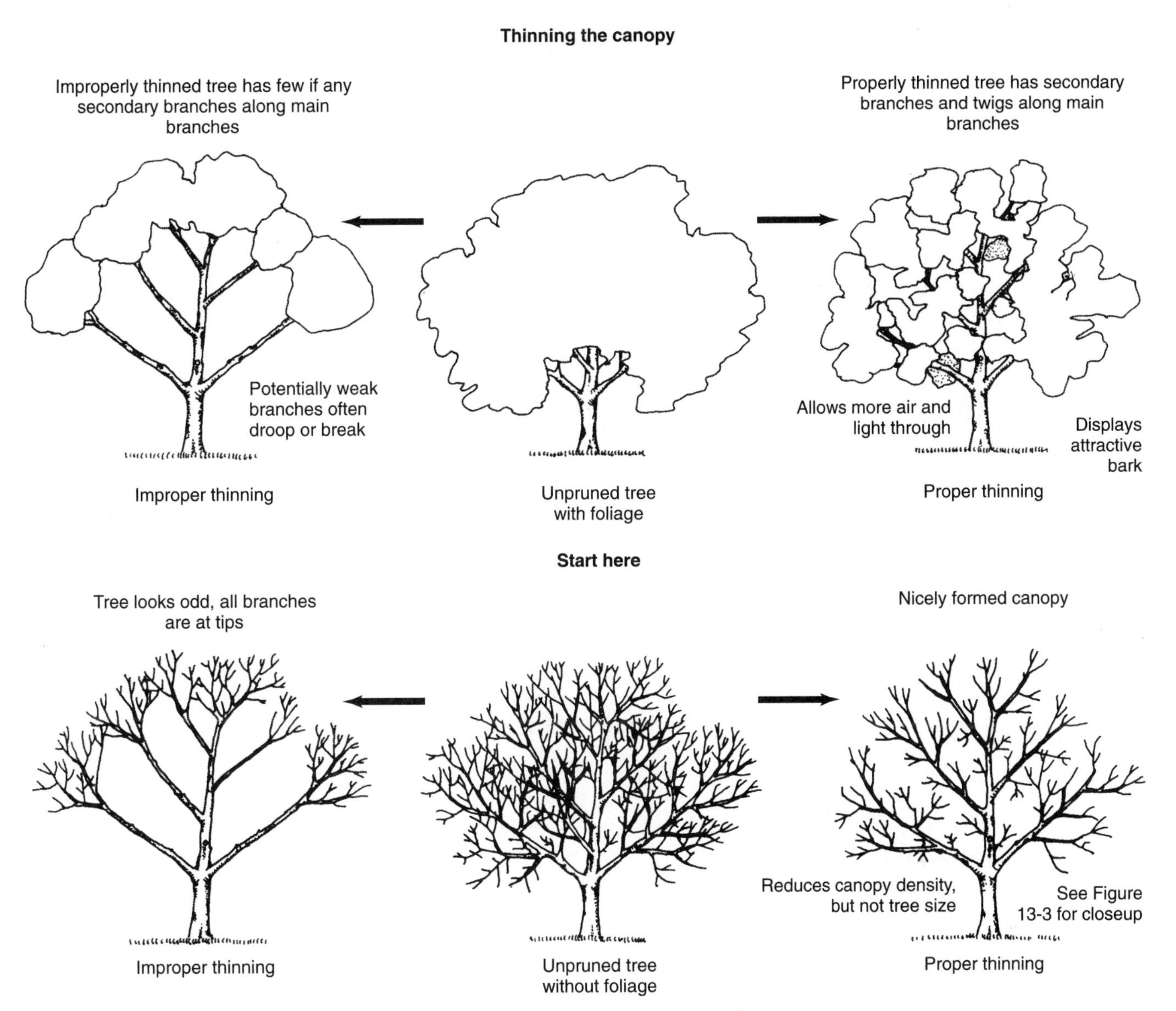

FIGURE 13-2. Appropriate thinning removes small branches from the edge of the canopy (right). Inappropriate thinning removes only interior and lower branches (left).

the end portion of limbs (Figure 13-3). It allows the wind to pass through the canopy, and might improve durability in a storm. Some researchers and practitioners are beginning to question whether thinning reduces damage from wind. Thinning is conducted on large or small trees for a number of other reasons. It is best to keep cuts less than 2 inches in diameter.

Thinning is regularly practiced by professional arborists and in Japanese gardens to create special effects (Table 13-4). The severity of some diseases and pest infestations can be reduced by thinning because more light and air move through a thinned canopy. This keeps the foliage drier, which discourages some pests. Thinning can also emphasize the beauty of the trunk and main branches and increases light penetration to the ground, which can enhance plant growth under the tree to a limited degree. Thinning is used to reduce limb weight on mature trees in order to reduce stress and strain on branches with structural defects such as cracks, hollows, overextended or long branches, and cavities.

TABLE 13-4. Reasons to thin the canopy.

Reasons
Reduce risk of storm damage
Improve growth on turf and other plants shaded by the canopy
Improve appearance
Open up a view
Show off attractive bark or trunk form
Reduce pests by improving air flow
Compensate for included bark and other defects

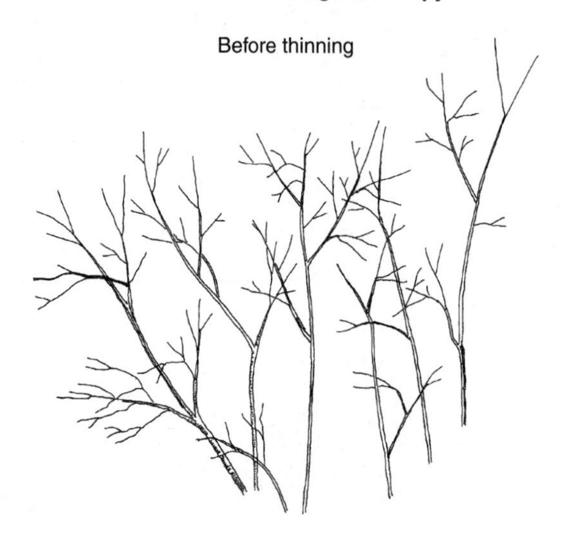

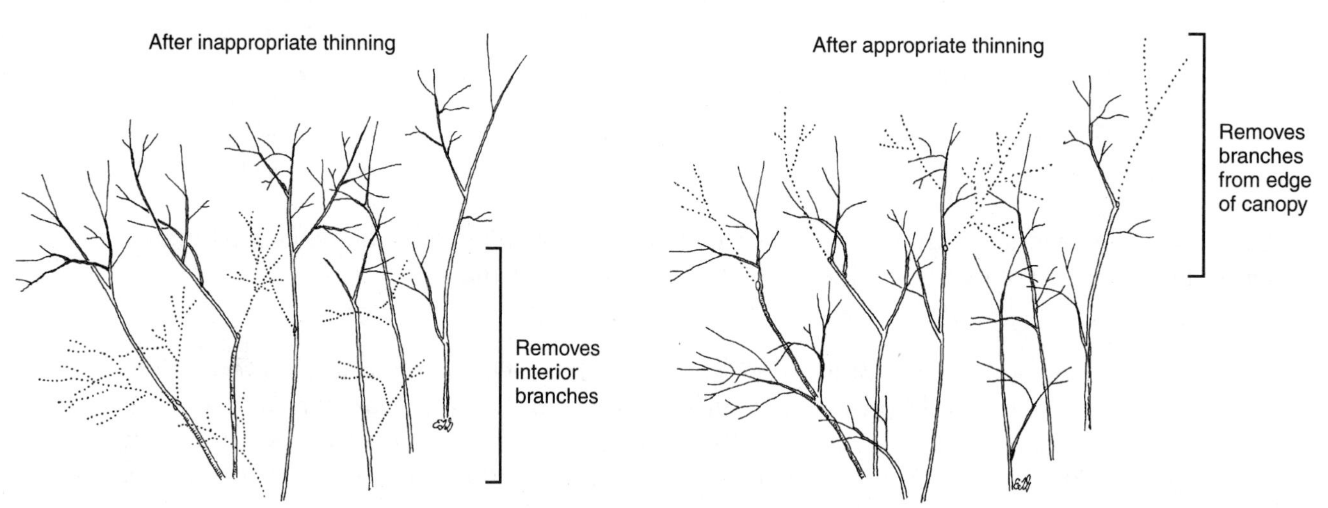

FIGURE 13-3. Thin a tree by removing small diameter branches primarily from the edge of the canopy (lower right). Removing branches only from the interior and lower portion of the tree is inappropriate (lower left).

Proper thinning does not change the overall size of the tree (Figure 13-3), which will remain about the same height and width as it was before thinning. Proper thinning is done on relatively small branches in the leafy area of the canopy toward the ends of the main branches (Figure 13-4). Remove branches growing parallel to each other attempting to occupy the same space (Figure 13-5).

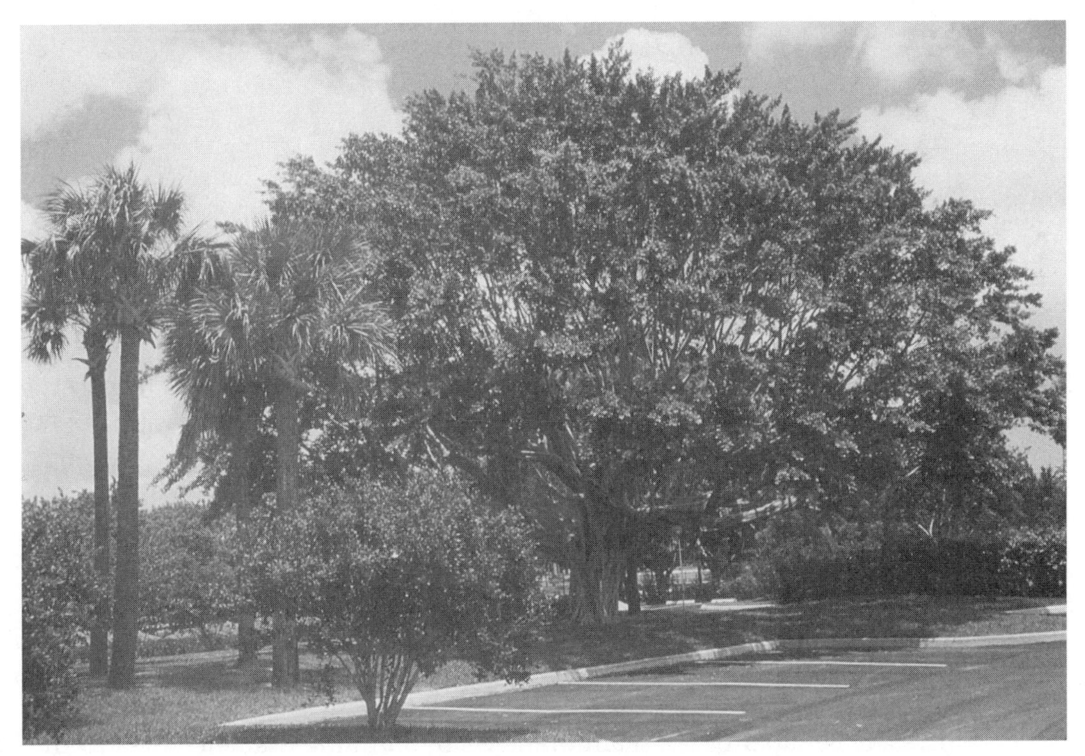

FIGURE 13-4. The canopy of this tree was dense before pruning (top). Properly thinned, the tree is the same size after thinning as before (bottom). The edge of the canopy is more jagged after pruning than before because branches were removed from the edge of the canopy. More air and light passes through the tree after pruning. (Photos courtesy of C. Way Hoyt)

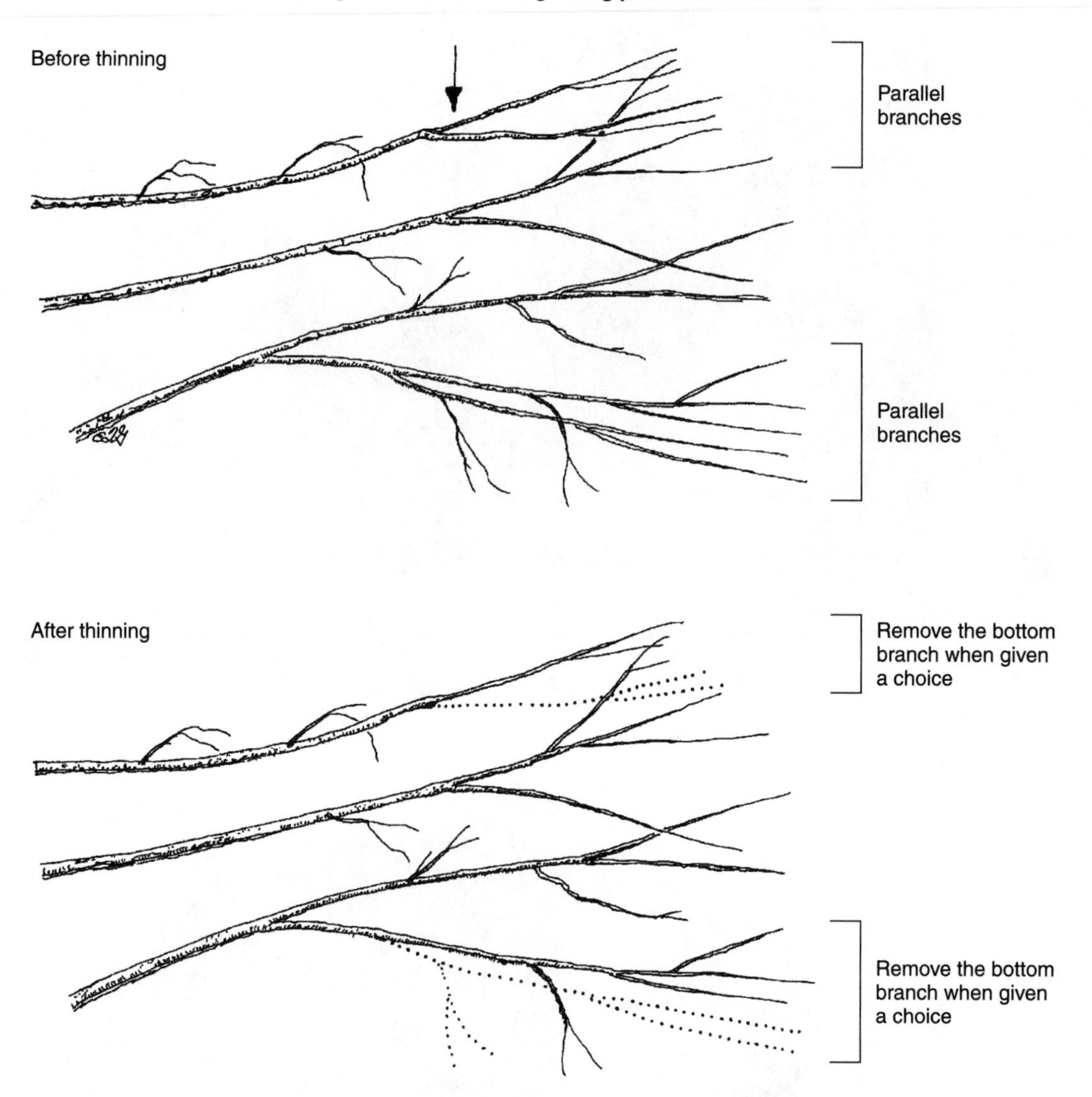

FIGURE 13-5. Problem: Branches are growing parallel to each other and are competing for the same space at the edge of the canopy. Before pruning there were seven branches at the position of the arrow. The three main branches are about 4 to 6 inches diameter. **Solution:** Two branches were removed (see dotted lines, bottom) resulting in more space between remaining branches. After pruning there were five branches at the position of the arrow. Keep cuts less than about two inches diameter.

To thin the canopy, remove some branches from the scaffold limbs, especially toward the ends. Removing foliage at the edge of the canopy allows light to penetrate inside, which encourages growth of existing small twigs in the interior of the tree. This helps increase the taper of the main branches by building diameter toward the base of the limbs, thus making the tree stronger.

When pruning young trees, remove no more than about ¼ of the foliage at one time. When pruning medium-aged trees, remove no more than about ⅕ of the live foliage at one time. Overthinning stresses the tree by reducing energy reserves (starch) and can initiate unwanted water sprouts on interior branches, creating an atypical look. Over-

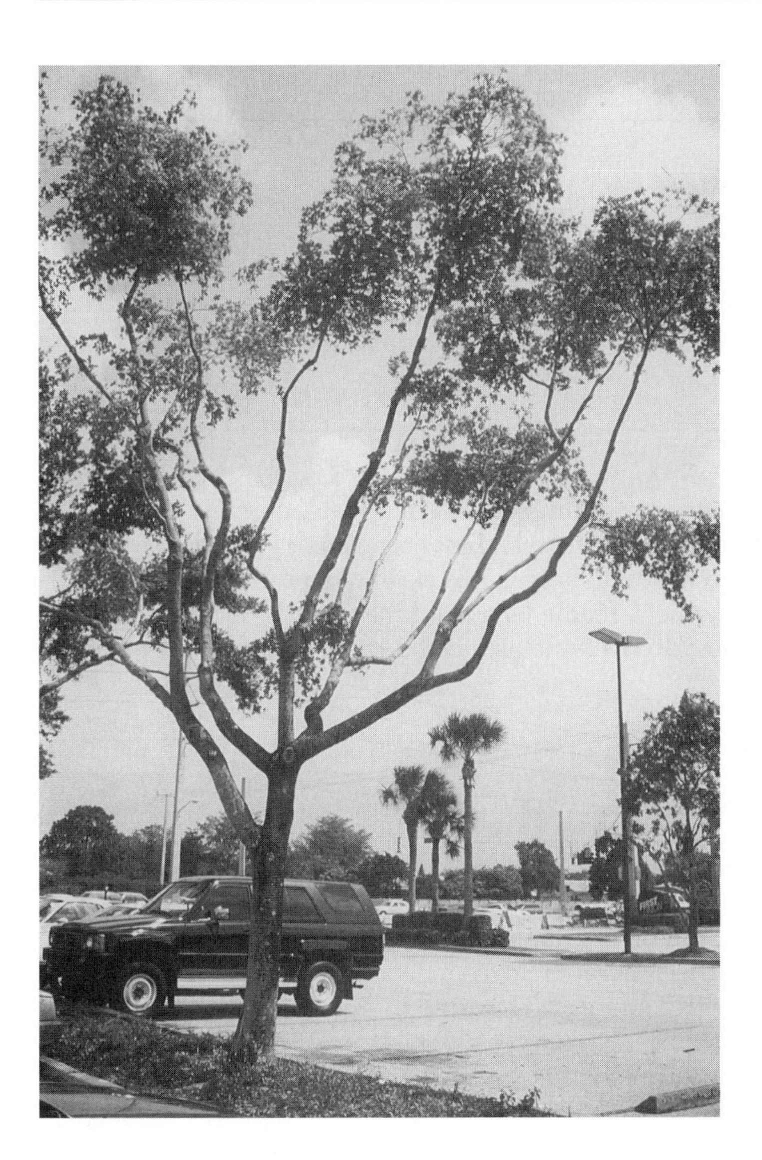

FIGURE 13-6. Improper thinning removes only interior branches and results in 'lions-tailing'. This is extremely damaging to trees and eliminates other pruning options later, such as reduction, because all interior branches are removed.

thinning reduces photosynthesis and enhances the likelihood of sunscald, cracks and death. Overthinning also forces the thinned branches to overelongate and changes the center of gravity creating substantially weakened branches, which may break easily in storms or under their own weight. Roots can also be negatively impacted by overthinning. If a tree sprouts vigorously after pruning it was probably overpruned. Few live branches should be removed from mature trees. Certainly no more than 15 percent of live foliage should be removed at one time.

Some (but not many) small-diameter interior branches can be removed to create a more open effect. *But proceed cautiously!* Removing *only* interior branches will not reduce the harmful effects of a storm and is inappropriate. Removing branches toward the edge of the canopy may provide some protection.

Unfortunately, many people misunderstand thinning and subscribe to the practice of only removing branches from the interior of the canopy. This is often referred to as "lions-tailing," overlifting, or overthinning (Figure 13-6). Little or nothing is removed from the ends of the limbs, and this is a mistake (Table 13-5). This leaves too much

TABLE 13-5. Stripping out interior growth and low branches has many negative effects.

Severely reduces stored energy reserves
Reduces valuable photosynthetic surface area
Changes wind loading patterns and makes the tree top heavy
Increases damage from storms
Increases number of entry points for pests
Spoils tree architecture and appearance
Causes branches to overelongate and droop
Can kill the tree
Can initiate cracks and decay

weight at the ends of branches and causes limbs to overelongate, possibly resulting in sunburn, water sprouts, or even death of the tree. Sometimes many water sprouts are produced, guaranteeing annual removal to keep limbs clean of sprouts. Limbs look terrible. Branches pruned in this manner become weak and may break, or they may have to be removed later because they droop too close to the ground. *Along with topping, "lions-tailing" is tree abuse at its worst!*

REDUCING (REDUCTION)

Too often, customers desire a tree to be smaller because it has grown too big. "Too big for what?" an arborist might ask. The customer's response is the tree has become dangerous and could blow over and fall on the house. After performing a risk analysis, an arborist may determine that the tree is acceptable for meeting owner objectives, yet the customer may still insist that the tree should be reduced in size. The most professional response an arborist can make is that this unnecessary task is not endorsed by professional arborists. In many cases the phobia of large trees has no basis in reality.

Reducing canopy size stresses the tree because many reduction cuts are required. Unlike a removal cut, reduction does not cut back to a natural boundary, which means that decay and other defects can spread inside cut branches especially in trees that compartmentalize decay poorly. For this reason, it is best not to perform canopy reduction if at all possible. However, reduction is a useful pruning type which does have applications (Table 13-6).

In many instances, canopies cannot be properly reduced in size to the extent desired by a customer. Certain species such as beech, birch, and other decay sensitive trees do not lend themselves to canopy reduction. Other trees are very susceptible to canker fungi infection following reduction pruning including maples, eucalyptus, rosacious plants, arborvitae, and falsecypress. The wishes of the customer often result in overpruning of the tree with heading and reduction cuts. This can initiate defects and decay in the trunk and branches and stimulate rapid sprout growth that fills in the canopy as it quickly grows to reestablish photosynthetic area.

It is very difficult to use canopy reduction to permanently maintain a tree at a smaller size without causing tree decline. Consider pollarding to reduce and maintain height if the

TABLE 13-6. Reasons to reduce the canopy with reduction cuts.

Make the tree smaller to reduce failure potential
Reduce weight at ends of long or defective branches to reduce likelihood of them breaking
Direct growth away from an object or structure
Open up a view
Bring one side of the canopy into balance with the other side

tree is young (see "Pollarding" in Chapter 10). However, tree removal and replacement with a smaller maturing plant may be the choice that minimizes the input of resources.

Some professionals consider it inappropriate to reduce the size of the canopy to open up a view, preferring instead vista pruning or raising the canopy (discussed later in this chapter). Because removal cuts are used in vista pruning and canopy raising, these techniques damage the tree far less than reduction cuts used in canopy reduction. However, when the customer wishes to reduce height, canopy reduction is much preferred over topping. Consider canopy reduction when the root system of a large-maturing tree is confined to a small soil space and it is in danger of falling over, or on a large tree that has substantial decay, making it at risk for catastrophic failure. Reduction can also be used on trees with height: trunk diameter ratio of 50 or greater (e.g., 50 ft. tall tree with a 1 ft. diameter trunk) to reduce risk of wind throw.

Before cutting any branches to reduce the size of the canopy, visualize the new canopy outline. The objective is to make reduction cuts so that branch tips are left intact on the outer edge of the new, smaller canopy (Figure 13-8). Ideally, pruning cuts should not be evident when you stand back from the tree after pruning. Heading (topping), shearing, tipping, lopping, or rounding over are not appropriate techniques for reducing the size of a tree (Figure 13-7) because they compromise the tree's structure and can cause defects and decay. If you plan on removing more than 25 to 30 percent of the foliage on anything but a young tree, consider dividing the job into two sessions, one growing season apart, to minimize sprouting and starch (energy) removal from the tree.

Reduce the size of a tree with reduction cuts, shortening those branches which extend beyond the edge of the new, smaller canopy (Figure 13-8). This maintains the approximate original shape of the tree. The tree is simply made smaller. The longest portion of the main branches is cut back to an existing, smaller, lateral branch that is large enough to assume the role of the branch. This is normally ⅓ (minimum) to ½ (preferred) the diameter of the removed branch. Excessive sprouting accompanied by dieback or decay can occur if you cut back to a branch that is too small. Many arborists think it is best to cut back to a branch that has a narrow angle of attachment to the cut stem than one with a wide angle (Figure 13-9). Wide angle branches are more likely to break later. Consider thinning the remaining branch (especially toward the end of the branch) if it is more than half the diameter of the cut stem or is very long. This could prevent it from breaking later.

It is probably unreasonable to expect more than about a 20 percent reduction in size of the canopy from a properly executed canopy reduction. Greater reduction results in

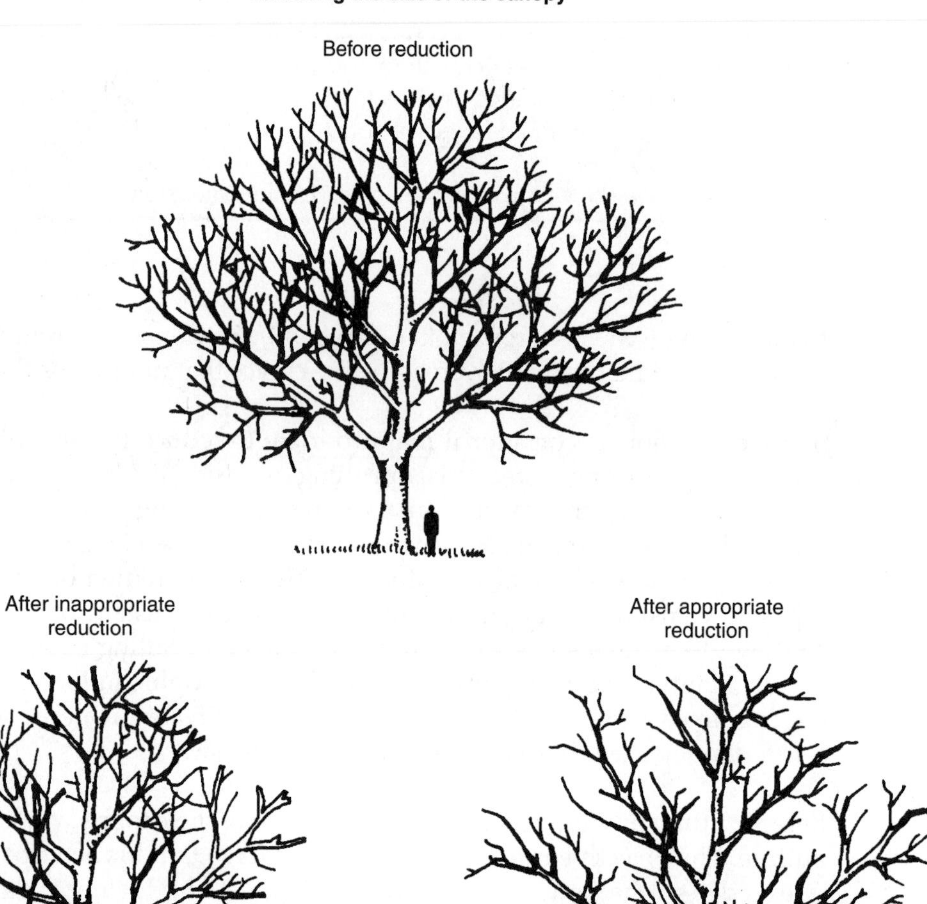

FIGURE 13-7. Canopy reduction makes a tree smaller by removing the end portion of branches with reduction cuts (lower right). Inappropriate reduction uses heading cuts and can result in more problems later (lower left).

larger pruning cuts and more potential decay and other defects. Sometimes as little as a 3 to 6 foot reduction in height can add a great amount of stability to a large tree. This time-consuming technique is more an art than a science. Professional arborists and horticulturists proficient at this technique can take an ordinary tree and create a unique specimen. It often requires substantial talent to perform this operation correctly and elegantly (Figure 13-10). Of course this is a temporary measure because medium-aged trees quickly grow to occupy available space. Mature trees respond slowly.

Pollarding (Chapter 10) can be used following canopy reduction to maintain a tree at a smaller size. This is needed if the decision to pollard the tree was not made at the out-

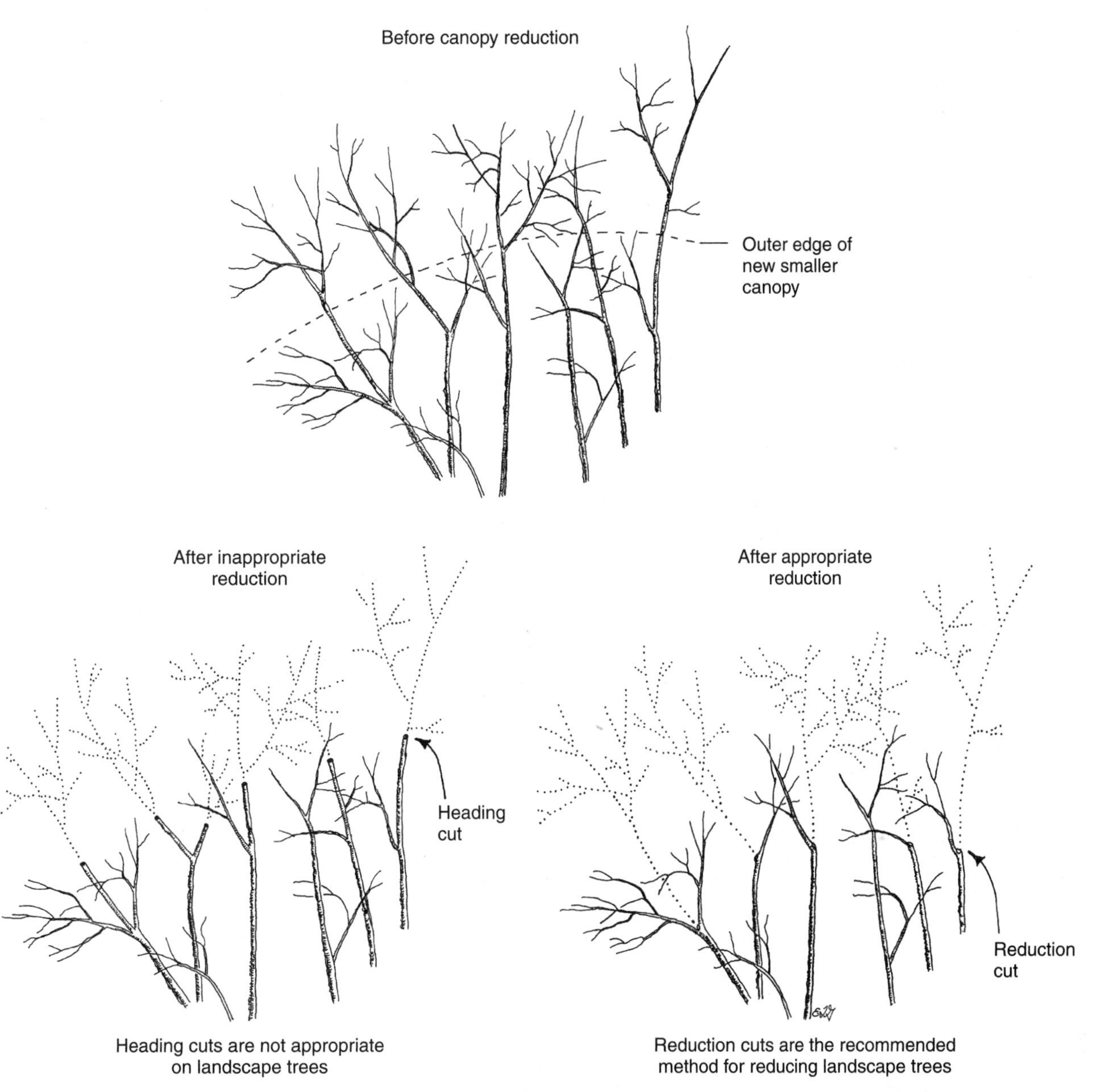

FIGURE 13-8. Reduce trees with reduction cuts by removing stems and branches back to live lateral branches that are at least one-third the diameter of the removed stem (lower right). Making heading cuts to reduce trees is topping and is not recommended (lower left).

set and the tree has more than several growth rings at the point you wish to maintain the pollard heads. Allow sprouts to grow from the reduction cuts made during canopy reduction. In the dormant season one or two years later, remove the residual lateral branch and all but one or two sprouts at each original pruning cut. Also remove sprouts originating lower on the cut branch. Continue the pollarding process, each year cutting back to the developing pollard head. Sprouts should always be cut back to the branch collar, not flush with the trunk.

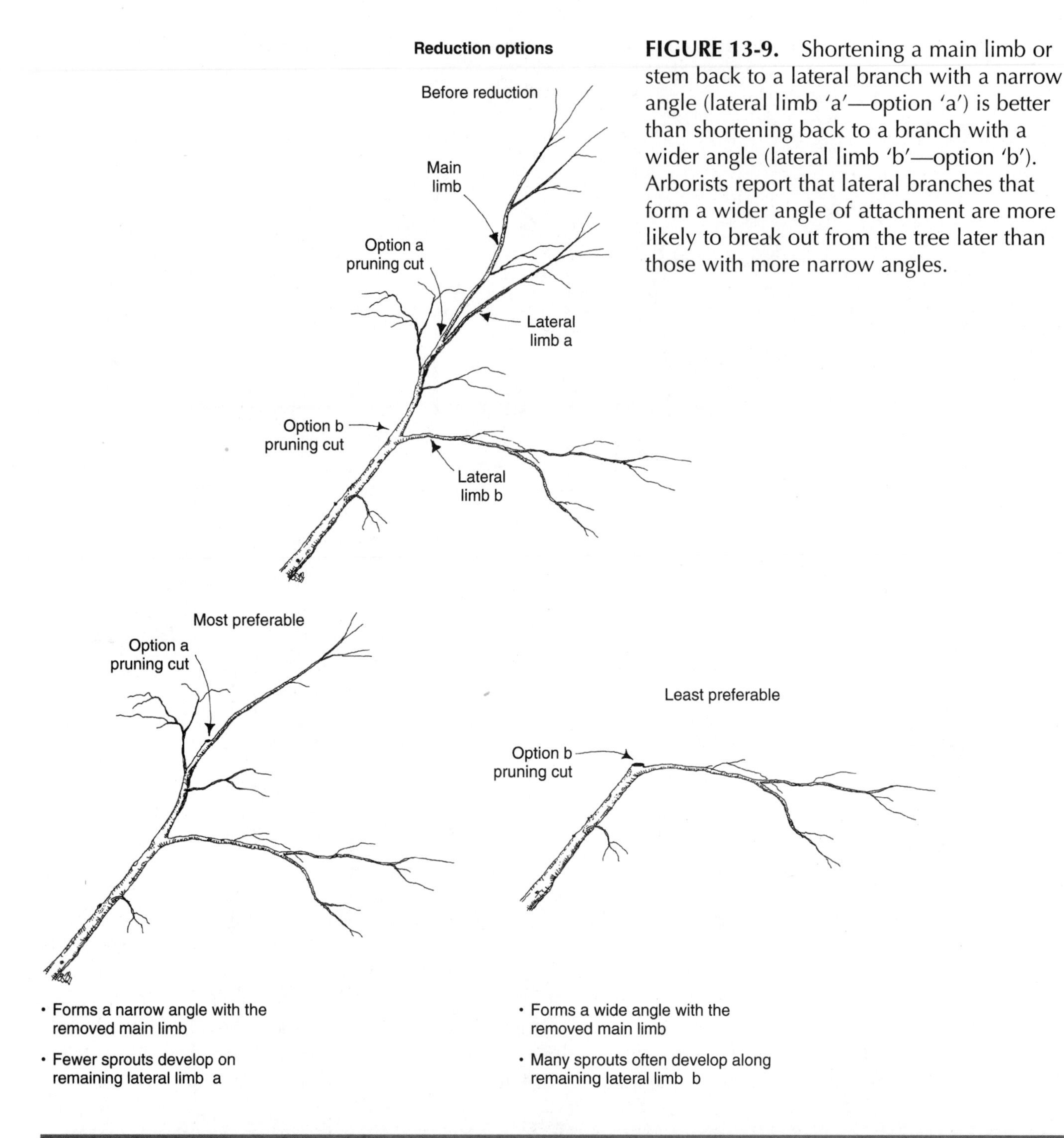

FIGURE 13-9. Shortening a main limb or stem back to a lateral branch with a narrow angle (lateral limb 'a'—option 'a') is better than shortening back to a branch with a wider angle (lateral limb 'b'—option 'b'). Arborists report that lateral branches that form a wider angle of attachment are more likely to break out from the tree later than those with more narrow angles.

RAISING

Lower branches often have to be removed to get them out of the way of traffic, keep them away from a building, make signs visible that were installed too far off the ground or open up a desirable view. This type of moderate pruning does much less damage to a tree than reducing the canopy size. To open up a desirable view, consider creating holes or windows in the canopy by removing selected branches within the canopy (thinning), or consider vista pruning (discussed later in this chapter) instead of raising it by removing only lower branches.

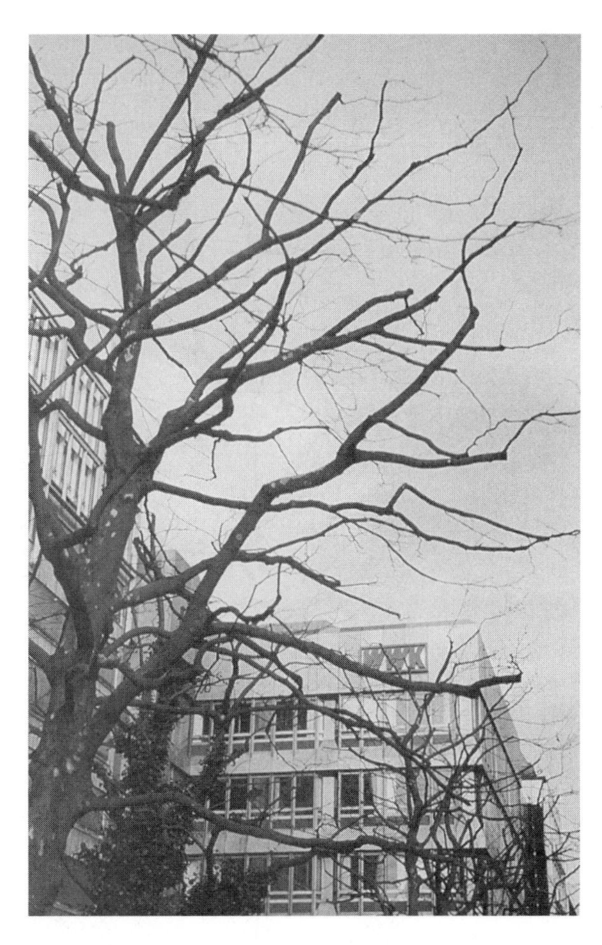

FIGURE 13-10. Proper canopy reduction makes the tree smaller but leaves no visible branch stubs. This orange tree was about 5 feet taller before the canopy was reduced in size (left). The canopy of this sycamore was reduced in size by about 20 percent (right). Some of the reduction cuts in the photo at right are inappropriate because they were made back to lateral branches that were too small.

Raising is best done gradually over a period of years. The trunk could be seriously injured if too many lower branches are removed at one time (Figure 13-11, top right). Discolored wood, cracks, and possibly decay may begin inside the trunk of an overpruned tree.

Removing too many lower branches can result in sunburn on the lower trunk, causes water sprouts, and forces the tree to grow taller. Some major branches should be left on the lower ½ of the trunk. Some foresters recommend that the live crown ratio should be at least 60 percent, meaning that there should be live branches along the upper 60 percent of the trunk to distribute wind stress and develop trunk taper. Similarly, half the foliage on scaffold limbs should originate from secondary branches on the lower ⅔ of these limbs. Try to leave small branches on the lower trunk intact for about a year (or more if possible) after removing large branches because they help minimize injury from sudden sun exposure by shading the trunk (Figure 13-11, bottom left). They also speed closure of the pruning wounds, could resist the spread of wood discoloration, and help build taper.

Raise the canopy in two or more steps if many branches need to be removed. First, thin or reduce the largest ones in the lower part of the tree, but leave the small ones intact (Figure 13-11, bottom right). This may provide for enough clearance for a year or so because the branch will often rise after removing branches from the tips. Figure 12-16 provides another example of reducing in lieu of removing low branches. If necessary, remove all branches back to the trunk one to several years later. Sometimes more than one large-diameter branch is located at the same point on the trunk, and both need to be removed to provide clearance. Since removing both at the same time could create a

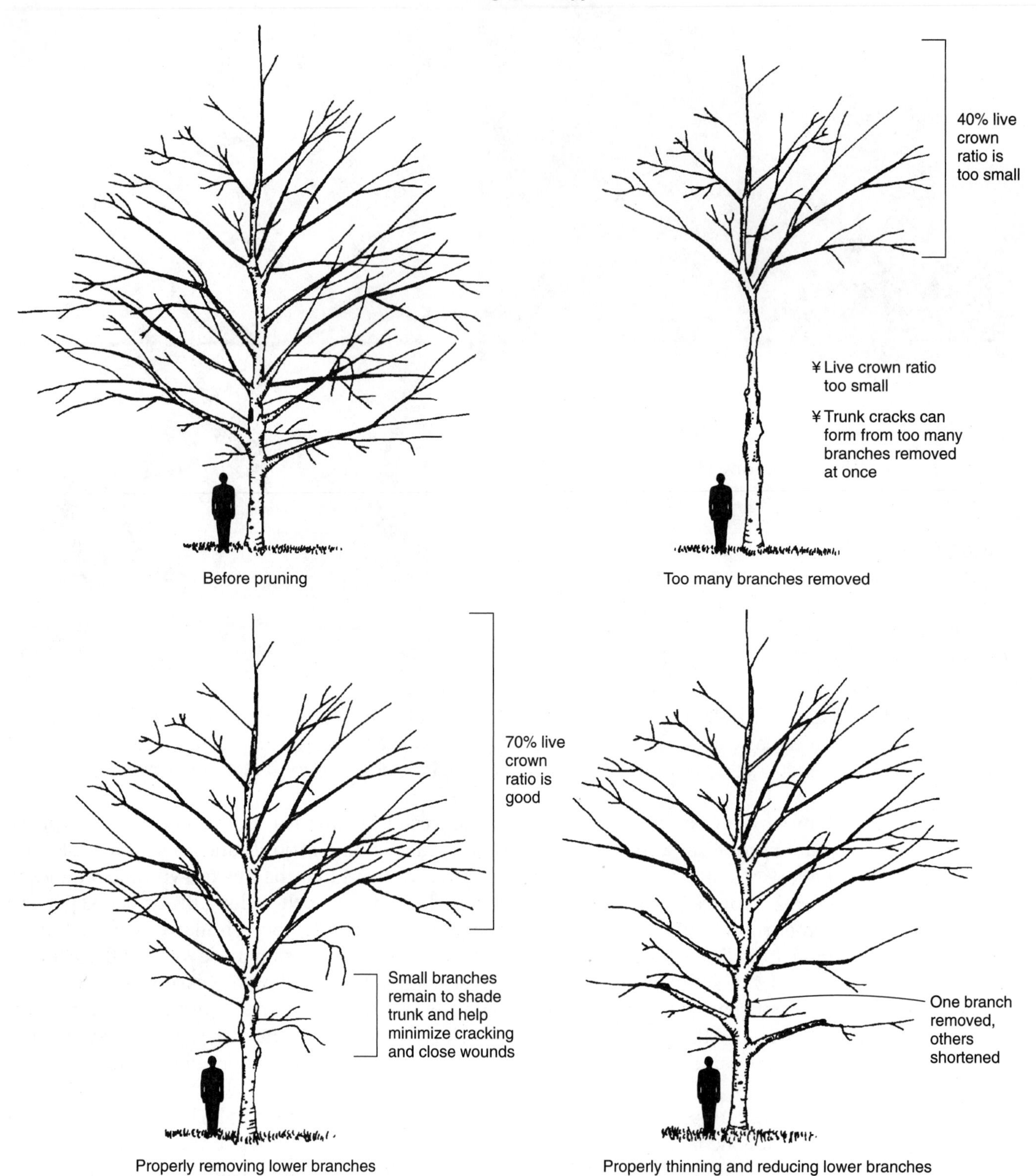

FIGURE 13-11. Many trees are overpruned when the canopy is raised (top right). After proper canopy raising, a good goal is to have foliage on branches in the upper ⅔ of the tree (bottom). Live crown ratio should be at least 60%. Small-diameter branches left on the lower trunk for about a year after pruning help close pruning wounds and protect the tree by providing shade to that region. They also help hide the pruning wounds.

fairly large pruning wound which could weaken the tree, consider removing one now and shortening or thinning the others. Come back a year or more later and remove some or all of the other branches. The objective is to only remove one large branch in a cluster on the trunk at a time. Of course it is best to prevent this clustered branch arrangement from occurring with regular subordination cuts on these lower branches.

Raising the canopy on trees with large-diameter (greater than ½ the trunk diameter measured directly above the branch) low branches can initiate trunk cracks and decay if these large branches are removed. To prevent this, consider thinning them (Figure 13-2) or reducing their length with reduction cuts instead of removing them. This will slow their growth rate, and eventually the trunk may grow to become much larger than the branch. This may give the tree an opportunity to form the branch protection zone at the base of the branch thus minimizing trunk decay once it is removed. When raising the canopy, do not forget to attend to any structural pruning that needs to be done to correct defects. *It is inappropriate to simply remove lower branches without correcting structural problems* (Figure 13-12).

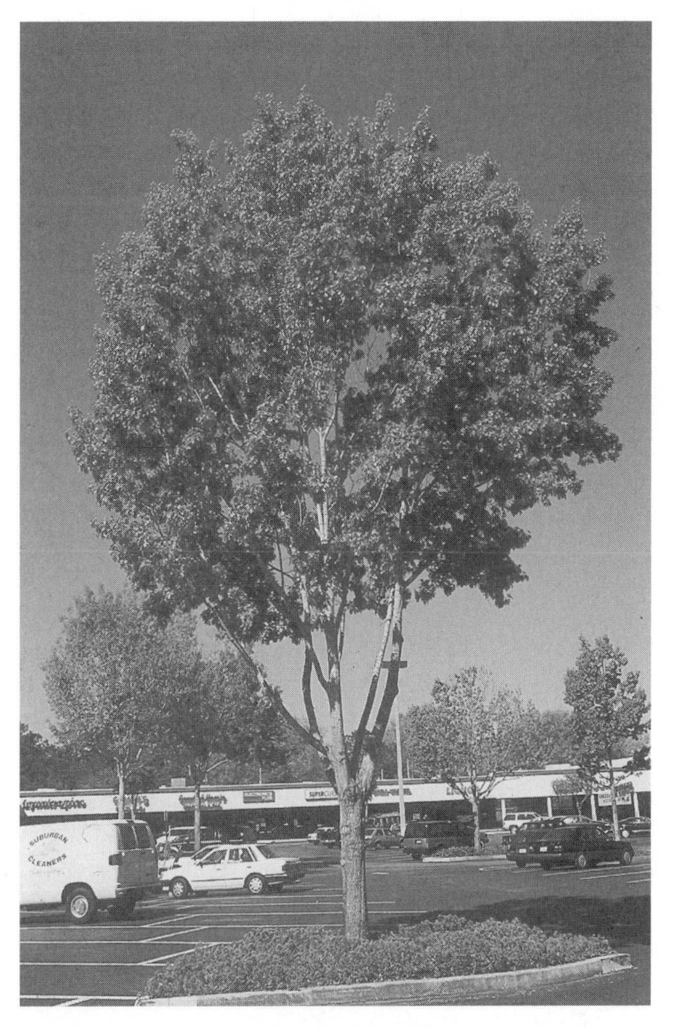

FIGURE 13-12. The canopy on this young oak was raised, but the needed structural pruning was not performed. The large-diameter stem on the right should have been subordinated to allow the slightly larger stem on the left to dominate the tree (left). Five years later, the tree has even worse form (right). The codominant stems have included bark in the union. The tree is taller placing more stress on this weak union.

BALANCING

A well-balanced tree is considered more stable, safer, and looks better than one with foliage and branches mostly missing from one side. It may be desirable to cut back aggressive, outwardly growing branches in order to make it easier to operate equipment in the landscape. Branches may need to be shortened to balance the canopy (Figure 13-13) or to prune it away from a structure or sidewalk. If the cut is made on a temporary branch on a young tree that will eventually be removed from the tree,

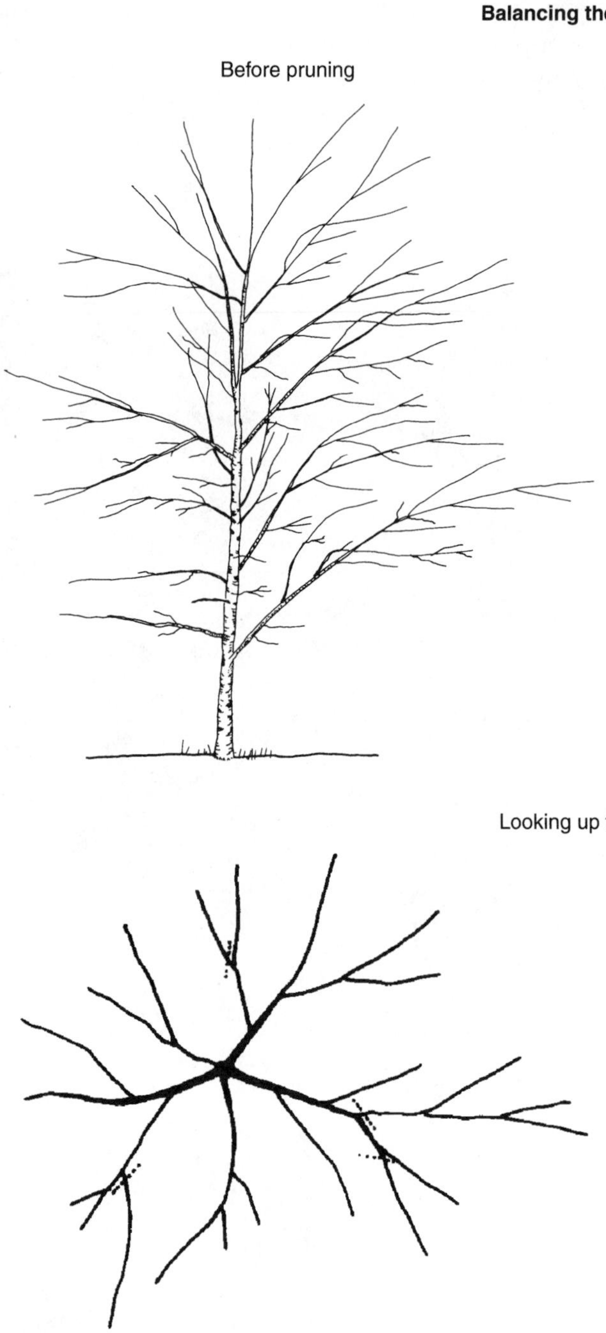

FIGURE 13-13. Reduction and removal cuts are used to balance the canopy. All cuts made on this young tree are reduction cuts (see arrows at top right). When deciding where to cut, it sometimes helps to stand under the tree and look up. Cut branches at the dotted lines (bottom).

the location is not as crucial as it would be if the offending branch were a permanent scaffold limb. When pruning a scaffold limb, keep in mind the future development of the branch. This could mean cutting back to a more upright-oriented branch along the scaffold to keep future growth out of the way. It might mean reducing to a lateral pointed in another direction.

Trees along the seacoast often develop an asymmetrical canopy sweeping away from the shore line (Figure 13-14). Trees in interior regions receiving winds from one direction also grow unbalanced canopies. Although these trees will never be symmetrical, they can be brought back into balance to a certain degree. Although unnecessary, you may also want to balance a tree simply to make it look more symmetrical (Figure 13-15).

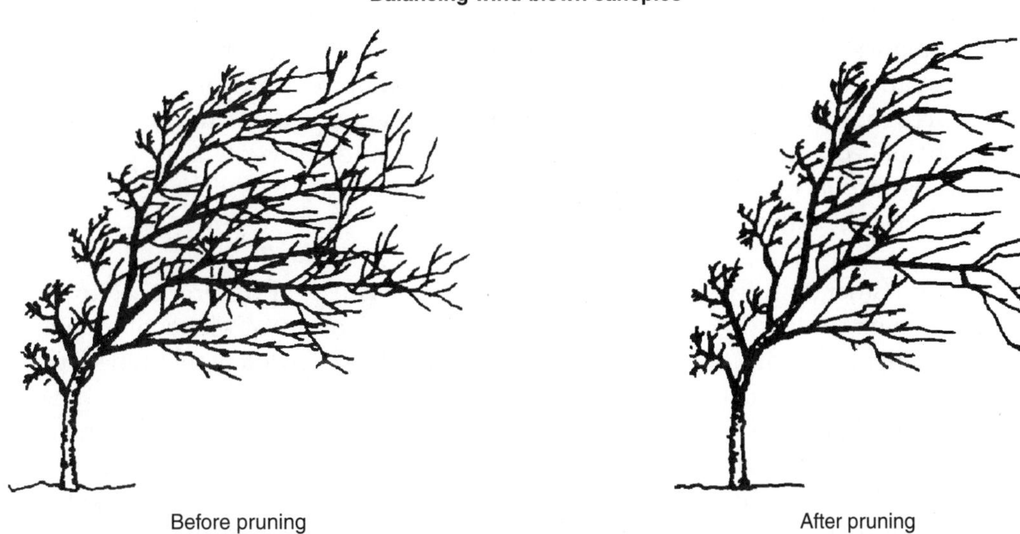

FIGURE 13-14. Trees growing to one side due to strong prevailing winds or salt exposure often become unbalanced. They can be brought into better balance with reduction and removal cuts.

FIGURE 13-15. Slight adjustments in the canopy can be made to create a better-balanced, more symmetrical canopy. Reduce back to the arrow at right. However, this is a cosmetic treatment and usually unnecessary.

Trees planted along a woodland or near a structure that shades one side of the tree often develop one-sided canopies, as branches bend toward the light and away from the shaded side. There are two solutions to help balance an asymmetrical canopy. It is best to start one or the other as soon as you recognize the problem because it is more difficult to balance older trees.

The first solution is to thin the canopy of the nearby trees if they are shading the unbalanced tree. This allows more light to reach the shaded tree and may help future growth proceed in a more balanced fashion. Then prune the shaded, lopsided tree to balance it. Remove some branches on the heavy side of the tree that are pointed toward the light and keep all branches (even small twigs) that are pointed in the direction of the shade. Use the reduction cut back to branches pointed toward the shade. If the tree has not been balanced for some time, large-diameter branches may have to be removed back to the trunk in order to create a symmetrical form. Although this treatment may balance the canopy, it may not be good for the tree. Do not expect miracles from this treatment; if you cannot provide the tree with more uniform sunlight after pruning, the tree may grow to be one-sided again after several years. Annual pruning to maintain a balanced canopy will keep it looking nice and prevent large pruning wounds and heartwood exposure. A tree receiving this treatment will have a more open canopy than one in the sun.

The other solution (on small trees) is to use wires to pull branches toward the shaded side of the tree in order to fill in a void in the canopy. This technique is used to bend branches toward the ground in Japanese gardens. To determine where to place the wire on the branch, use your hand to pull on the branch, bending it into the desirable position. Your hand will usually be located ½ to ¾ of the way up the branch. Screw a small eyelet into the side of the leaning branch at the point where you are holding it. Tie a wire or line to the eyelet and pull it until the branch reaches the desired position. Secure the wire to a small stake driven into the ground or to another eyelet on a different part of the tree. You can wire several branches on the tree at the same time. The wire may have to remain on the tree for two or more years until sufficient wood has developed to hold the branch in the new position. Remove the wire and cut the eyelet flush with the bark using a hacksaw. Unscrewing the eyelet may do more damage.

REDUCING RISK

Evaluating a tree for potential risk from structural failure is a complex process not covered in this book. Other books provide a guide to hazard tree evaluation (e.g., Lonsdale, 1999; Matheny and Clark, 1996). However, one of the many potential risks trees can pose, which can be partially reduced by pruning, is due to bark inclusions in the unions of large-diameter limbs. Managing trees to minimize hazards also includes several other strategies (Table 13-7). Chapters 12 and 13 are really all about risk reduction.

Trees at the edge of woods or trees near a home built on a wooded lot often grow one-sided (as in Figure 13-16). Branches formerly suppressed by surrounding trees are suddenly exposed to more sunlight, and they can respond by growing vigorously. Stems and branches with included bark elongate and bend down, increasing the likelihood of splitting from the trunk, especially if bark is included in the union. Thinning

TABLE 13-7. Pruning to reduce risk from structural failure can involve at least six strategies.*

Attempt to maintain or develop a dominant leader
Shorten or remove branches that will become (or are now) too long and heavy
Shorten branches with excessive end weight or bark inclusions
Remove dead or detached branches or those with cracks
Balance or secure trees deformed by a storm or leaning trees
Consider reducing trees with extensive decay or hollows, or those in restricted soil spaces or those with structural root loss

*All strategies may not be applicable to all trees.

and reducing the branch can help prevent if from falling. Other measures, including cabling, or totally removing the branch or the tree, might also be recommended by a professional arborist. There are many other situations that can make trees potentially harardous (see Tables 1-2 and 1-3) (Matheny and Clark, 1996), that can be partially corrected by pruning.

Another common situation encountered in residential and commercial landscapes carved out of forests is very close spacing between trunks. This is especially common in a young successional forest where pioneer species dominate the canopy. Two or more trees grow very close to one another forming a clump (Figure 13-17). Trunks may eventually touch, or they may be touching already. Although the clump appears to have a uniform, well-balanced canopy, each tree is severely one-sided. As trees begin to grow and lean away from each other, increased stress is placed on the lower trunk and root systems. Touching trunks can also exert a force pushing each other apart.

Tip: *Reduce risk by implementing a tree pruning plan.*

The root system on the side of the tree away from the lean plays a significant role in holding trees erect. Roots cannot develop on the side of the tree away from the lean because the trunks of the other trees are in the way (Figure 13-17). If these roots do not develop appropriately, trees could fall over. When one of the trees in a clump falls over or is cut down, the remaining trees are even more susceptible to falling because of increased exposure. Reduction pruning to shorten the tree could minimize the chances of the trees breaking or falling. Tall, slender trees with a tree height: trunk diameter ratio (both measured in the same units) of 50 or more are blown over more often than those with a smaller ratio. Consider reduction or removal of the tree to minimize risk of these failure-prone trees.

Large reduction cuts can cause serious decay problems on certain trees. If the objectives cannot be adequately met by reduction pruning, or if the cuts required to minimize the risk of failure will be very large, consider cabling and bracing techniques. Combined with pruning, arborists often use these two techniques in an attempt to reduce the risk of a tree failing. Implement a management program to evaluate reduced trees at regular intervals to check for decay, cracks, and other possible defects that could result from severe reduction.

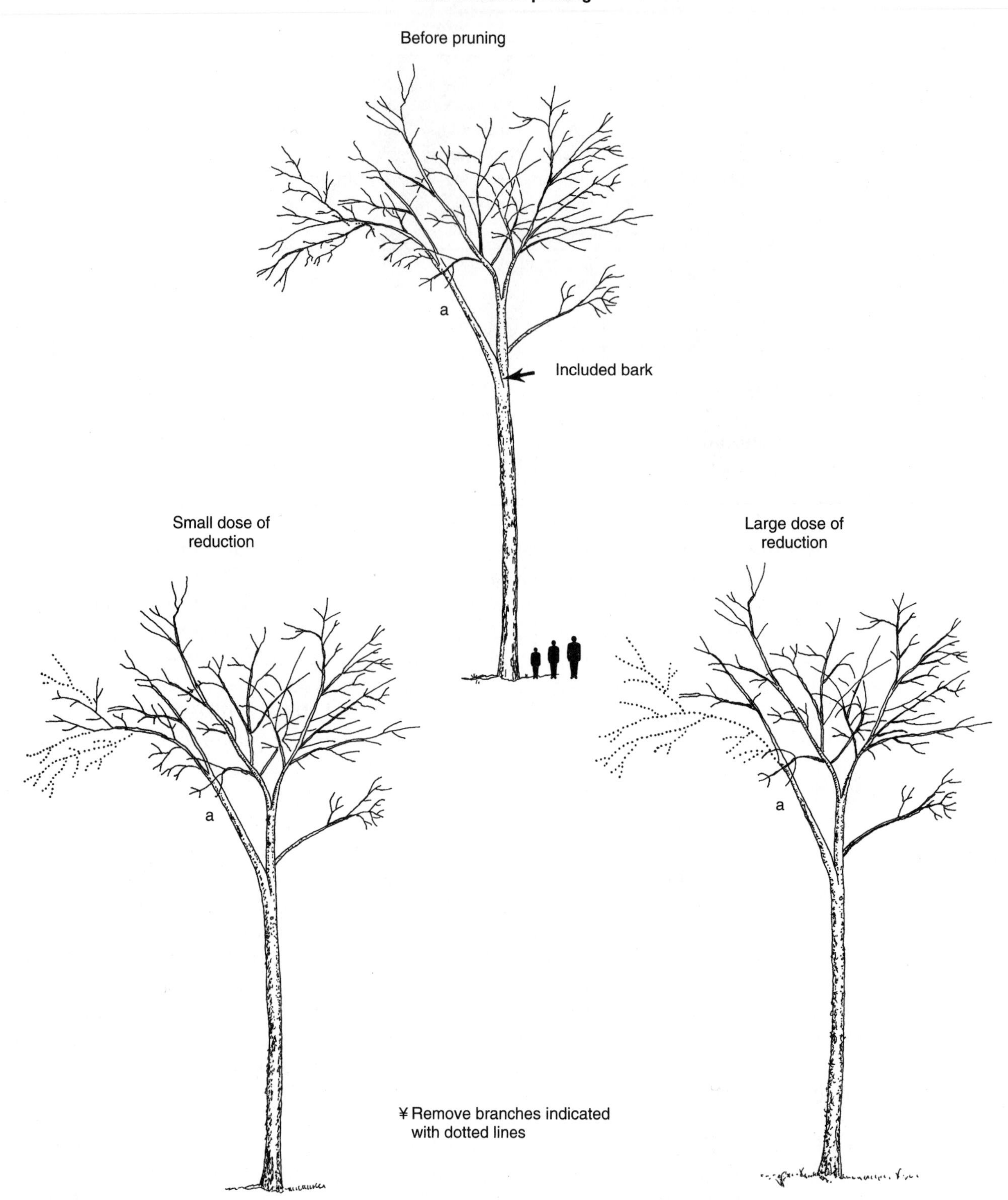

FIGURE 13-16. Reducing long, over-extended branches (branch 'a'), especially if they have bark inclusions, reduces risk by reducing end weight. A small (lower left) or large (lower right) dose may be appropriate depending on the situation.

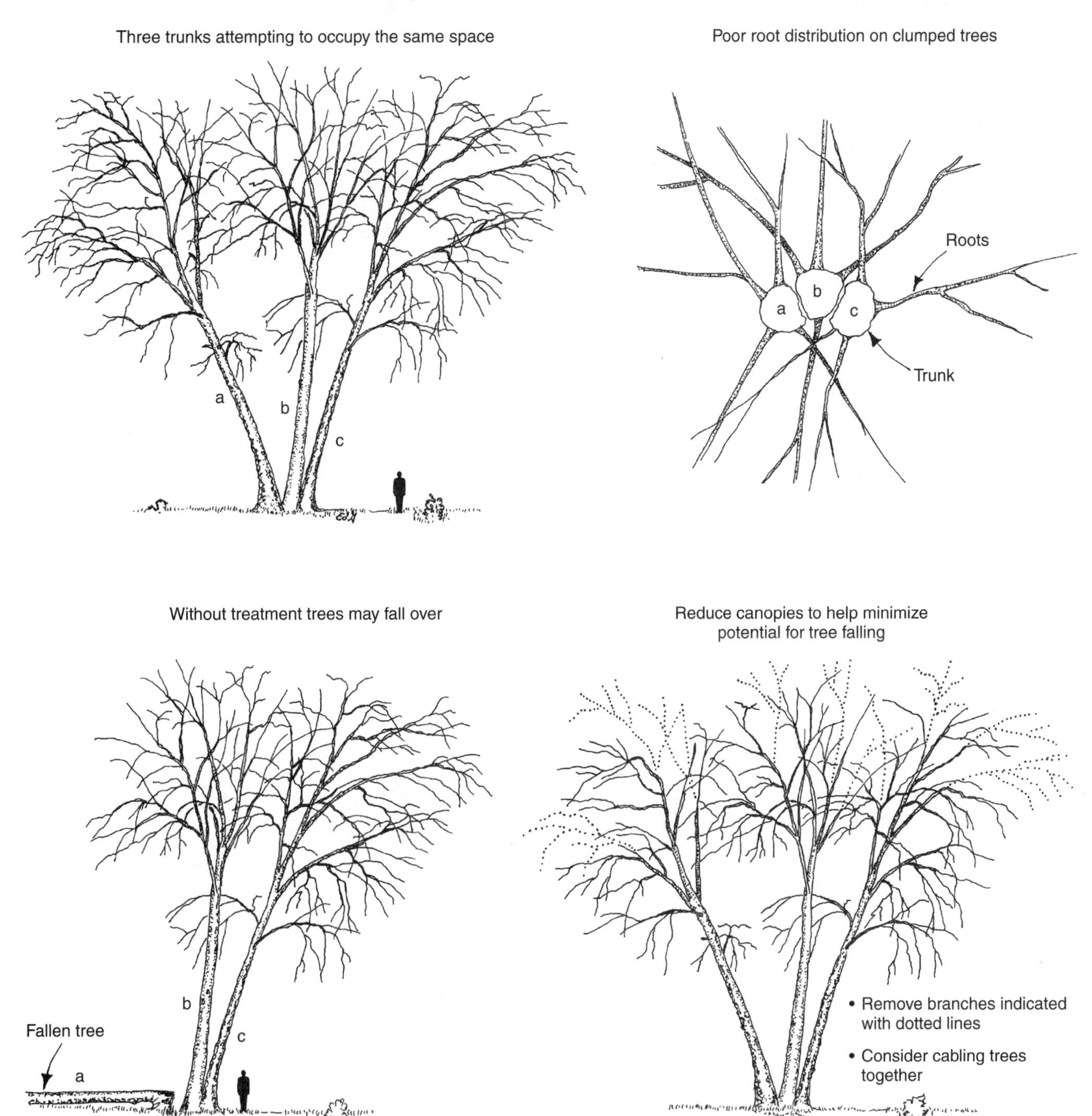

FIGURE 13-17. Several trees clumped together often form a nice symmetrical canopy but each individual tree is very one-sided (top left). Tree 'a' is beginning to lean to the left and tree 'c' to the right. A close look at the root system (top right) of tree 'a' shows that there are no roots growing toward the right which is in the direction away from the lean. Roots on the side of a tree away from a lean play a large role in holding the tree up. Since these roots are missing, it is not surprising that tree 'a' is prone to falling over (lower left). Reducing the canopy could help prevent the trees from separating (lower right).

BEFORE A STORM

Executing a planned tree management program that includes pruning is the best method of reducing tree damage resulting from a storm. Once a storm is predicted, it is really too late to perform meaningful treatments. However, if a super-cell, hurricane, or other major storm is expected soon and you have the opportunity to apply some last-minute treatments, consider removing dead and detached branches (Table 13-8). These can become missiles if they break free from the canopy. Large fruit such as coconuts, walnuts, and osage-orange fruits can also become hazardous in strong storms. Treatments such as thinning or reducing the canopy may reduce damage in a storm but take a great deal of time to perform. Customers might not be able to find a tree service to perform this treatment if the storm is eminent.

TABLE 13-8. Last-minute preparations before a hurricane or large storm.

Remove dead limbs
Remove large fruit
Thin or shorten large branches with included bark
Turn off irrigation systems to keep soil as dry as possible—wet soil makes trees less stable

AFTER A STORM

It comes as no surprise to most people that storms, especially hurricanes and tornadoes, cause severe damage to trees. Damage to trees from storms can be difficult to evaluate. Some is easy to see, such as defoliation; broken branches; split branch unions and trunks; large, broken surface roots; and leaning trees. Defoliated trees that were healthy before a storm often leaf out quickly following a hurricane. If this is the only damage to the tree, no special treatments are needed. Some trees, such as mango and others, are damaged by sudden exposure to sunlight. There is no need to apply fertilizers or other chemicals or substances. Time and irrigation (if soil becomes dry in hot weather after the storm) are the best treatments for this type of damage.

Broken branches should be pruned back to existing, intact lateral or parent branches with removal and reduction cuts (Figure 13-18). In most circumstances pruning paint is not needed and will do nothing for the tree. If there is a crack or split evident where a major branch meets the trunk and the crack goes into the trunk, it is probably best to remove the limb. Limbs with this type of damage may not be well secured to the trunk. If the limb is very large and the crack extends well into the trunk, consideration should be given to removing the tree. Because professional arborists are best qualified to evaluate trees, it is best to have one to look at the damaged trees.

In the days and weeks following the storm, small-sized leaning trees can be righted and staked as you would a transplanted tree. Treat these as if transplanted by watering regularly. If the tree is not too large and the area receives adequate rainfall or irrigation in the next several months, the tree has a good chance of recovering. Some large trees can be righted and secured with cables and other structures, but recovery to prestorm

FIGURE 13-18. After a storm has damaged trees and broken branches, make pruning cuts such as those shown by the dashed lines at the arrows.

conditions depends on many factors, some of which are unpredictable and difficult to evaluate. Large trees could require many years to develop a root system capable of supporting the tree. Roots may regenerate and eventually be able to hold the tree erect, but if large roots close to the trunk were broken or severely damaged, they may eventually rot, causing the tree to fall over in subsequent years. Guying to right storm-tipped trees is a long-term treatment.

Less obvious damage includes cracks or splits in the trunk and major limbs and hidden breakage of the root system. Trunks with cracks are dangerous, and serious consideration should be given to removing the tree if it poses a potential risk. These cracks will not heal; they remain for the life of the tree. If there is any question as to the safety and health of the tree that is close to a building, schoolyard, park, parking lot, or other place where people live, work, or play, consult a consulting arborist (contact American Society of Consulting Arborists) or a certified arborist (contact the International Society of Arboriculture in Champaign, IL, for a list.)

RESTORATION

Topped trees

Topping, hat-racking, stag heading, de-horning, lopping, rounding over—all refer to the same damaging malpractice. Unfortunately, these techniques are practiced worldwide. Several to many sprouts grow quickly from the cut end of a topped branch. There is not enough room on the cut branch to allow for secure attachment of all the sprouts to the tree, and they are held on the cut branch only by a thin layer of new wood generated after pruning. The new growth is so dense that few side twigs are produced on the new sprouts. As a result, new sprouts are tapered poorly, which means that they are about the same diameter at the base of the branch as they are near the top. This makes them weak.

Topped branches can also decay rapidly which increases their susceptibility to storm damage. Topped trees also regain their original height quickly, sometimes in a year or two depending on the severity of pruning, soil characteristics, and the species. Topped or rounded-over trees often have structural defects other than those created by topping, which remain within the framework of the trunk and main branches. Unfortunately, these defects are often ignored when the tree is pruned. Cold injury, storm damage, flooding, and drought can cause the same devastating topping effect on trees.

Tip: *Topped trees can be hazardous to your health.*

Keep in mind that it may not be possible to restore a topped tree completely to the structure it had before topping. Canopy restoration attempts to improve structure more than restore it. To improve structure on a small-diameter topped stem or branch, you might choose to remove all sprouts except two or three. Head back (or reduce if sprouts have branches) the sprouts that remain except the one you choose as the new leader (Figure 13-19, top). The new leader may need to be reduced (or headed if there are no laterals) if it is long, with few side branches and poor taper. Use this technique on all topped branches throughout the tree.

On larger diameter topping cuts, there may be more sprouts growing from each cut (Figure 13-19, bottom). Remove several and shorten all others except the one you want to become the new leader. Again, the goal is to prune so that eventually there is only one main leader at the location of the topping cut. You want to make room for growth and development of lateral branches from the new leader. Even when the correct sprouts are removed to improve the structure of a topped tree, the sprout left as the new leader may be susceptible to breakage in a storm. The connection between the cut branch and the new leader is weak because the sprout is attached only to the thin cambium layer on the topped branch. The sprouts eventually become somewhat better secured to the cut branch. Do not forget to evaluate the scaffold branches lower on the tree as well. There may be defects, such as included bark in the unions at the main trunk, that should be attended to. Subordination cuts on some branches might be used to improve structure, or some topped branches may need to be removed from the tree entirely, back to existing laterals.

Lions-tailed trees

To restore a tree that has been overthinned or **lions-tailed,** allow water sprouts to develop along the branches for a few years. Your goal is to eventually develop a few of the water sprouts into permanent branches. This will help the tree rebuild its energy reserves that were severely depleted due to overthinning. Remove some of the sprouts and shorten others two or three years after the tree was overthinned, making certain that the ones you leave on the tree are more or less evenly distributed along the branches (Figure 13-20). Make a special effort to keep sprouts toward the base of the main scaffold limbs. This will help develop taper on the scaffold limbs making them stronger.

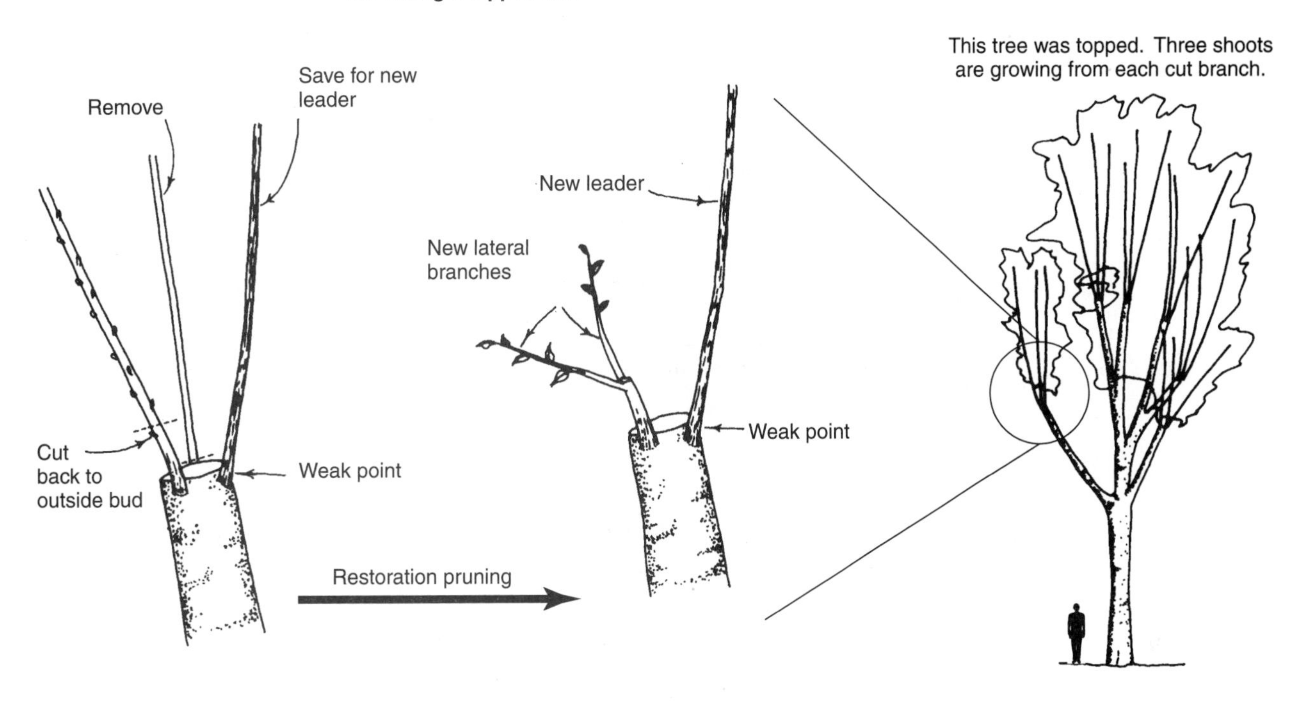

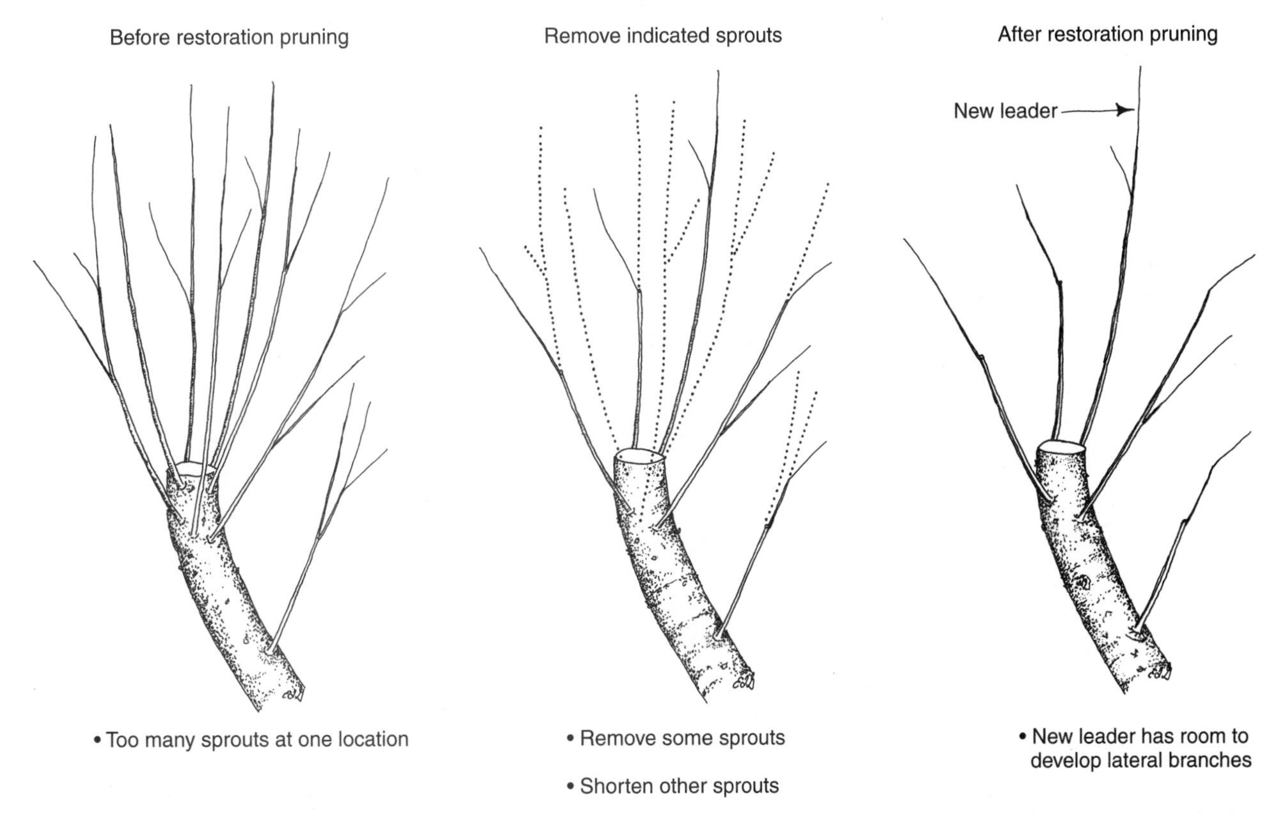

FIGURE 13-19. Problem: Too many sprouts occupy the same space on decaying topping cuts. **Solution:** Pruning to help restore structure in a topped tree removes and shortens most or all sprouts except the one that should become the new leader. There will be a weak point at the base of the sprouts for many years to come. Even when the sprout eventually closes over the topping wound, there is a crack in the tree at this point forever.

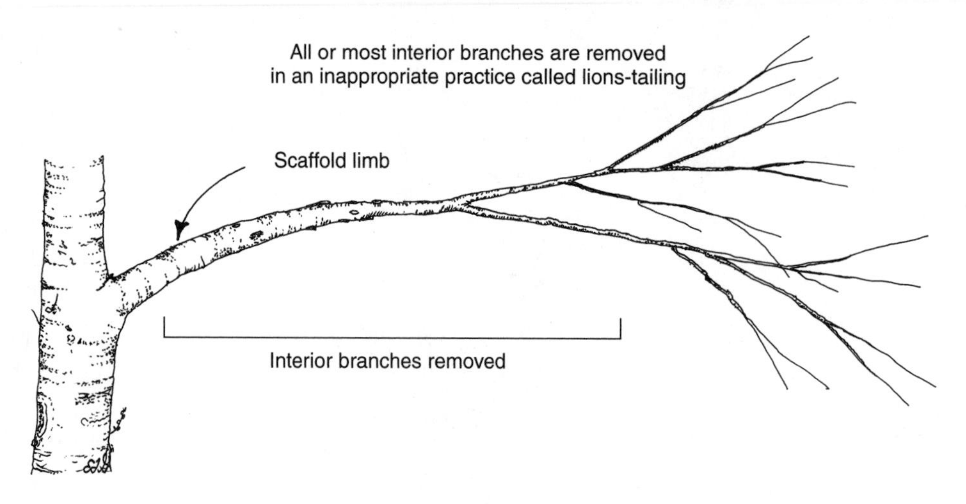

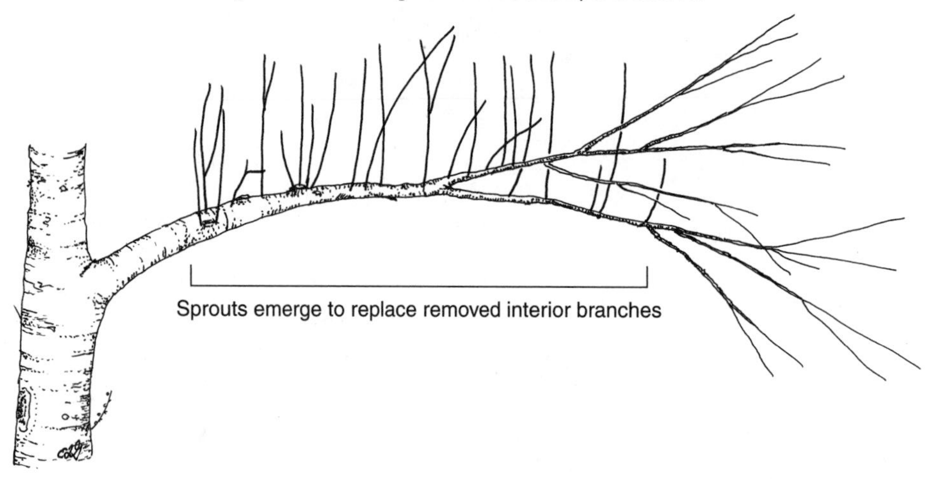

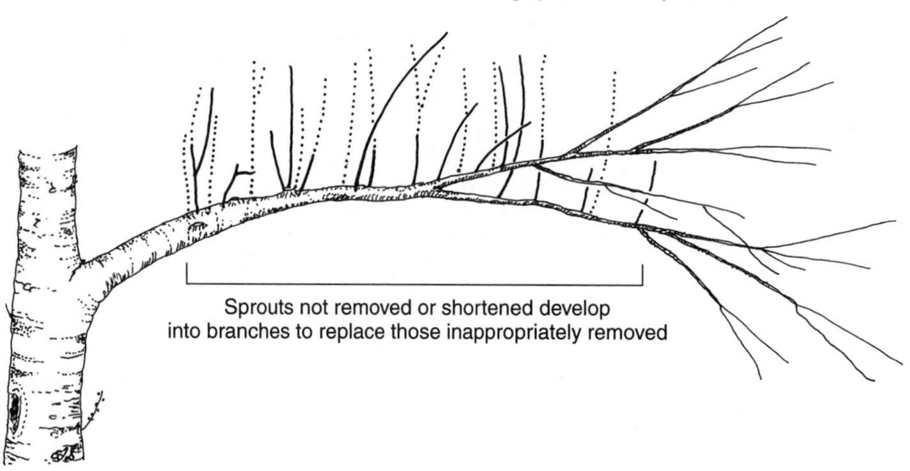

FIGURE 13-20. Problem: Interior branches were removed leaving most branch weight at the ends (top). **Solution:** To help a tree recover from overthinning or lions-tailing, first *do nothing at all.* Allow the tree to generate sprouts and grow for at least two to three years to regain energy supplies (middle). Then remove some sprouts and shorten others to encourage those that remain to grow (bottom).

A few years later, some of the sprouts left intact in the first restoration pruning might require shortening to allow the remaining ones to develop further. This process might have to be repeated for about ten years until remaining branches are spaced along the scaffold limbs. As sprouts slow down their growth rate and begin acting more like branches, you might consider shortening the scaffold limbs with a reduction cut. This would only be recommended if the limb was becoming too long, or began to droop, or was getting in the way. Do not reduce the scaffold limb too soon after the tree was over-thinned because this could cause further stress and more sprouting.

Tipped trees

Tipping trees is like topping except that the pruning cuts are smaller in diameter. Many people refer to this as rounding over. This practice essentially creates a "lollypop" effect and spoils the structure of the tree. This vision of a tree might come from the drawings we did when we were in elementary school.

Tipping creates a unique structure in trees that is challenging to correct because many branches end up the same diameter (Figure 13-21, left). In some ways, restoration is similar to the procedure for topped trees. Your goal is to (1) remove all but one sprout from

Restoring a tipped or rounded-over tree

Before restoration pruning

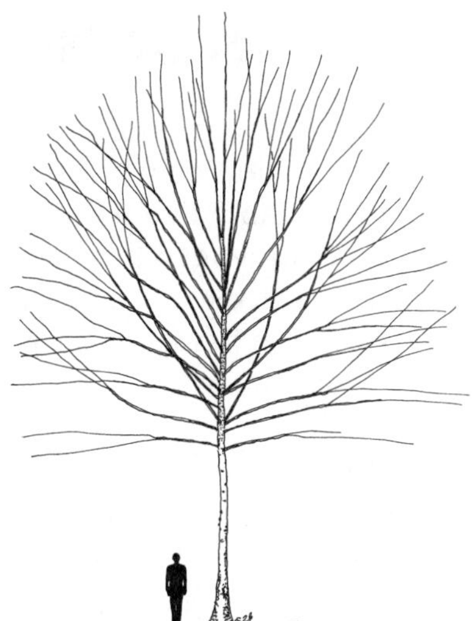

- Too many branches crowded together
- Branches are forced to grow too long

Note: Closely-spaced, small diameter branches are normal on excurrent trees and usually require no treatment.

Remove indicated branches

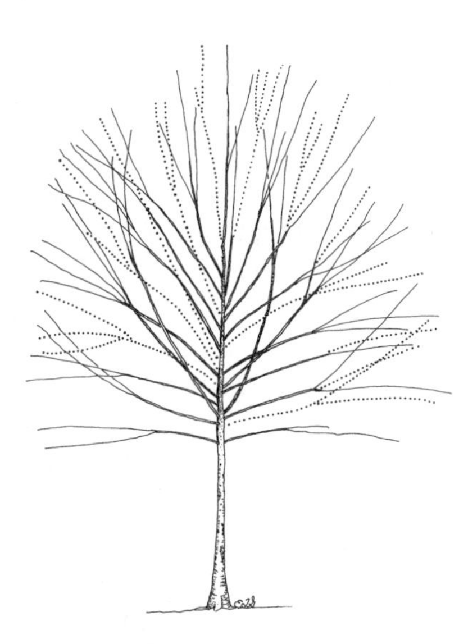

- Remove and shorten some so remaining ones develop lateral branches
- Do not remove too many back to the trunk — trunk cracks could develop

After restoration pruning

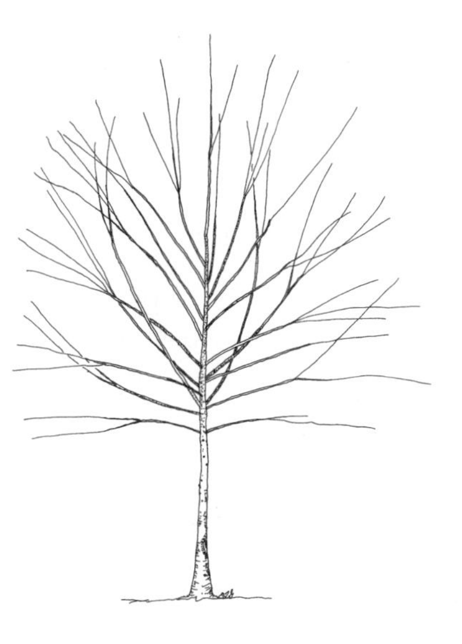

- Main limbs are now spaced apart
- A second and third pruning may be needed in the next few years to help scaffold limbs develop appropriate spacing

FIGURE 13-21. Problem: Tipped tree with many small diameter branches mostly originating from heading cuts; too many equal-sized scaffold branches (left). **Solution:** Remove and shorten branches so those that remain can generate lateral branches (center). Main limbs are spaced apart. This will help prevent each from over-extending and growing too long (right).

some of the tipping cuts (Figure 13-19), (2) remove some branches that were tipped, back to the trunk or to lateral branches inside the canopy and (3) space scaffold branches far enough apart so they can develop lateral branches (Figure 13-21, middle and right). Several prunings over ten years may be required to put good structure into a tipped tree.

Storm-damaged trees

Like topped and lions-tailed trees, those recently damaged by storms will develop water sprouts. This is an attempt by the tree to replace the photosynthesis-generating foliage that was removed suddenly. This process requires expenditure of stored starch. Sprouts should be allowed to grow for several years so energy reserves can be replaced. Once sprouts have grown for several years (or sooner if sprouts are extremely vigorous) they begin competing for the same space and should be reduced in number. Broken upright stems may look like those in Figure 13-19, whereas broken horizontal branches often generate sprouts all along their length (Figure 13-22, top).

Choose to save several sprouts spaced apart from each other that appear to be capable of growing into strong limbs with plenty of lateral branches. Remove ⅓ of the sprouts and shorten another ⅓ to allow the chosen ones to develop lateral branches and good taper. This prevents all the sprouts from growing too long and becoming weak. You may have to return several times during a ten-year period to put good structure back into the tree.

NEGLECTED TREES

Considerable pruning may be needed to improve structure and appearance on some trees. Subordination, reduction, thinning, cleaning, and risk reduction may be used on the same tree to meet your objectives. Remember to remove no more live tissue than the recommended maximum (young tree—25 to 30 percent; medium-aged trees—20 to 25 percent; mature trees—10 to 15 percent). It may be best for the tree to remove living tissue over a period of two or three years instead of all at once. The following provide examples of real-life pruning situations taken from around the country.

Branches may be growing into a street light or traffic light (Figure 13-23, top left). Remove branches A and C to allow the light to shine through the tree to the ground. Thin branch B. This tree also has included bark in the union marked with an arrow. To prevent this from getting much worse, subordinate stem D by thinning and perhaps reduction to slow growth rate and reduce weight on this stem, allowing stem E to dominate. In addition to allowing the street light to penetrate the tree, pruning branches B and C will subordinate stem F. This will help encourage the main trunk, E, to dominate the tree. Consider removing stem F as the tree grows to eliminate regular repruning to clear branches from the security light. Since this will leave a large wound consider all other options first.

The tree could be pruned differently if the light was not next to the tree. Instead of removing branches A, B, and C, (Figure 13-23) the top portion of stem F could be reduced back to branch B. In this case, stem F could remain on the tree, perhaps for a very long time. Several reduction cuts over a period of years may subordinate stem F enough to prevent it from growing larger than about ½ the size of the trunk.

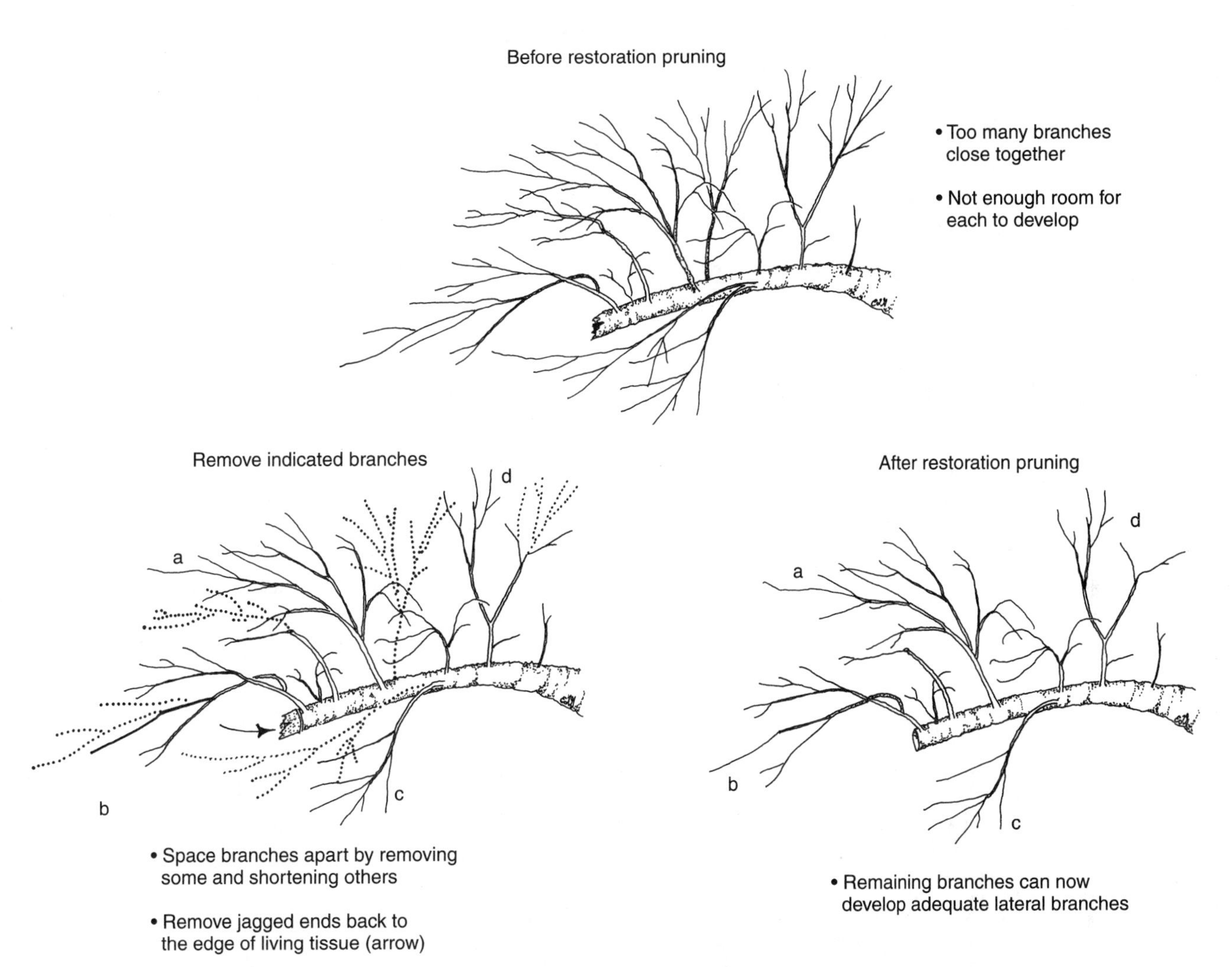

FIGURE 13-22. Problem: Too many closely-spaced sprouts are attempting to grow into branches; there is not enough room for all to develop appropriately (top). **Solution:** Shorten and remove some so the remaining ones ('a' through 'd') develop lateral branches; remove dead, jagged branch end (see arrow) back to living tissue (lower left). Branches 'a' through 'd' can now develop lateral branches and become secured better to the damaged branch. This process usually takes several to many years to correct; additional pruning may be required every year or two for ten years to develop appropriate structure (lower right).

Established, moderately sized (12–20 inch trunk diameter) trees sometimes need a small number of pruning cuts to make a dramatic improvement in structure. For example, you might remove one branch at the center of the tree in Figure 13-23 (top right) and thin one other branch. One reduction cut is required on the tree at left center to help prevent the branch from splitting from the trunk. One branch could be removed on the tree at center because it is nearly touching another larger branch. The tree on the far right center needs several cuts on the branch with included bark (indicated with an arrow) to slow its growth rate. This, perhaps combined with a cable, will help keep the branch from splitting from the trunk. The tree at the bottom center has a fairly strong branch union because there is no included bark, but thinning some of the branches on the left-hand stem every year for several years will allow the right-hand stem to grow larger and perhaps dominate the tree.

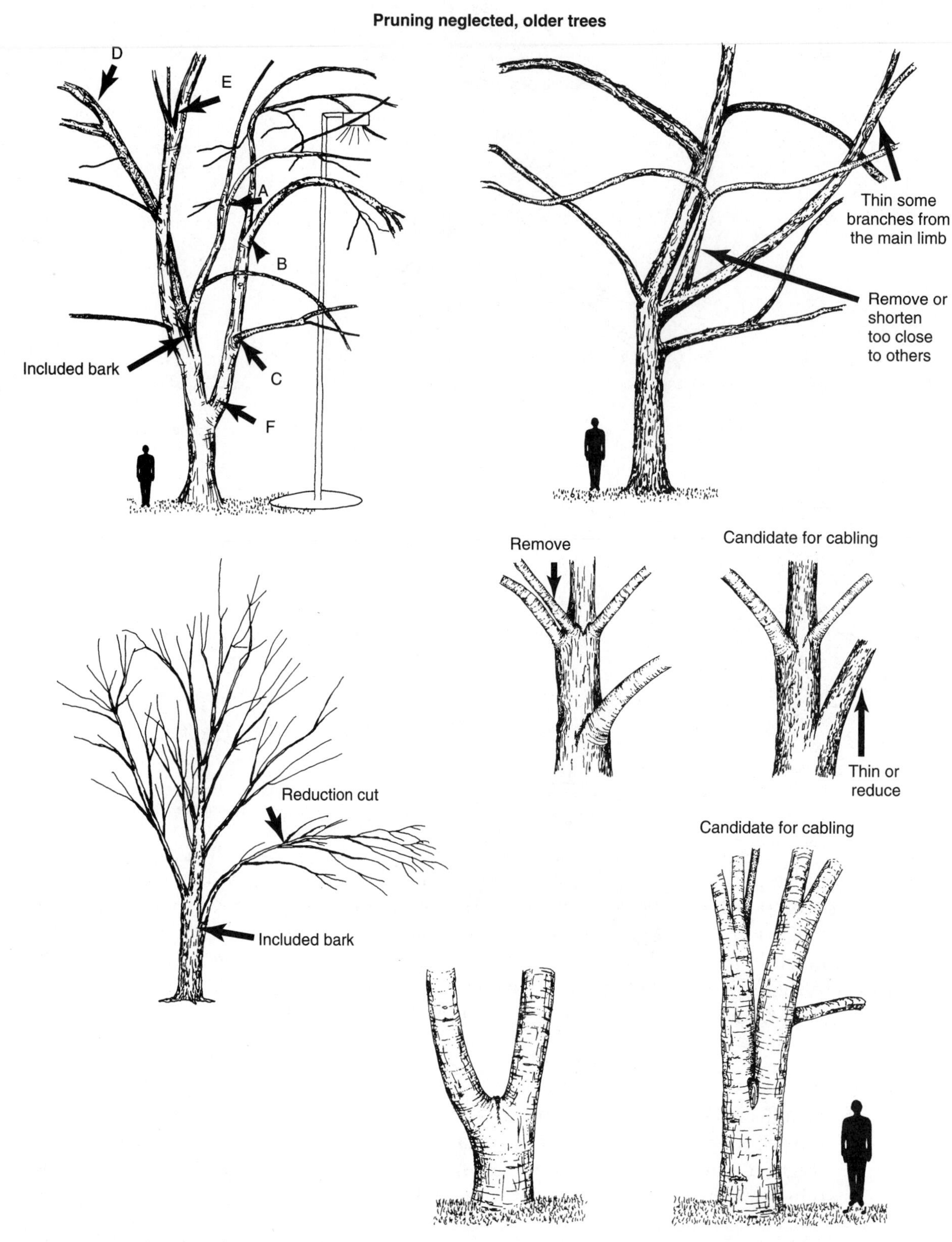

FIGURE 13-23. Remove branches A and C and thin branches B and D to allow stem E to dominate the tree. Other secondary branches on stem F can be thinned in order to allow the security light to shine to the ground (top left). Remove unwanted stems and branches that are too close to others (marked with an arrow, top right and center). Reduce a drooping branch at the arrow (middle left). Thin a branch that has included bark (middle right). Thin some branches from the left-hand stems to slow growth rate and minimize the potential for splitting at the union (bottom).

This will improve the tree structure because the stem at left will eventually be smaller than the stem at right. A collar may even develop around the base of the branch, which is a sign that the stem is becoming more like a branch.

The tree on the bottom right has a 3-foot length of included bark between the main stems. This is a serious condition and indicates a very weak union. The tree may need to be cabled or a threaded rod may need to be installed just above the union by a qualified arborist to help prevent it from splitting. Thinning some branches from the left stem each year to slow its growth rate will not hurt the tree but may not improve its structure very much immediately because the tree is so committed to this form. A tree with this poor a structure should have been attended to many years before. Figure 13-24 shows how just two reduction cuts can improve structure on a 20-year-old tree that was never pruned before. Some trees are difficult to prune effectively because form is poor and defective or misplaced limbs are so large. Pruning won't create miracles for these trees. See restoration for more on pruning neglected trees.

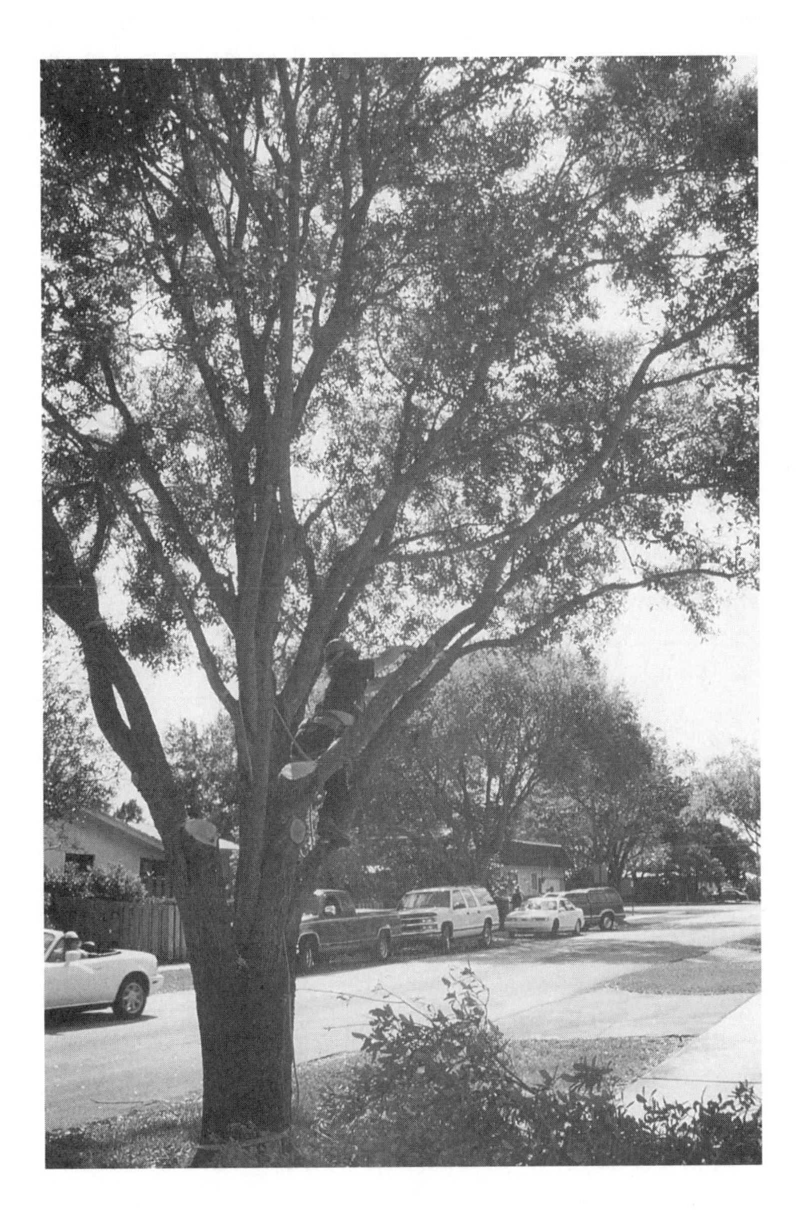

FIGURE 13-24. The codominant stems on the right and left were subordinated (note the light-colored pruning cuts) in order to allow the trunk in the center of the tree to dominate the structure. This is likely to increase the tree's longevity. Both branch unions have included bark, indicating poor attachment with the trunk.

Without care, older fruit trees often develop a dense, bushy habit and produce poor-quality fruit. Thin and clean the canopy and make certain that vigorous upright sprouts are removed or cut back with reduction cuts. Remove root sprouts back to their point of origin. Space major limbs apart by removing or shortening other large-diameter branches that are closer than 4 inches. If fruit production is poor remove or shorten some upright limbs and shorten major limbs slightly with reduction cuts to stimulate new shoot production. Do not remove more than 10 percent of the foliage, as this stimulates excessive sprouting. If more needs to be removed, do it over a period of two or three years to prevent the formation of unwanted sprouts, which can spoil your efforts.

PRUNE FOR PEST CONTROL

Early disease infections or infestations of pests can sometimes be picked off or pruned away before they become a major problem (Table 13-9). This keeps the tree healthy and reduces the use of pesticides. Regular monitoring is essential for this to work effectively. If you wait until more than about 5 percent of the foliage or branches are involved, you have waited too long, and more traditional chemical control may be necessary. Due to the labor-intensive nature of this strategy, it is most effective for residential or small landscapes or in a landscape with a high maintenance budget. This is more easily performed on small trees than on large ones.

DIRECTIONAL PRUNING

Directional pruning (also called lateral pruning) guides the tree to grow in a certain direction by removing live branches from other portions of the tree. It is really a version of subordination. Reduction and removal cuts are typically employed in directional pruning (Figure 13-25). This is especially useful near buildings, roads, walks, streetlights, and overhead utility wires. Directional pruning usually provides for only

TABLE 13-9. Disease and pest problems that may be reduced by pruning.

Pest	Action*
Sphaeropsis (Diplodia)	Remove infected branches (preferably in dry weather)
Cytospora canker	Remove infected branches back to trunk immediately
Anthracnose	Remove dead branches
Rust diseases	Remove affected branches on coniferous hosts
Tip blights	Remove infected and dead branches
Dutch elm disease	Prune out flagged or symptomatic branches immediately
Black knot	Remove branches with knots before new growth emerges
Fireblight	Remove infected branches (preferably in dry weather)
Leaf spots	Thin branches from the edge of the canopy, raise canopy
Tent caterpillars	Remove egg masses
Sawfly	Remove infested branches immediately
Mistletoe	Remove branches with mistletoe back to next branch union
Webworm	Remove infested foliage and twigs

*Dispose of plant material away from susceptible plants to reduce likelihood of re-infestation.

a temporary solution. Ample access to sunlight usually means that branches will return to the portion of the tree that was reduced. More discussion of directional pruning near utility lines can be found in Chapter 14.

VISTA PRUNING

Trees often block views to a pleasing landscape such as a body of water, valley, or mountain range. Although many people choose only to raise the canopy or top the tree to reduce its height to open a view, a combination of moderate canopy raising and thinning often looks best and is best for the tree (Figure 13-26). Special thinning, which opens small windows in the canopy, may also be done to open a view to a desirable landscape. Remove no more than 10 to 15 percent of the canopy on a mature tree. Up to about ¼ can be removed on a younger tree. If height must be reduced, make reduction cuts, not heading cuts.

SLOWING THE GROWTH RATE

Pruning a tree will not prevent it from getting big because genetics and available resources at the site control tree size. However, removing branches from a tree slows its growth rate slightly. Although root growth slows as the tree replaces the removed branches, young and medium-aged trees rapidly outgrow the slowing effect of prun-

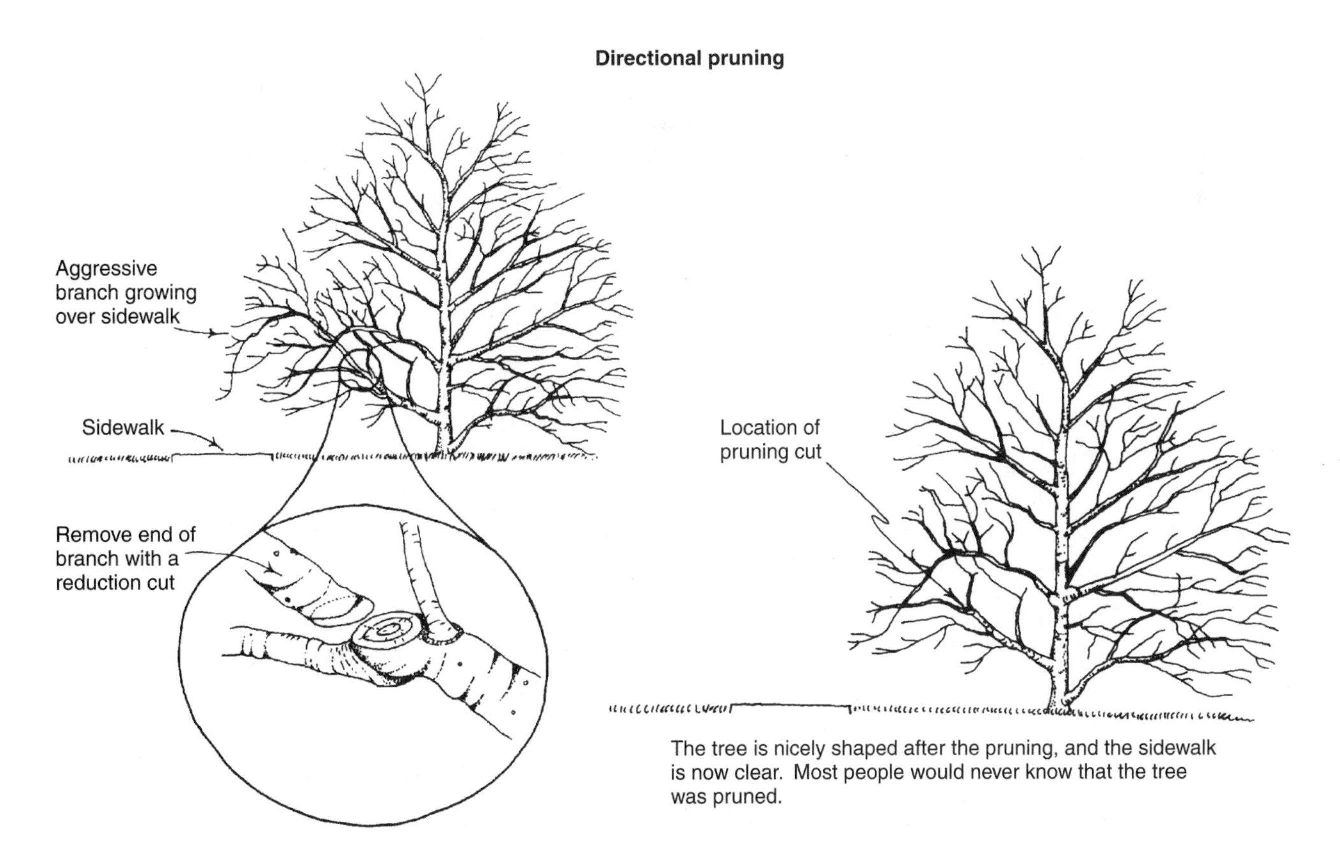

FIGURE 13-25. Reducing the length of a branch with a reduction cut directs and slows its growth.

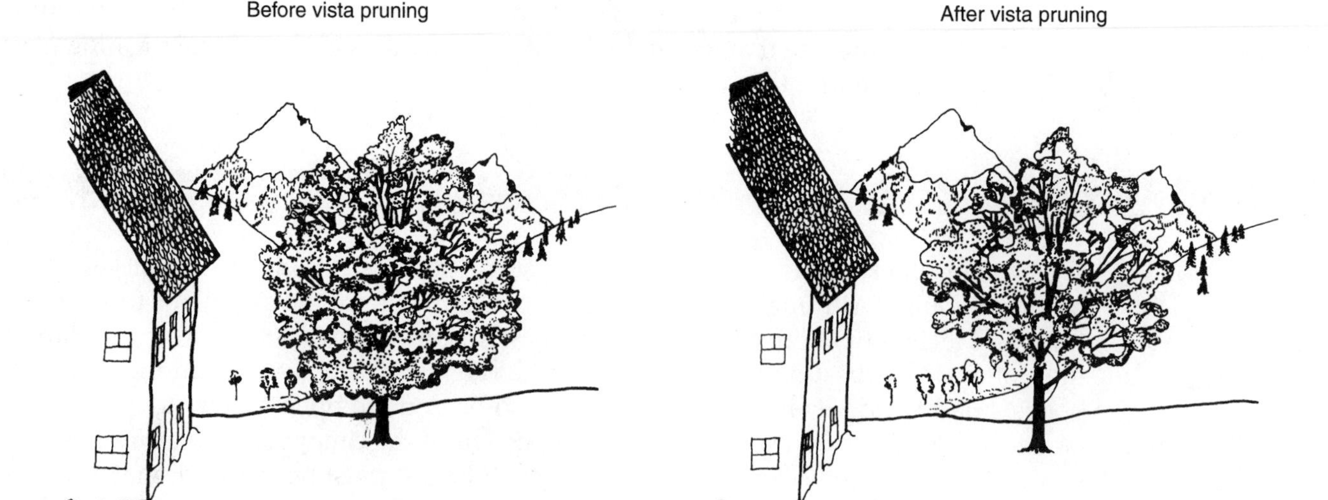

FIGURE 13-26. Thinning and raising can open up a view blocked by a tree canopy. Remove no more than ¼ of the foliage on young and medium-aged trees, less on mature trees.

ing. Remove, reduce, or head (head only on branches 1 to 2 years old) to slow diameter growth on a main branch. Remove main branches at the trunk to slow trunk diameter growth. While increasing total growth on the tree, leaving laterals on a leader or branch will slow length growth of that leader or branch. Conversely, overthinning by removing only lower or interior branches increases length growth of the parent branches. It is not appropriate to prune older, mature trees to slow their growth rate as this can starve the tree by removing branches with valuable stored energy.

CHECK YOUR KNOWLEDGE

1) The most appropriate cut types for reducing tree height are:
   a. heading cuts.
   b. reduction cuts.
   c. removal cuts.
   d. All are equally appropriate.
2) Reducing the length of a scaffold limb with included bark:
   a. slows growth on the limb.
   b. raises the limb slightly.
   c. reduces the likelihood of the limb splitting from the tree.
   d. all of the above
3) If the shade cast by the canopy of a mature tree is causing the turf beneath the canopy to decline, which is the BEST option to maintain tree health?
   a. raise the canopy and reduce the amount of foliage in the canopy with removal cuts to allow more light to reach the turf
   b. raise the canopy and reduce the amount of foliage in the canopy with reduction cuts to allow more light to reach the turf

c. suggest replacing the turf with shade tolerant ground cover
d. top the tree to allow more light to reach the turf

4) Which pruning type encourages growth on interior lateral branches and promotes branch taper?

a. cleaning
b. raising
c. restoration
d. thinning

5) In many instances, sprouts:

a. indicate that the tree was overpruned.
b. should be removed.
c. indicate a healthy tree.
d. indicate a nutrient deficiency.

6) It is usually most appropriate to:

a. install cables and insert screw rods in branches with included bark.
b. remove branches with included bark.
c. let branches develop with included bark.
d. subordinate or shorten branches with included bark and consider cabling on heavy limbs.

7) Tree A was pruned by removing only interior lateral branches creating lions-tailing; tree B was thinned by removing some lateral branches mostly from the ends of limbs. Which of the following is true?

a. Tree A will grow taller and/or wider than tree B.
b. Tree B will grow taller and/or wider than tree A.
c. The thinning technique has no impact on tree size.
d. Tree B will sprout more than tree A.

8) In order to raise the end of a low drooping main branch:

a. remove secondary branches from all along the main branch.
b. remove most interior secondary branches.
c. remove secondary branches mostly from the end of the branch.
d. a branch usually cannot be raised with pruning.

9) If tree height must be reduced to clear trees from wires or other overhead obstructions, the best method is called canopy:

a. thinning.
b. cleaning.
c. restoration.
d. reduction.

10) Lions-tailing is common because it:

a. provides for a strong-structured tree.
b. is easy to perform this technique.
c. reduces the likelihood of damage in a storm.
d. reduces the chances of decay in main branches.

11) Where trees have been previously topped and have resprouted, thin the tree to remove:

    a. the most vigorous sprouts.
    b. the least vigorous sprouts.
    c. some of the stubbed heads plus shorten and remove some sprouts from those that remain.
    d. none of the above

Answers: b, d, c, d, a, d, a, c, d, b, c

CHALLENGE QUESTIONS

1) Describe a pruning plan for a medium-aged tree to reduce the likelihood of it failing in a windstorm.
2) Let's say there is a large-maturing oak tree 10 feet from a two-story building. The tree is about 30 feet tall. What could you do now and for the next ten years to help direct growth away from the building?
3) Compare the pluses and minuses of removing a low 6-inch diameter branch from a 14-inch trunk versus shortening now and removing several years later.

SUGGESTED EXERCISES

1) Bring a group to a mature tree with some dieback in the top of the canopy, trunk cavities, or other defects. Divide into subgroups of two to ten people. Ask each subgroup to devise a plan to manage the tree over a period of ten years. Have each subgroup present and discuss their plans to the entire group.
2) Find a tree that does not require pruning. Discuss why no pruning is needed. Ask other members of your group if they agree with your evaluation.

CHAPTER 14

CONSIDERATIONS FOR MAINTAINING SPECIAL SITES AND TREES

OBJECTIVES

1) Present precautions for pruning mature trees.
2) Contrast pruning strategies for street trees with park and landscape trees.
3) Present special considerations for pruning certain genera and tree forms.
4) Develop an understanding of the challenges of pruning near utilities.

KEY WORDS

Edge trees
Feature trees
Lateral pruning
Mop tops
Ornamental trees
Primary branches
Quaternary branches
Secondary branches
Tertiary branches

MATURE TREES

Pruning large and mature trees focuses on ensuring human safety and passage, minimizing limb failure or total-tree failure near targets such as buildings and cars, and maintaining tree health and vigor. This means (1) minimizing hazardous conditions by cleaning and reducing weight where needed, (2) canopy raising where needed, and (3) maintaining small-diameter interior branches. It may be too late to make meaningful structural changes in trunk and scaffold limb architecture on mature and over-mature trees.

Remove live foliage from a mature or over-mature tree only for good reason! Refrain from removing any live foliage from a stressed tree because they need as much sugar-generating capacity as possible. Removing live tissue on a mature tree removes stored energy and forces the tree to react and expend energy unnecessarily causing many potential problems (Table 14-1). ANSI A300 pruning standards allow up to 25 percent of live foliage removal on mature trees, but this is too much in many circumstances. If you decide to thin an old tree, make cuts primarily on **tertiary branches, quaternary branches,** and even smaller branches toward the canopy edge only (Figure 14-1). Removing **primary branches** such as scaffold limbs and more than just a few **secondary branches** growing from scaffold limbs may leave large pruning wounds and remove too much live tissue. Branches that are more than about a third the diameter of the trunk, and those that are more than about 15 years old (depending on species and other factors) may have poor ability to restrict spread of decay following removal. Consider shortening or thinning the limb by removing tertiary and smaller branches instead of removing the branch entirely.

A common malpractice on large trees is removing many or all interior low branches less than about 4 inches diameter (Figure 14-2). Industry professionals and many educators consider this overthinning or lions-tailing to be no more than an income-generating scheme practiced by uninformed, untrained people practicing arboriculture. Unfortunately, this practice has become commonplace in many communities throughout the United States (see Chapter 13 for more details). When people prune trees in this abusive manner, excessive live tissue is removed from the tree and no structural pruning is performed. This creates poor form and numerous wounds, and the tree becomes more prone to failure especially if there are few trees nearby. Interior branches also provide a local source of carbohydrates needed by the tree to carry on normal defense and other functions. Old trees can decline as a result of removing too much live tissue. Only a small amount of live tissue may be too much to remove on a mature or over-mature tree. There should be a very good reason, such as to reduce likelihood of branch failure, when removing more than 10 percent of the live foliage.

TABLE 14-1. Risks associated with removing live tissue from old trees.

Forces energy expenditure by initiating compartmentalization
Can cause cracks
Removes energy reserves
Can cause sprouting that further reduces energy reserves
Can cause branch death
Can cause tree death
Reduces available energy storage space

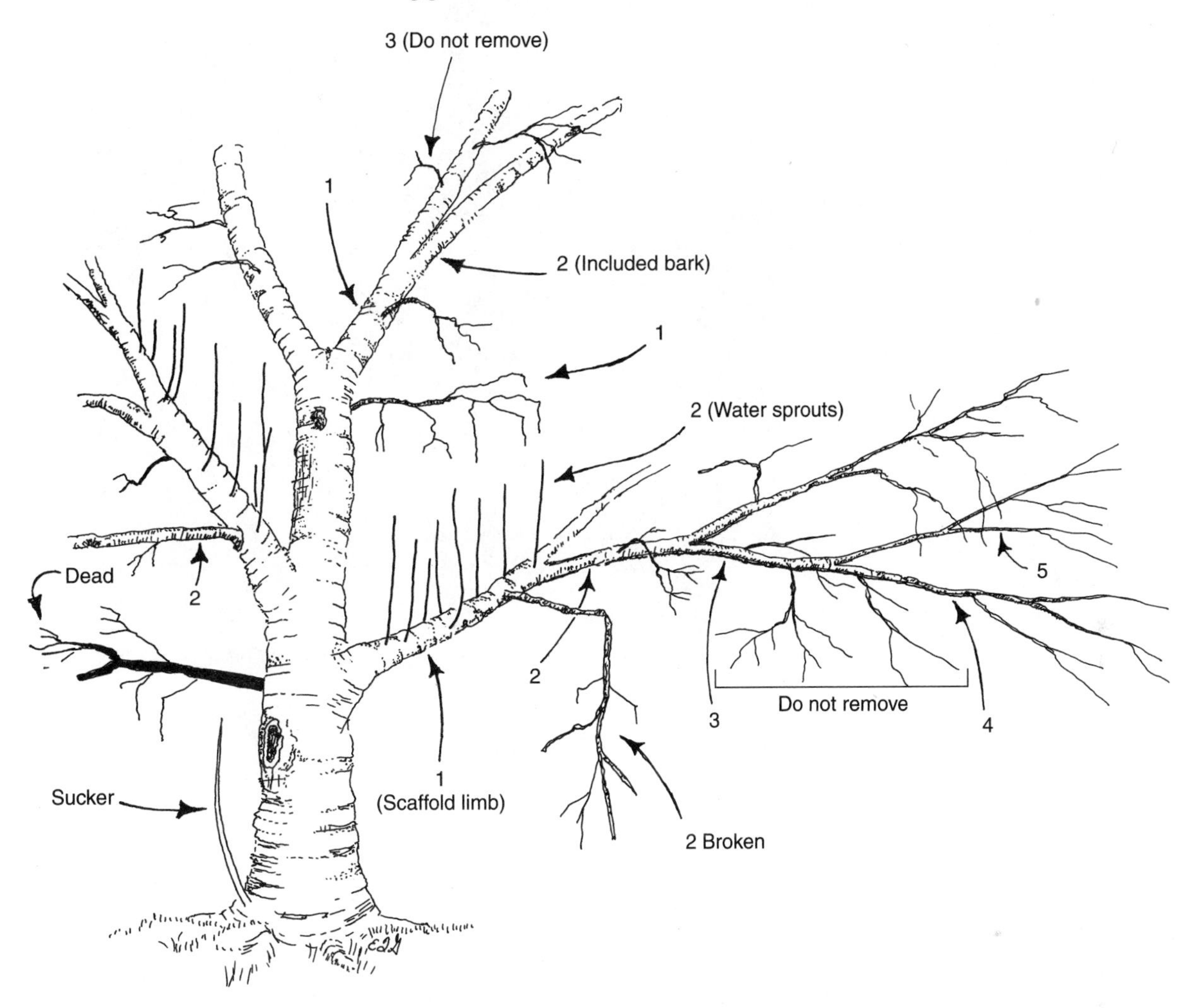

FIGURE 14-1. (Only the lower right branch on this tree is drawn to completion. The rest of the tree has been truncated for illustration purposes). Removing primary and secondary branches (those marked with a 1 or 2) even small-diameter ones, from mature trees can begin a downward health spiral for the tree. Notice the decay beginning on the lower left trunk where a large primary branch was removed several years earlier. Remove live primary and secondary branches only as the last option. If live tissue is to be removed from a mature tree, cut small-diameter branches toward the edge of the canopy (branches marked with a 4 or a 5). There may be instances where removing a secondary branch is appropriate such as when there is a severe bark inclusion or broken branches (see upper right and lower right canopy). Suckers and dead branches (lower left) can be removed. Some water sprouts could be shortened or removed under certain circumstances. Never clean out or remove all live interior growth from the canopy.

FIGURE 14-2. One of the most common abuses of medium-aged and mature trees is over-thinning or over-lifting the canopy. Too many interior and low branches are removed in the abusive treatment leaving foliage only at the edge of the canopy. This practice has no place in tree care. This practice should not be prescribed by professional arborists.

 **Tip:** *Remove live foliage from mature trees only for good reason.*

Older trees that have not been pruned for some time, or those that were never pruned correctly, may have bark inclusions in unions of large branches. It may be difficult to see this from the ground in some trees. Although small branches with included bark are also poorly connected to the tree they may be of lesser priority than the heavier limbs that carry more weight. In addition to a poor structural connection with the trunk, large limbs with included bark can cause unseen injury to the trunk below the union. This occurs because the bark and cambium of both rub against one another as the branch moves in the wind. Look for subtle outward signs of this problem in the union such as bleeding and excessive callus growth at the base of the union. Consider reducing these limbs to lessen the weight toward their ends. This could be combined with cabling and/or bracing systems (see ANSI A300, part 3 Support systems). Many professionals consider this combination the appropriate treatment to lessen likelihood of breakage on large trees.

It is most difficult, lacking any meaningful research in this area, to determine how much to remove (i.e., the dose) when reducing a limb that is potentially at risk of failing. Some refer to this as a limb at risk. Several principles should be considered when making this decision. We know that shortening a limb with a reduction cut reduces end weight, but this can introduce decay at the cut because reduction does not cut back to a natural boundary. Only weak wall 1 retards the spread of decay. We really do not know how much to remove to appropriately reduce an at risk tree by a known amount.

Species play a role in this as well because certain trees are more resistant to decay following reduction cuts than others. It may be reasonable to assume that larger reduction cuts can be made in the trees that compartmentalize decay well (see Table 4-1). Reducing the overall size of the canopy using reduction cuts may be recommended occasionally to clear the tree of structures such as buildings, wires, and lights. Since this can cause severe injury and decline on certain mature trees, *consider all other options first!*

STREET AND PARKING LOT TREES

Some urban foresters, arborists, and other people familiar with street tree management know that shade trees perform best with one dominant trunk for the first 30 feet or more (Figure 14-3, lower left). Trees trained this way are easier to manage than those with several low leaders or large low branches. Those with large low stems and branches require massive pruning cuts to remove them once they droop in the way of traffic and people (Figure 14-3, top and right). This can lead to the decline of the tree and probably contributes to the short life reported for street trees in many communities.

The best management practices for street tree care include a program that develops a dominant leader high up into the canopy and keeps low branches small (Figure 14-4). This is accomplished by regular pruning in the first fifteen to twenty-five years after planting (see Chapters 11 and 12 for details). Branches above the 30-foot-tall single trunk can be allowed to spread and grow to a larger diameter because they are high enough in many situations to remain clear of vehicles and other obstacles. Even some branches that originate 30 feet from the ground may eventually droop to get in the way on certain species. If this will be the case, shorten them as they grow to prepare them for coming off the tree eventually or to prevent them from getting in the way. Limbs this far up the tree can be guided to grow upward by removing lower branches from the end of the limb. This may allow them to remain on the tree indefinitely. Branches and stems on upright trees can originate low on the trunk, but keep them less than half the trunk diameter with regular pruning.

GOLF COURSE AND PARK TREES

Trees in golf courses, municipal parks, theme parks, cemeteries, and other areas where tall clearance is not required under the canopy can be managed differently than street trees. In some instances, large branches can be left on a lower position in the tree because only lawn mowing equipment needs to pass under them. Management of trees in golf courses and parks depends on where the trees are located.

Trees at the edge of woods

Edge trees have access to sunlight from only one side so they often grow more on that side. This can create an unbalanced situation by encouraging several branches on the sun side to overextend. As these branches grow longer, much of the weight is located far from the trunk so they begin to droop. Look for subtle longitudinal cracks in the branch close to where it meets the trunk. These indicate weakness that could result in breakage. Branches on edge trees might also be blocking play.

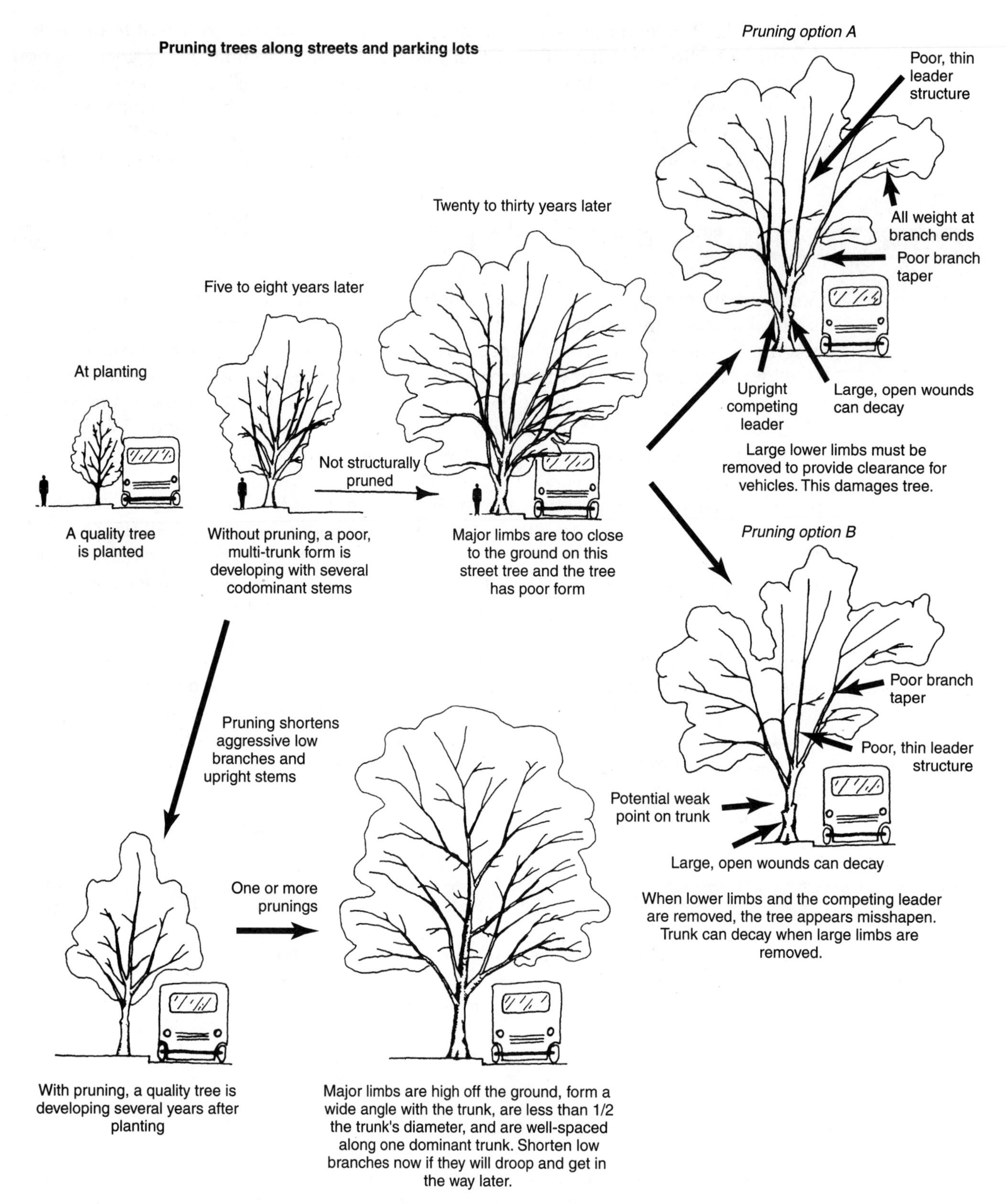

FIGURE 14-3. Planning ahead helps to maintain the health of trees planted along streets and in parking lots.

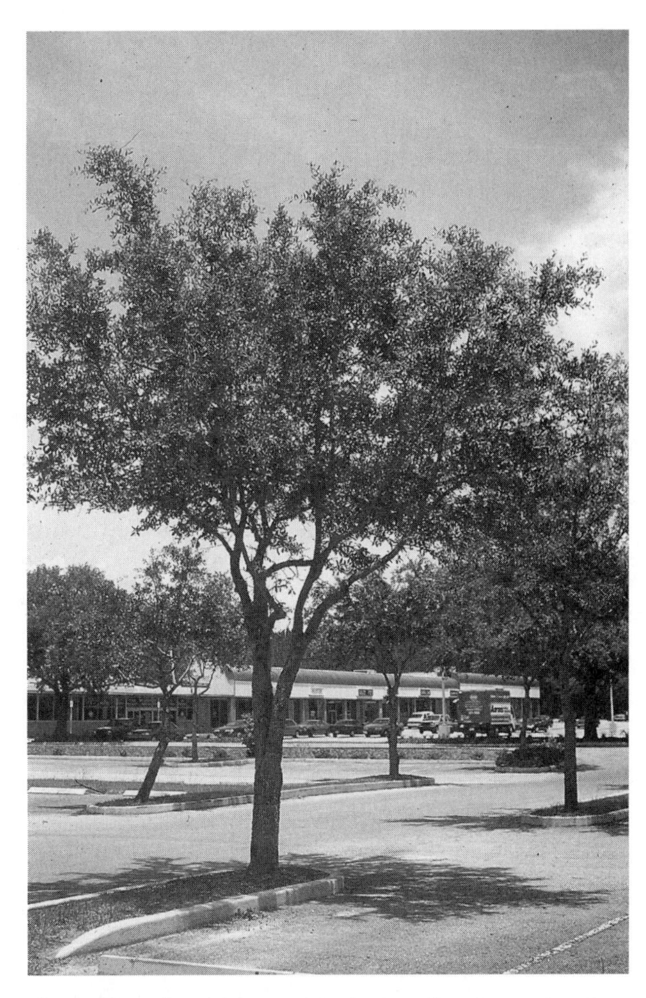

FIGURE 14-4. Shade trees planted along streets and in parking lots should be trained to a dominant leader with small diameter branches (left). When lower branches are removed, as in this photograph, to provide clearance under the canopy only small-diameter wounds are created. Even unskilled labor end up pruning the trees correctly. Shade trees with several trunks are typically pruned inappropriately by unskilled labor (right). They do not know to make the appropriate subordination cuts in the upper canopy. Instead, they remove only low drooping branches as shown in this photograph. This will encourage the poor, codominant leader form to continue developing.

Consider reduction to balance the canopies and to shorten overextended limbs with defects. This often raises the branch and can lessen the likelihood of breakage. Also watch for trees that begin to lean out away from adjacent trees. Root systems or trunks might have been damaged during construction creating an at risk tree—the tree may be at risk of falling over. Consider removing, reducing, cabling, or bracing these trees.

Feature trees

Trees located by themselves surrounded by turf, ground cover, and shrubs are often referred to as **feature trees.** It is often desirable to allow plenty of sunlight to reach the ground under the tree. Do not be tempted to remove only low branches as this often reduces live crown ratio below 60 percent resulting in a top-heavy condition, excess weight at the ends of limbs, tree decline, and can create a potentially at risk tree condition (Figure 14-1). The tree could blow over.

There are two ways to provide sunlight under a tree canopy. One is to space scaffold limbs far apart creating an open canopy. Instead of scaffold limbs several feet apart, your goal might be to space them 10 feet apart. Trees in a forest often grow this way. Spacing scaffold limbs is easier to do if you begin when trees are young. If the tree is already large and has not received this type of pruning before, proceed toward your goal slowly, over a period of years. You cannot make changes quickly in large trees and maintain the tree in a healthy condition. Start by reducing live branches between scaffold limbs you want to eventually retain (Figure 12-10). Reduce the same branches each year until the desired form and canopy density is reached. Trees tend to fill in holes you create by pruning so this process will have to be an ongoing one for a long time. Trees will remain healthiest if light pruning is conducted each year rather than heavier pruning less often; remove no more than about 5 to 10 percent of the foliage when pruning mature trees annually. The second way to provide more sunlight under the canopy is to perform a traditional thinning operation (Figures 13-2 and 13-3).

Drooping branches on feature trees often require lifting to allow for passage of lawn maintenance and other equipment. Low branches could also be blocking a view or restricting play. Again, be cautious not to overlift or overthin the canopy. Raising can be done by thinning low branches instead of removing them. This makes it difficult to tell the tree was pruned (Figures 13-11 and 13-12) and is considered a more professional approach. Basically, remove the lowest branch, reduce the next one, leave the next one, alternating up the lower third of the tree. This thins the lower canopy and creates a gradually lifting. Removing too many low branches can cause problems.

Trees separating fairways or in clusters

Trees in the middle of a cluster may have a fine form but those on the outside edge often grow as described for edge trees. Plan to manage these as you would edge trees watching for drooping, overextended branches, and unbalanced canopies. The canopies might also have to be raised eventually. Shorten low branches early in their life if they will eventually be removed from the tree (see Chapters 11 and 12).

Root management

Roots sometimes grow into high-value turf areas such as greens, tees, and other athletic turf areas. Tree roots can steal some of the water and nutrients meant for the turf and cause surface irregularities. Roots also can grow under and raise pavement and sidewalks. Pruning roots within about 10 feet of a large tree will certainly stress it and could kill it. If the tree survives it could fall over from lack of ample support roots, especially if soil drainage is poor and all roots are shallow. Roots could rot after root pruning, eventually causing instability. Root barriers near newly installed trees could provide a temporary solution, but some roots eventually grow under many barriers. If roots were pruned close to the tree, consider removing or stabilizing the tree. Some people might recommend canopy reduction and/or thinning to reduce weight in an attempt to keep it from falling.

ORNAMENTAL CHERRIES, CRABAPPLES, AND OTHER ORNAMENTAL TREES

Some of the best-looking ornamental cherries, crabapples, and plums have never been pruned, or only dead branches have been removed. Most people remove too much live tissue from these trees by overthinning or topping them. This causes many water sprouts to generate which quickly grow back to fill in the canopy. Removing excessive live tissue can also lead to decline and eventual death of the tree.

In short, the best strategy is to remove all dead branches including dead stubs, then step back from the tree to determine if you really want more tissue removed. If so, remove no more than 10 to 15 percent of live tissue. This usually means that all crossed or rubbing branches cannot be removed. Remove the rubbing ones that will give you the best aesthetic improvement. Leave as many small twigs as possible along the main branches and trunk since these bear many flowers.

Avoid heading and reduction cuts where possible. These cuts cause great stress on plums and cherries because they decay rapidly. Make only removal cuts on these trees if at all possible, and perform canopy reduction pruning with caution. Do not "clean" the inner portion of the canopy by removing small-diameter branches. A properly thinned tree will have fewer branches than before thinning but every space in the canopy will still be filled with branches. Crabapples and some other small trees often develop one to several branches that shoot out beyond the edge of the canopy. Resist removing or cutting on these since sprouts are sure to follow and fill in to ruin the looks of the tree. At most, shorten them back to the edge of the rest of the canopy. Remove suckers from the roots, graft union, and lower trunk.

If your tree was overpruned, the best course of action is probably to do nothing for two to four years. If you remove sprouts that form the first year or two after overpruning, they are sure to grow back. You will have a yearly pruning job on your hands that will only end when either the tree or you die. Begin removing some of the sprouts when the tree reaches its original size and sprouts begin acting like branches.

FRUIT TREES

In addition to the themes presented in the previous section on ornamental cherries, consider the following. Once major scaffold limbs have been established and dominate the tree structure, live branch pruning can be conducted once or twice each year to clean the canopy. Pruning less often could result in excessive sprouting because too much would be removed at one time. In addition, remove shoots that contribute to crowding (i.e., thin the canopy).

Keeping the center of the tree open by removing upright branches back to more horizontal laterals encourages flowering, helps keep branches low, and makes it easier to pick fruit. Peaches and apples color best in this system because they receive direct sun exposure. If more than 10 percent of live foliage needs to be removed, do it over a period of two years. If you absolutely have to remove more and can live with excessive sprouting, do it late in the dormant season to minimize formation of sprouts and to speed wound closure. Be prepared to fight sprouts for a very long time.

Light summer thinning by removing some live twigs from the edge of the canopy can enhance color on apples and perhaps other fruits but could reduce fruit size. This happens because photosynthetic capacity was removed. Do not remove too much or sunscald on branches could result. On the other hand, thinning the edge of the canopy on citrus enhances fruit color *and* fruit size, but reduces number of fruit. Remove some young developing fruits on peaches, apples, and other fruit trees to increase the size of the remaining fruits.

Without care, older fruit trees often develop a dense, bushy habit and produce poor-quality fruit. Thin and clean the canopy and make certain that vigorous upright sprouts are removed or cut back with reduction cuts. Remove root sprouts back to their point of origin. Space major limbs apart by removing or shortening other large-diameter branches that are closer than 4 inches. If fruit production is poor remove or shorten upright limbs and shorten major limbs slightly with reduction cuts to stimulate new shoot production. As with most other mature or over-mature trees, do not remove more than 10 percent of the foliage, as this stimulates excessive sprouting. If more needs to be removed, do it over a period of two or three years to prevent the formation of unwanted sprouts, which can spoil your efforts.

CRAPE-MYRTLE AND OTHER SUMMER-FLOWERING TREES

Crape-myrtle and other trees that flower on current season's growth are pruned in a variety of ways. There is little research on the effect of pruning type on growth and flower development. Most has been performed on crape-myrtle. Crape-myrtle pruned hard sprout the most; those that are not pruned sprout very little (Table 14-2). Sprouts on this and other small ornamental trees are a perennial nuisance. Synthetic auxins applied to fresh cuts can slow development of sprouts on some trees. Flower number was greatest when trees were not pruned and professional horticulturists preferred the look of the non-pruned trees. Progressive topping (Figure 10-7) is probably the most common method of maintaining crape-myrtle trees, but it turns out that the resulting flower display and form of the tree is the least preferred.

PALMS AND CYCADS

Palms are pruned to remove dead, diseased, infested, or chlorotic lower leaves; eliminate flowers and fruit; remove leaf bases (boots); or remove sprouts or a stem from a cluster palm for aesthetic reasons. Palms with large dead leaves or fruit left unpruned can pose

TABLE 14-2. Response of crape-myrtle to pruning types.[1]

	Pruning Type			
Response	**None**	**One-year-old Stems Headed Annually**	**Pollarding**	**Progressive Topping**
Sprouting	minimal	some	some to moderate	heavy
Flower no.	most	moderate	moderate	least
Panicle size	smallest	intermediate	intermediate	largest
Aesthetics	most preferred	acceptable	acceptable	least preferred

[1]Knox and Gilman, unpublished.

a potential risk. Removing all leaves from cabbage and Pygmy date palms may reduce transplant shock on those receiving infrequent irrigation after planting. *There is no need to remove leaves in an irrigated landscape.* The trunk often develops a slight permanent bottleneck in response to this treatment meaning that it will have a smaller diameter than immediately above and below this point. Removing the bud at the top of the trunk will kill a palm. Do not injure the trunk or petioles of remaining leaves with the pruning saw or knife. Spurs used to climb palms are not recommended because they create wounds that do not callus over. Wounds could become entry points for decay organisms and trunk disease (e.g., *Thalaviopsis*), which compromises strength and health.

Palms are often overpruned (Figure 14-5, right). There is no need to remove green leaves from a palm. Removing green leaves slows growth rate and can cause a narrowing of the trunk and root problems. On some dates and other palms, fusarium wilt can follow and eventually kill those that are overpruned. Insect pests could be attracted to palms following fruit, flower, or leaf removal. Some arborists report that overpruning of green foliage increases the amount of foliage the next season.

With the introduction of the lethal fusarium wilt disease on Canary Island date palm to California and Florida, some professionals recommend not pruning these palms as a

FIGURE 14-5. It is rarely necessary to remove green leaves from a palm. However, if you wish to do so, only remove those drooping below an imaginary horizontal line drawn through the bottom of the canopy (center). Growth will be slowed and the palm can be damaged and attract pests and diseases when green leaves are removed from above this imaginary line (right).

measure to prevent spreading the disease. This may not be practical because old leaves that naturally die and turn brown at the base of the canopy remain on the palm for several months or longer. Since this looks unslightly, these are traditionally pruned off to improve appearance. Presently, there is no good solution for this devastating disease. Diseased palms are removed from the site to help prevent the disease from spreading. It has already eliminated this palm from the planting palette of some regions of southern California.

PRUNING NEAR UTILITY LINES

Electric utilities prune trees to ensure safe, reliable service to their customers and to gain access to utility structures. This practice is referred to as line clearance. Homeowners, horticulturists, and other persons without Electrical Hazard Awareness training must leave this to a line clearance tree trimmer or trainee. Call the utility company or a utility arborist to do this hazardous work. Never prune within 10 feet of a utility conductor unless you have the appropriate training.

Removal of a medium- or large-maturing tree and replanting with a small tree or large shrub is the best method of minimizing pruning requirements near wires. However, this is not viewed as practical in many instances due to the high cost of removing trees. Pruning them is viewed as more economical than removing them. A general rule used by the industry is, unless imminently hazardous, trees are not removed unless the cost of removal is less than 2 ½ to 3 times the cost of pruning. Most large trees do not meet this criterion because they are expensive to remove. In some instances, wires can be placed underground or moved away from the trees to allow trees to develop properly.

Ice, snow, and wind can break branches growing above or among wires and cause the wires to malfunction or break. The number one cause of outage is branches over wires falling down on the wires. Therefore, medium- and large-maturing trees inappropriately located under or near wires need regular pruning to keep them away. Utility companies often hire specialized line clearance contractors to perform this type of tree pruning. The utility company typically writes the pruning specifications and the contractor prunes trees according to these specifications. Controversy sometimes arises when property owners feel that their trees have been pruned inappropriately or too severely by the line clearance contractor. Citizens should contact the forester at the utility company or the person in charge of line clearance pruning to voice their concerns. Though sometimes mistakes are made, the contractor is almost always following contract specifications as directed by the utility.

Trees pruned with heading cuts generate sprouts that quickly grow back into the wires (Figure 14-6, right). Branches on trees pruned with reduction and remove cuts can be encouraged to grow away from wires (Figure 14-6, left). This is called directional or lateral pruning. In addition to increasing service reliability, directional pruning could reduce pruning costs by increasing the interval between prunings. Pollarding and other forms of architectural pruning are high-maintenance options (see Chapter 10) requiring regular pruning. Pollarding has not been tried on a regular basis in the United States, but deserves consideration in certain instances.

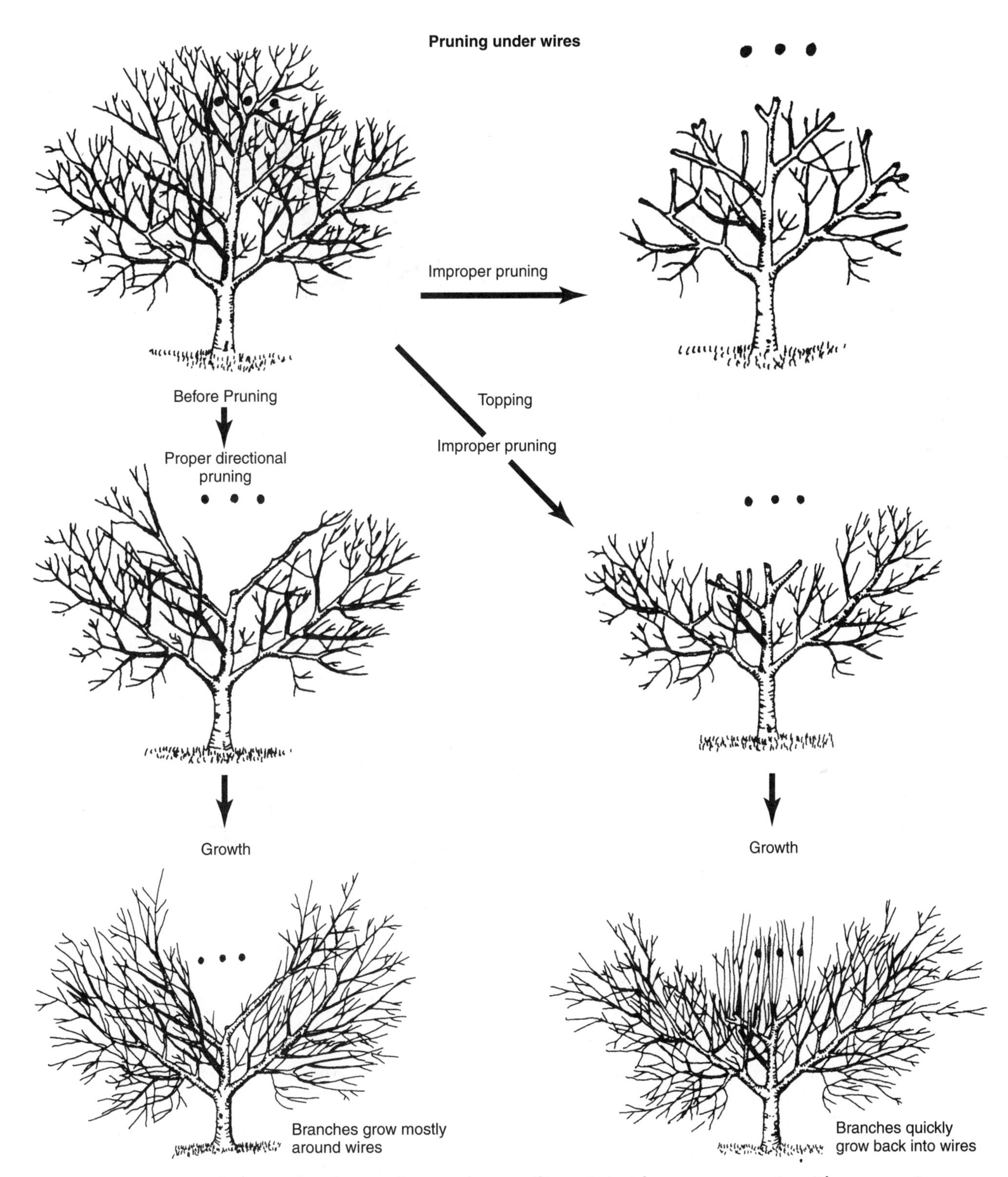

FIGURE 14-6. Branches can be directed away from utility wires (shown as • • •), with appropriate directional pruning (left). However, topping encourages sprouts to quickly grow back into the wires and damages the tree (right).

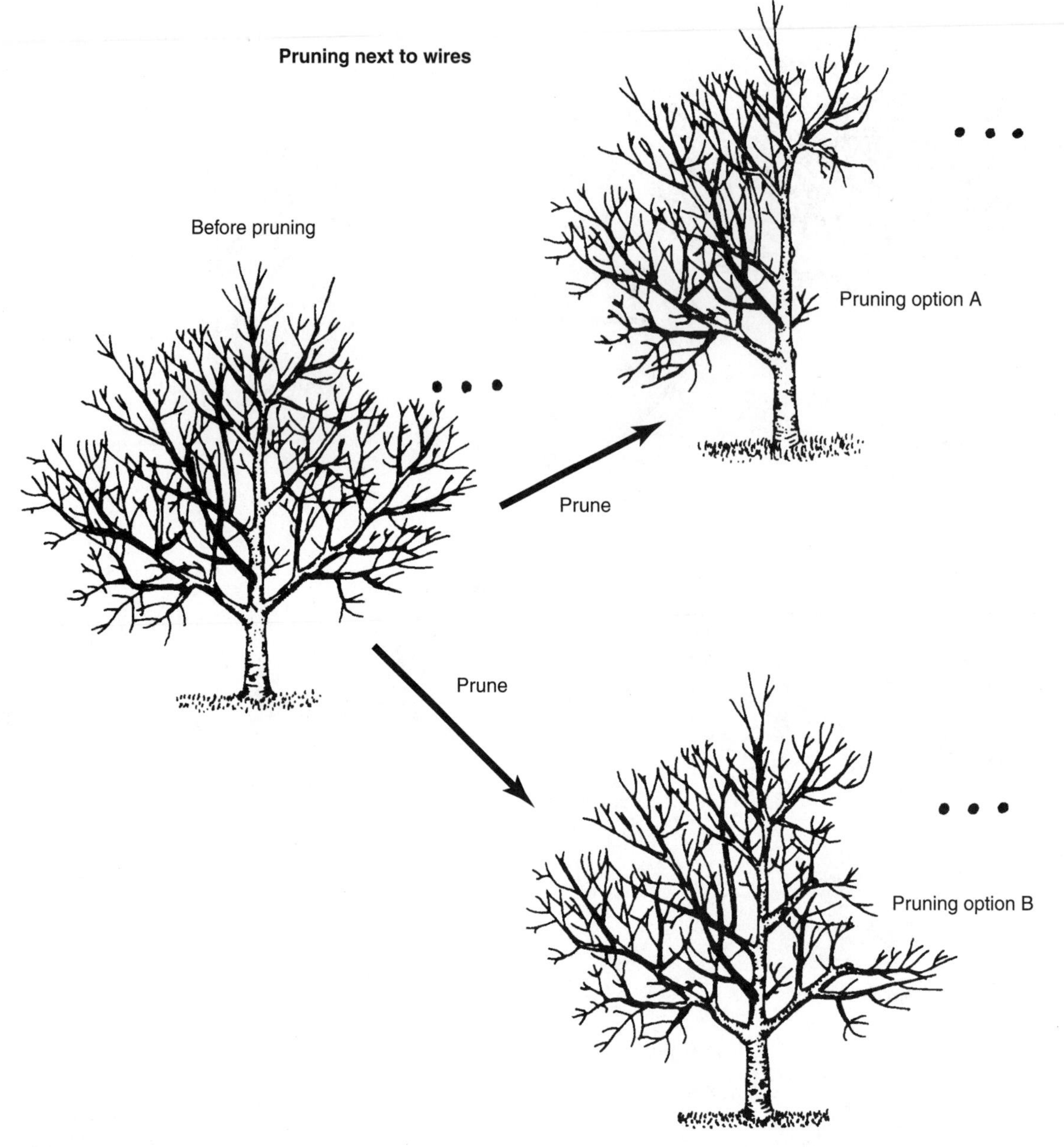

FIGURE 14-7. Option A provides more clearance from the utility wires (shown as • • •), but option B looks better. B is also more expensive because the trees will require pruning sooner than those shown in A.

Trees growing into wires from the side are pruned differently from those under wires (Figure 14-7). Some utility companies prune as shown in option A; others prune less drastically as in option B. Only two pruning cuts were made in each option. The tree in option B may need pruning again sooner than that in option A, but option B results in less debris removal and less trunk decay, and it looks better. Many people would hardly recognize the tree was pruned. Option A can cause more trunk decay because large branches or codominant stems were removed back to the trunk. Large branches and codominant stems lack a branch protection zone at their base. Maintaining a tree as in option B may require more frequent pruning so it could cost more. However, the long-run cost of option A could surpass option B's costs if a severely decayed option A tree fails and damages the line—or, worse, injures someone. Since this is likely to result in litigation against the utility, customers could end up paying more for option A.

In lieu of option A in Figure 14-7, if all lower branches must be removed back to the trunk on the wire side of the tree, reduce the branches as in option B and remove them over two or three pruning cycles, not all at once. If more than half of the foliage must be removed from a branch, or if the sprouts will grow back into the wires quickly, plan on removing the entire branch from the tree. Although there are many exceptions, some utility arborists attempt to make no more than three cuts to remove most of the foliage required to clear the lines of branches. This is not always possible, but provides a guideline to work from.

Lateral branches on trees near wires are often headed to provide clearance (Figure 14-8, option A). This is inappropriate and usually is not in conformance with the ANSI A300 pruning standard. Not only will sprouts quickly grow back into the wires, they are poorly attached to the tree because the headed branches often crack and begin to decay. A much better alternative is a combination of reducing some offending branches and removing others back to the trunk (Figure 14-8, option B).

Depending on the importance of the line, the tree species and condition, utility companies sometimes choose to remove all overhanging branches from one side of the tree. Although this provides the most clearance, it raises aesthetic concerns, and may be extremely damaging to the tree (Figure 14-8, option C). Many trees cannot be expected to recover gracefully from this type of pruning. Removing all branches from one side of the tree at one time is likely to initiate severe trunk decay and cracks and may predispose the tree to sunscald, wind throw, or trunk failure. Bark often dies on the side of the tree facing the wires and roots can decline. If this type of pruning is required to provide clearance consider removing the tree. As an alternative, reduce the offending branches (see Figure 14-8, option B) and remove only one or two back to the trunk at the first pruning. Remove another one back to the trunk at the next pruning cycle and a third one the following time. This will stress the tree less than removing all at once.

It should be apparent that line clearance is controversial. All techniques have serious draw-backs—some worse than others. Moving the wires to a location away from the trees is a possible solution in some cases. Removing large trees near wires and replacing them with small-maturing ones is another good low-maintenance alternative. Communities and utilities will continue to evolve programs to provide for reliable delivery of electricity with minimum impact on trees.

CONIFERS AND OTHER EVERGREENS

Pruning is conducted primarily to change or control the shape, density, or form of the plant or to remove dead, diseased, dying, damaged, or hazardous branches. It is best to locate conifers in the landscape so they will not have to be pruned, because many look odd and are severely injured when extensively pruned. One of the most common types of pruning on large conifers is the removal of the lower branches to create room beneath the canopy for passage of people or vehicles. Branches are removed back to the trunk or, less often, back to a living side branch. Reduction and thinning are usually not appropriate on conifers.

Most conifers grow with one central leader and small-diameter branches, and have a structurally sound form. Little structural pruning is needed. If a double leader develops on a young tree, or in the top portion of an older tree, shorten or remove one of

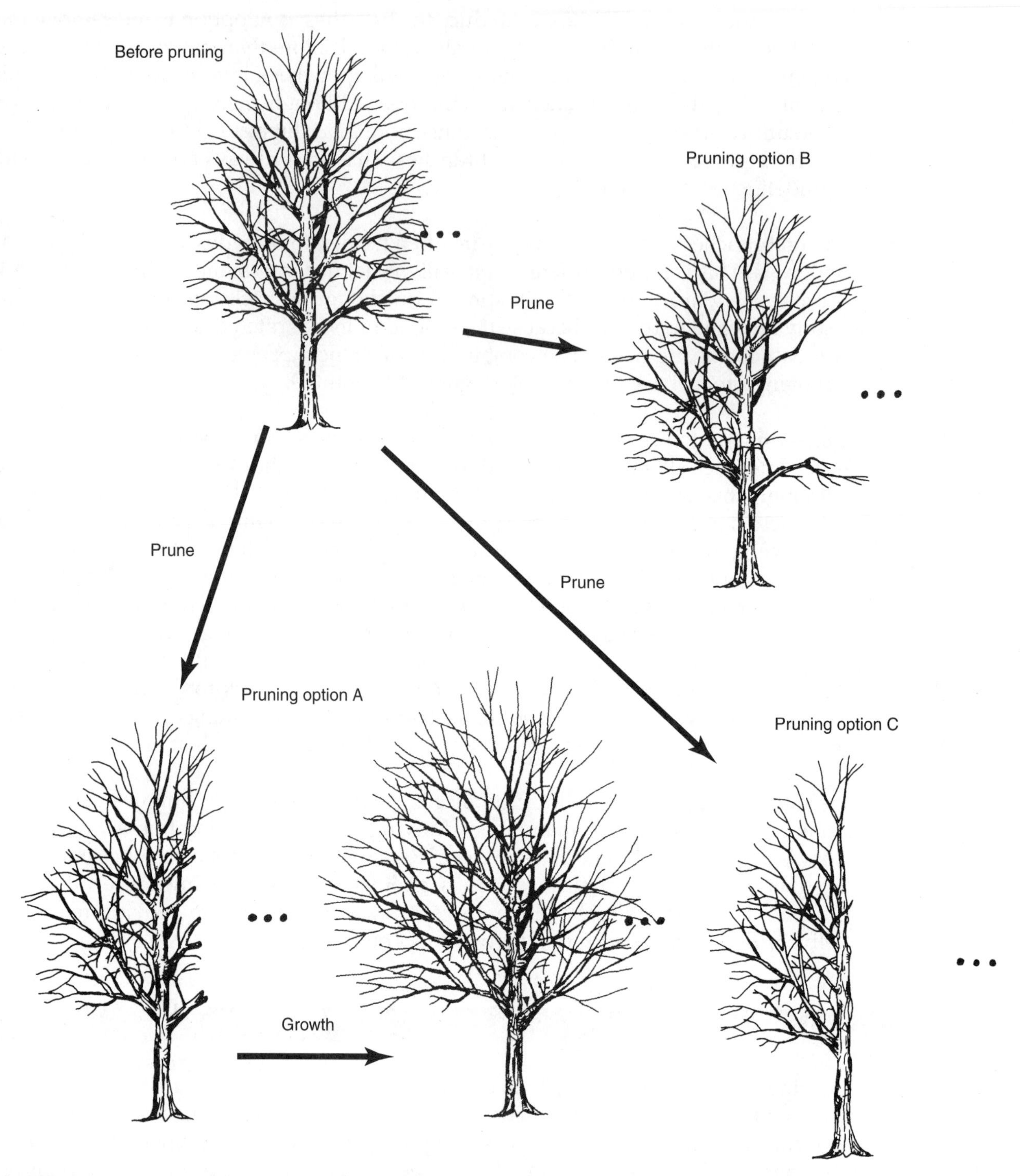

FIGURE 14-8. Option A is a poor choice because sprouts from heading cuts quickly grow back into the wires (shown as • • •), and the headed branches then decay. For a lower-maintenance alternative, reduce and then at a later date remove the three branches indicated with arrows at the trunk (bottom center). Some utility companies choose between options B and C because they provide clearance for a longer time than option A. Option C provides the best clearance but is over-pruning resulting in sunscald, cracks and decay in the trunk and roots.

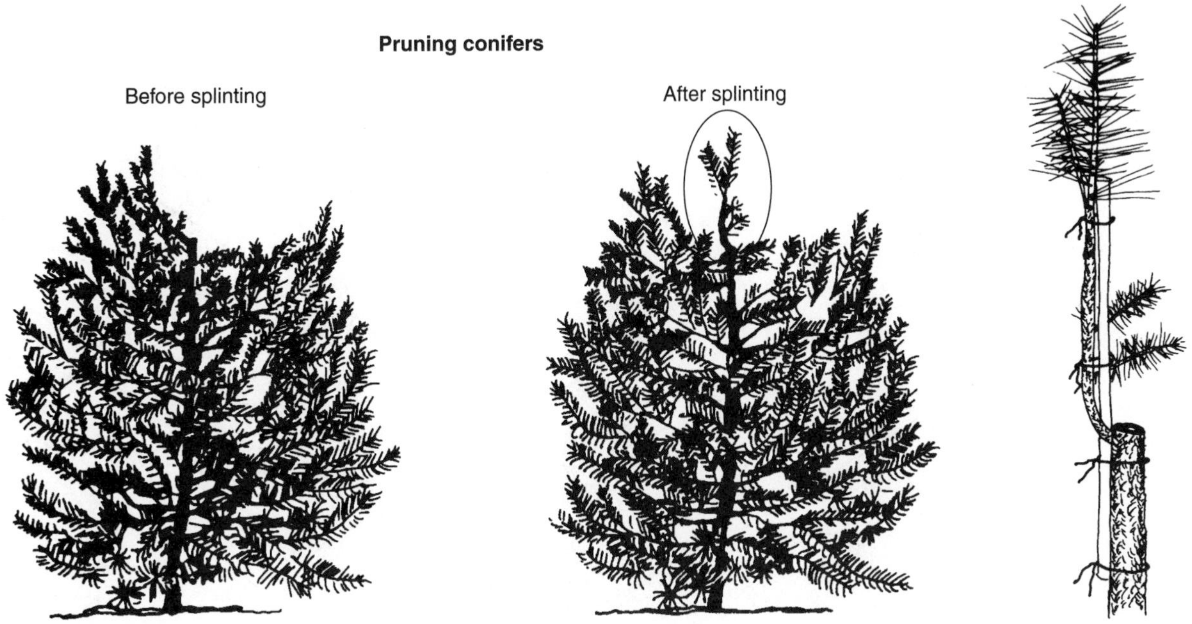

Pruning conifers

Before splinting

After splinting

FIGURE 14-9. Fixing a damaged or missing leader on a conifer is easy. Simply attach the upper-most branch to a short splint made from a piece of wood or metal, which will help direct this limb to become the new central leader. (The close-up drawing at right provides a detailed view of the circled area of the full drawing.)

them to prevent the tree from splitting at this point. You may need to install a cable in an older tree if the double leader formed many years ago. If the leader is lost or badly damaged, a branch can be trained to assume the role of the leader (Figure 14-9), but some trees restore themselves without intervention.

Young arborvitae, falsecypress, Leyland cypress, yew, and some junipers can be cut back into wood 1 or 2 years old that does not have living foliage on it. The best time to do this type of pruning is in the spring, just before new growth emerges. Dormant buds will usually generate foliage and fill in the plant. They can also be sheared on the outside of the canopy to create and maintain a certain shape or size. But remember, shearing could shorten the life of plants.

Because spruces, firs, and most pines (except Canary Island pine and a few others) do not produce shoots from older wood, when pruned they should not be cut back past the current-year twig unless the entire branch is removed back to a lateral branch or the trunk. A branch with all foliage removed will probably die. Emerging candles on pines can be pinched to create more branches or to slow the growth rate (Figure 14-10). Candles are pinched in the spring just as the needles can be seen emerging from the candle. This encourages several new buds to form at the pinch and creates a more compact plant. Christmas tree growers practice this technique to form symmetrical, dense canopies. The entire candle can be removed directing growth into lateral shoots that were not pinched. Growth can be guided in this manner around a structure or to form an unusual shape. Pines pinched at other times of the year will not form new shoots behind the pinch, and the terminal portion of the branch eventually dies. Spruces and firs can be cut back to a living lateral branch. They can be headed back to any position on the current season's growth because buds exist all along the twig. New growth emerges next spring from buds along last year's twig.

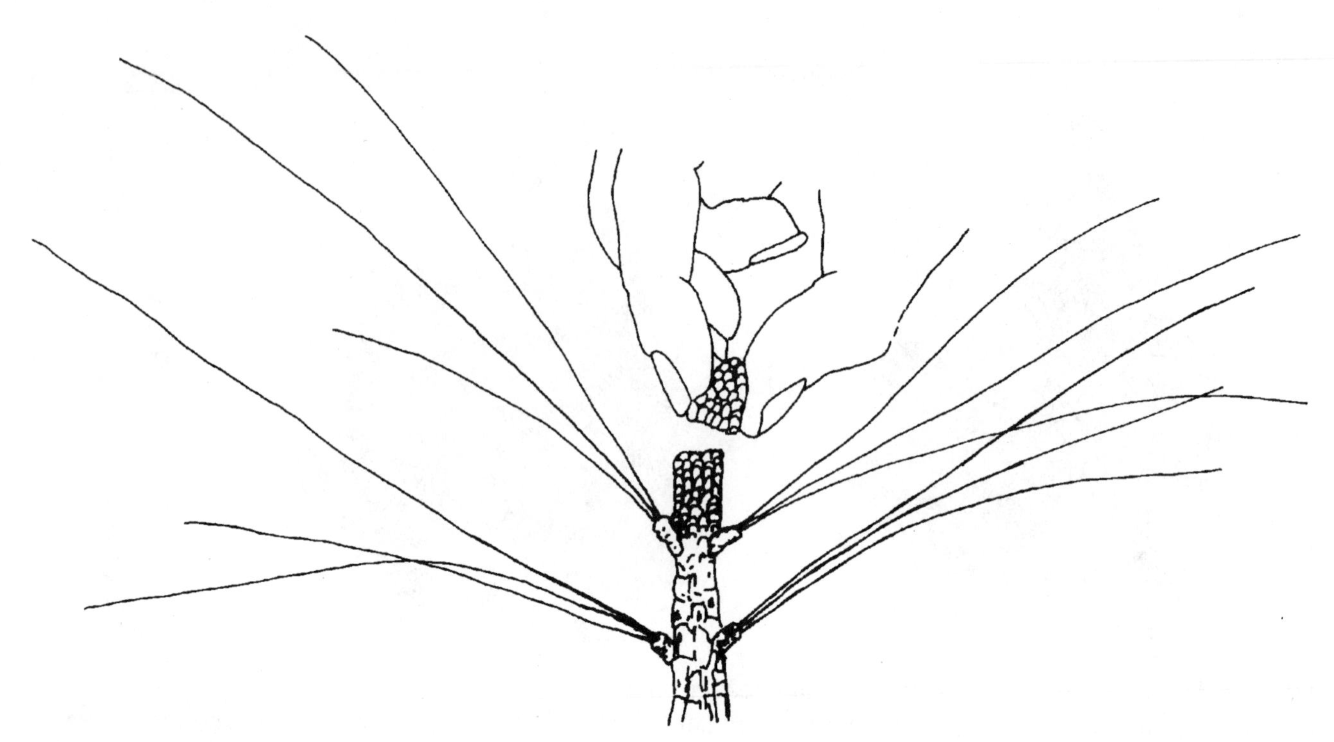

FIGURE 14-10. Prune pines by pinching a portion of the new shoot just before the needles begin to elongate in order to minimize shoot length growth. New buds will form behind the pinch.

Branches (other than emerging shoots) on pines may be best pruned during the dormant season so that beetle pests will not be unduly attracted to the tree. Some beetles are attracted to pruned trees in summer droughts and can kill the tree. Most other conifers can be pruned at any time of year, but for maximum tree growth prune just before new growth begins in the spring.

Most small maturing evergreens, such as podocarpus, holly, privet, juniper, viburnums, and wax myrtle, can be pruned anytime. However, in USDA hardiness zones 8 and 9, moderate to severe pruning in the fall through mid-winter could stimulate growth in the dormant season if a warm spell occurs. The tree or shrub could be damaged by subsequent cold weather.

WEEPING TREES

Many weeping trees will grow as a ground cover or sprawling or mounded shrub if not pruned initially to an upright trunk (Table 14-3, right). Some people refer to these as **mop tops** because their form reminds them of a mop. Even many that are initially trained to one trunk in the nursery develop branches that eventually reach the ground. Branches will grow along the ground unless pruned.

You can enhance growth in the upper portion of the tree and discourage the canopy from sweeping the ground. Encourage growth in branches that are more upright and those at the top of the canopy by shortening branches that will soon reach the ground. Use reduction cuts, where possible, to shorten. This technique is similar to pruning

TABLE 14-3. Some[1] weeping trees form a billowing or upright canopy, others form a low, mounded canopy.

Billowing/Upright Canopy	Low, Mounded Canopy
Acacia pendula	*Acer palmatum* 'Dissectum'
Acer saccharinum 'Beebe'	*Betula pendula* 'Youngii'
Betula pendula	*Caragana arborescens* 'Pendula'
Callistemon viminalis	*Cedrus libani* 'Glauca Pendula'
Chamaecyparis nootkatensis 'Pendula'	*Gleditsia triacanthos* 'Emerald Kascade'
Fagus sylvatica 'Pendula'	*Larix decidua* 'Pendula'
Ilex vomitoria 'Pendula'	*Morus alba* 'Chaparral'
Juniperus rigida 'Pendula'	*Pinus strobus* 'Pendula'
J. scopulorum 'Pendula'	*Prunus subhirtella* 'Pendula'
Picea omorika	*Salix caprea* 'Pendula'
Pinus densiflora	*Sophora japonica* 'Pendula'
Salix babylonica	*Tsuga canadensis* 'Sargentii'
Syringa pekinensis 'Pendula'	*Ulmus alata* 'Lace Parasol'
Ulmus parvifolia 'Drake'	*Ulmus* x *vegeta* 'Camperdownii'

[1]This is a partial list of plants. See Horticopia Professional (*www.horticopia.com*) for more plants.

shade trees in that lower branches are shortened to force growth in the upper part of the canopy creating a billowing canopy effect. This technique will help prevent formation of large low branches that might get in the way later. It would be a shame if these attractive, large low branches had to removed later. Some young stems can be headed at the top of the canopy (see Figure 10-9). Do not cut all stems at the top like this; only cut a few scattered in the top. Several sprouts may develop from the heading cuts and grow upright, at least for a short period. Notching a branch or stem above a bud (on the side of the bud away from the trunk) could also encourage a sprout to emerge there.

Other weeping trees generate upright shoots in addition to weeping branches (Table 14-3, left). Several weeping trees such as cultivars of *Acer, Picea,* some *Juniperus,* and others normally develop weeping branches from a more or less dominant trunk. To help prevent breakage later, this form should be encouraged by subordinating codominant stems as they develop. Weeping trees that grow to be a large size might also benefit from a dominant leader form.

COLUMNAR AND UPRIGHT TREES

Columnar and upright shaped trees often form a symmetrical, dense canopy. The canopy consists of many small-diameter branches growing from several to many upright stems from the lower, middle, and upper canopy (Table 14-4). Those originating from the lower and middle canopy can grow to be quite large as the trees age. Some become codominant stems with included bark. Despite the bark inclusions, these are normally adequately secured to the tree provided they remain small and do not begin extending outside the edge of the rest of the canopy. When this happens, they become wind loaded and can accumulate ice and snow in storms. Even tight, symmetrical canopies can fail if the larger stems and branches become laden with ice and snow. Small-diameter branches usually have no problem securing themselves to the tree.

TABLE 14-4. Some columnar and upright non-coniferous trees that may benefit from reducing the length* of codominant stems.

Acer x *freemanii* 'Armstrong' and 'Armstrong II' and Scarlet Sentinel™
Acer platinoides 'Columnar' and 'Olmsted'
Acer rubrum 'Bowhall' and 'Columnar'
Acer saccharum 'Endowment', 'Goldspire', and 'Newton Sentry'
Agathis robusta
Carpinus betulus 'Fastigiata'
Erythrina variegata var. *orientalis* 'Tropic Coral'
Fagus sylvatica 'Dawyck'
Gordonia lasianthus
Ginkgo biloba 'Fastigiata' and Princeton Sentry™
Ilex x *attenuata* (most cultivars)
Magnolia grandiflora (most cultivars)
Magnolia virginiana (cultivars)
Populus alba 'Pyramidalis'
Populus nigra 'Italica'
Prunus sargentii 'Columnaris'
Quercus petrea 'Columna'
Quercus robur 'Skymaster'

*Subordination should begin when the tree is small. Subordinating these trees later can deform the canopy.

If upright stems and branches are regularly subordinated as the tree grows they can be kept in check and remain well secured to the trunk. This is most effective and aesthetically acceptable if the program begins when trees are young. If a large pruning dose is applied for the first time when the tree is in middle age, the canopy might be misshapen beyond what most people will accept. Apply a light dose frequently on trees to prevent large voids. On young trees, it might be best to remove entire stems back to the trunk where possible, especially if there are many smaller-diameter branches near the cut (Figure 14-11). Shortening aggressive stems would also be acceptable. On cultivars with *all* stems and branches in the upright position, it will be difficult to impossible to apply this treatment.

CHECK YOUR KNOWLEDGE

1) No more than about 10 percent of living foliage should be removed from fruit trees and small flowering trees such as cherries, peaches, and crabapple because:
   a. many sprouts will appear along the branches.
   b. pruning wounds will close slowly.
   c. growth rate will be slowed substantially.
   d. the tree canopy will be too open.
2) Only small amounts of live foliage should be removed from mature trees because:
   a. it is difficult to reach the ends of the branches where most of the foliage is located.
   b. if overpruned, mature trees can become sunscalded, sprout excessively, decay, or roots could decline.
   c. mature trees have more dead branches which should be removed instead of live foliage.
   d. overpruning can cause formation of wall 4 and a branch protection zone.

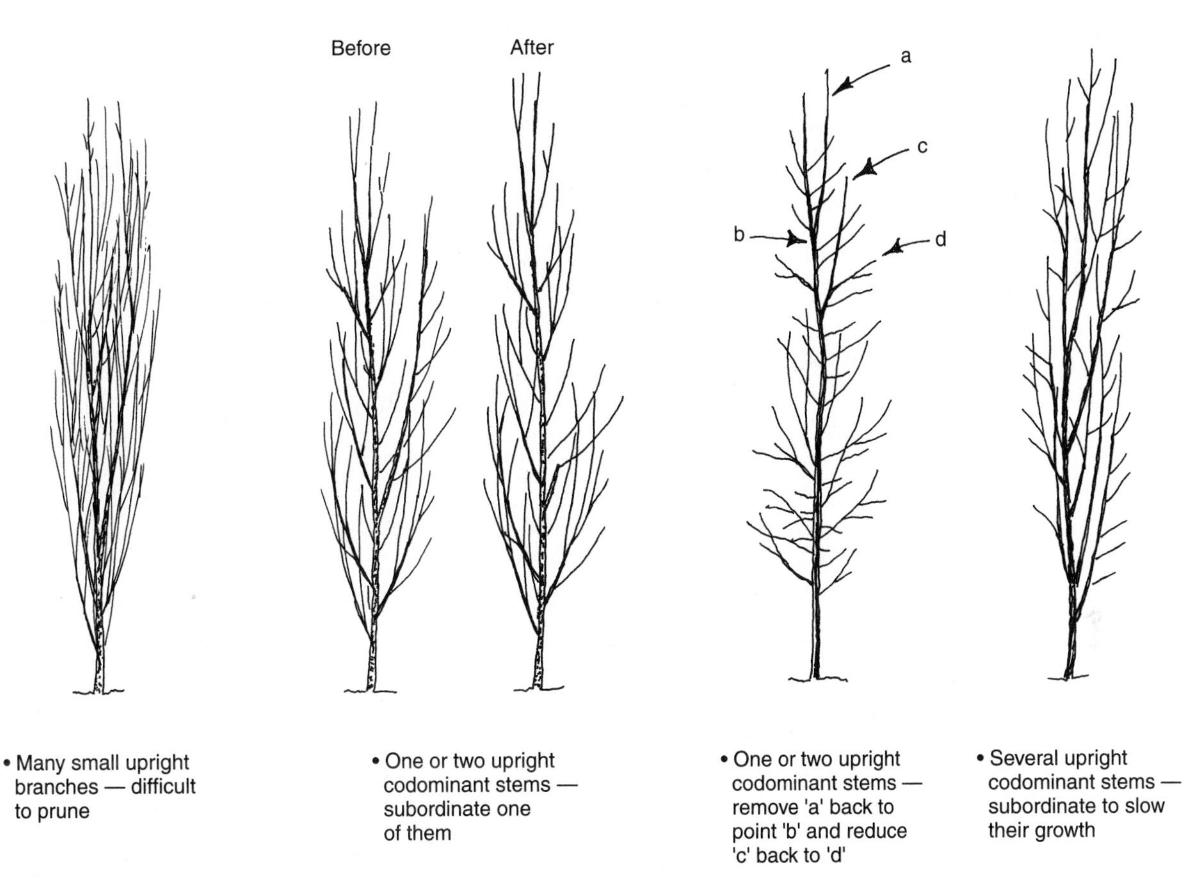

FIGURE 14-11. Upright trees with numerous small-diameter, upright stems and branches are difficult to prune because voids in the canopy become very noticeable (left). Trees with a few upright, but larger stems can be meaningfully pruned if this is done when trees are young (center and right).

3) What is the maximum amount of live foliage that can be removed from a mature tree in one pruning operation according to the ANSI A300 pruning standard?
   a. 15 percent
   b. 25 percent
   c. 35 percent
   d. 40 percent
4) Which pruning cut is LEAST appropriate for mature trees?
   a. heading
   b. reduction
   c. removal
   d. pollarding
5) Order the pruning cuts from most appropriate to least appropriate for mature trees.
   a. heading, removal, reduction
   b. removal, reduction, heading
   c. reduction, removal, heading
   d. removal, heading, reduction

6) When pruning fruit trees it is best to remove what first?
   a. sprouts
   b. drooping branches
   c. upright branches
   d. dead branches

7) Assuming the same amount of foliage is removed from the tree in each of the following pruning types, which is likely to cause the greatest harm to a mature tree?
   a. canopy raising
   b. canopy cleaning
   c. canopy thinning
   d. canopy reduction

8) Which is the BEST strategy for pruning a medium-aged or mature tree?
   a. remove no more than 20 percent of the live foliage
   b. remove dead branches first, then live foliage if there is good reason
   c. remove no more than 30 percent of the live foliage
   d. remove only small-diameter branches

9) What is the BEST method of minimizing the pruning requirement near utility lines?
   a. reduction cuts
   b. removal and reduction cuts
   c. proper directional pruning
   d. removing medium- and large-maturing trees

10) Which of the following should not be cut back into 2-year-old wood?
   a. arborvitae (*Thuja*)
   b. Leyland cypress (*Cupressocyparis*)
   c. yew (*Taxus*)
   d. fir (*Abies*)

Answers: a, b, b, a, b, d, d, b, d, d

CHALLENGE QUESTIONS

1) What do the strategies for pruning palms and pruning shade trees have in common?
2) Describe a strategy for allowing more sunlight to reach under the canopy of a mature oak tree to stimulate better turf growth.

SUGGESTED EXERCISES

1) Find several mature trees on a site. Divide a group into subgroups of about five to seven people. Have each subgroup devise a plan for pruning each tree. Give each subgroup about 15 to 20 minutes for this exercise. Each subgroup will present their plan to the entire group. Be sure the subgroups are instructed to look for all faults and to provide appropriate treatments.
2) Overprune a cherry, plum, or other fruit tree by thinning to remove about 20 to 25 percent of the live foliage. Thin another one by removing about 10 percent. Come back in the growing season a few months later to see the amount of sprouting that occurred. Compare this to a tree that was not pruned.

CHAPTER 15

STANDARDS AND SPECIFICATIONS

OBJECTIVES

1) Describe the function of a national standard.
2) Compare a standard with a specification.
3) Provide guidelines for writing good pruning specifications.

KEY WORDS

ANSI A300
ANSI Z60.1
ANSI Z133.1
Specifications
Standards

STANDARDS

The American National Standards Institute (ANSI) in Washington, D.C. publishes standards for many industries across America including the green industry. Each standard is updated about every five years. Standards do not provide prescriptions. That is not their intention. **Standards** are a set of definitions and guidelines to assist in writing good specifications. Specifications include prescriptions. Although they were never meant to be specifications themselves, many people use the ANSI standards this way. This is not appropriate and represents misuse because it results in ambiguity, confusion, and wildly varying bids from contractors. Using a standard as a specification would be like saying to your home builder, "Build me a house." Obviously, this does not communicate enough information for anyone to proceed.

The nursery and tree care professions have each developed standards for pruning trees. The "American Standard for Nursery Stock" was sponsored by the American Association of Nurserymen (Washington, D.C.) and approved in 2002 by the American National Standards Institute as **ANSI Z60.1.** It presents standards for growing nursery trees. Various states such as California (Department of Transportation) and Florida (Department of Agriculture) adopted standards for nursery stock. Standards for pruning young and mature trees in the landscape are included in American National Standards Institute **ANSI A300.** The National Arborist Association pruning standards have been superceded by the A300 standards. The ANSI A300 standard was not intended to encompass all the detail required to prune trees. Europe and Australia have also developed a guide and standards, respectively, for pruning landscape trees.

Many terms are defined in the A300 pruning standard. This allows for better communication among parties in a contract. If the industry uses different terms for the same technique, then we have a failure to communicate. A good example of how standard language works best is the worldwide recognition of Latin names for plants. Many people refer to the same plant using a different common name but there is usually only one Latin name. For instance *Nyssa sylvatica* is called tupelo, black gum, black tupelo, swamp tupelo, and other common names. Using the Latin name is usually the best way to avoid confusion as to what plant we are referring to. Using standard language and definitions contained in A300 is the best way to avoid confusion in pruning specifications and contracts.

Although the ANSI A300 standard is not meant to be prescriptive, it includes several practices that should be followed or avoided (Table 15-1). You will note that some contain the word "shall"; this is a requirement. You must adhere to the requirement to be working within the standard. Others contain the word "should"; this is a recommendation and is a proper work procedure that is *strongly recommended* by the ANSI A300 standard. If a specification does not follow a "should" recommendation, the specification writer needs to be able to justify the reason why the recommendation was not followed. It is a good idea to write the reason for not following a "should" recommendation on the work specification.

SPECIFICATIONS

Most tree pruning is performed without written specifications or the instructions or specifications are inadequate and vague.

TABLE 15-1. An arborist pruning according to the ANSI A300 pruning standard.

- Should not climb a tree to be pruned with tree climbing spikes or spurs except in case of emergency to rescue an injured climber
- Should use removal (thinning) cuts instead of heading cuts
- Shall not cut off the branch collar (not make a *flush* cut)
- Shall use sharp pruning tools so as to not leave jagged, rough, or torn bark around cuts
- Shall not *top* or *lions-tail* trees
- Should not remove more than 25% of the *live* foliage of a single limb on mature trees
- Should not remove more than 25% of the total-tree *live* foliage in a single year on mature trees
- Should leave 50% of the foliage evenly distributed in the lower 66% of the canopy on mature trees
- Should not use equipment or practices that would damage bark or cambium beyond the scope of the work
- Shall not leave cut limbs in the crown of a tree upon completion of pruning, at times when the tree would be left unattended, or at the end of the workday
- Should not use wound paint

Note: *Should* refers to recommended practices; *shall* refers to mandatory practices.

Arborists often receive requests for a proposal or bid or receive a specification from a municipal, commercial, or even residential job where the request or contract calls for "All pruning to be done according to the ANSI A300 standard." That's it. That is all the information given! They contain few or no additional **specifications,** no pruning type is called for, no pruning objective, nothing. Specifications describe what pruning types should be performed and provide other details of the pruning required for the job. With only this stated, the arborist has no way of knowing what type and amount of work is being requested. How can an arborist possibly bid on this? Yet too many pruning specifications are written in this oversimplified fashion. This is hurting the professional image of the tree care industry.

Tip: *Arborists should write tree pruning specifications and charge for them.*

Without good specifications, each arborist bidding on a pruning job will be bidding on the work they think should be done and this could vary widely among arborists. Since each arborist will be bidding on what is essentially different work, the estimates or bids are likely to be very different from one another. Just like a builder would not even dream about bidding on constructing a home without a set of plans, why should arborists be asked to bid on pruning a tree without a set of plans (i.e., specifications)?

Unfortunately, most arborists bid on pruning jobs that do not have written specifications. If there are specifications, they are often vague and provide little guidance. When asked to bid on pruning jobs like this, point out the value of well-written specs. The biggest advantage is that all bidders will know exactly what is to be done so they all bid on the same amount and type of work. Arborists should offer to write the specifications, and *charge for them.* The person that writes specifications may or may not bid on the work.

A specification writer should understand that A300 sets some guidelines for basic pruning practices that arborists should follow (summarized in Table 15-1). It is always a good idea to include the phrase "Prune according to the ANSI A300 pruning standard." The

TABLE 15-2. Properly written specifications should include, at a bare minimum, the following information.

- Statement that says all work will be performed in accordance with ANSI A300 pruning standard and ANSI Z133.1 safety standard
- Clearly defined pruning objectives
- Pruning type to be performed to meet the objectives
- Size specifications of the minimum and/or maximum branch size to be removed
- Work should be performed by an International Society of Arboriculture (ISA) certified arborist

specification writer might want to stipulate some or all the items listed in Table 15-1 on the contract, since some bidders might not understand all the implications of the sentence, "Prune according to ANSI A300 standard." A well-written pruning specification should include at least the information listed in Table 15-2. Other information, such as a drawing, is also appropriate and is recommended under many circumstances.

The specifications in Appendix 5 are only examples to be used as models. Each job will require different specifications that should vary depending on the species; region of the country; size, age, condition, and location of the trees; objectives and budget of the customer; and other factors. Also, with some minor wording changes, you can use the sample specifications on residential and other small jobs, where the customer does not offer written specs. This helps ensure that customers will have a good idea of what work is to be performed on the trees. They can also use your specifications to compare to other tree care company estimates to see that your company offers more quality per dollar and a more professional image.

Specifications must state the objectives of the pruning that needs to be performed. For example, on young shade trees (maturing at greater than 35 to 40 feet tall) with less than about a 15-inch trunk diameter, specifications should usually include structural pruning with the objective of developing and maintaining a dominant leader and well-spaced scaffold limbs. While important, removing dead branches is usually of secondary importance on trees this age. On large or mature trees, removing dead branches and reducing at risk in the tree is of prime importance. The maximum diameter of a dead branch that can be left in the tree at the completion of the job should be specified. Also specify the minimum diameter (measured at the base of the branch) of live branches that will be removed. For example, specifying 1-inch-minimum branch size means that for the stated pruning objective branches smaller than 1 inch in diameter will not be pruned. This helps prevent lions-tailing. In order to prevent the removal of large-diameter branches on medium-aged and mature trees, some specifications might state a maximum branch size that can be removed, such as 2 or 3 inches.

Be sure to specify which pruning procedure or type is to be conducted, including subordination of codominant stems, cleaning, thinning, raising, reduction, restoration, risk reduction, or vista pruning. Specify the pruning cut types that can or should be used, such as removal or reduction. For example, on a particular job site, you might not want reduction cuts made, only removal cuts. If this is the case, state that in the pruning specifications. In some instances, you might want to specify the approximate maximum amount of foliage to be removed (e.g., 20 percent).

It might be useful to make it clear to the bidders that all structural defects may not be correctable with one pruning. A second pruning several years later may be necessary to bring the tree to a suitably strong structure. You might require the arborist performing the work to indicate which trees need this follow-up pruning. This will inform your decisions for the next pruning cycle and will help prevent the trees from being overpruned.

Writing good specifications is the only way to help ensure that both customer and arborist understand what is to be accomplished by pruning. Poorly written specifications can result in a dissatisfied customer. If you feel uncomfortable writing pruning specifications, hire someone to help you write them.

HOW TO HIRE AN ARBORIST

Arborists make a career of caring for trees in urban and suburban landscapes. They work for companies commonly referred to as tree experts, tree services, tree care, arborists, tree specialists, and others. Here are a few tips for selecting a company to work with your trees:

- Avoid arborists who advertise they top trees. Ask the arborists if they will top your trees. If they say yes, do not use them.
- Have more than one arborist look at the job, and get a written proposal specifying the work to be done. *Consider paying an arborist to write specifications so you can give them to the prospective companies.*
- Ask for and check local references.
- Be sure the company has the appropriate licenses, insurance, and certifications. Some communities require special permits, insurance, or certifications for all arborists to practice in the community.
- Ask them what the ANSI A300 pruning standard and the ANSI Z133.1 safety standards are. Ask if their practices will be in compliance with these two standards.
- Ask them what tree book they have read most recently.
- Ask them what seminar they attended most recently and when was it held.
- Beware of an arborist who suggests removal of living trees. Removal of living trees is sometimes necessary, but should usually be considered the last resort after all other options have been considered.
- Determine if the arborist is certified with the International Society of Arboriculture in Champaign, Illinois. Certified arborists must pass a written test and maintain certification by regularly attending classes. Except for registered consulting arborists with the American Society of Consulting Arborists, membership in other organizations is useful, but no tests or training are required for membership. Members simply pay dues to belong.
- Ask for verification of personal and property liability insurance and worker's compensation (or a waiver of worker's compensation).

- Low price is a poor gauge of a quality arborist. Often the better ones are more expensive because of more specialized equipment, more professional training, and insurance costs.
- Know if your state or municipality requires a specific license or certification for providing these services.

SUMMARY OF TREE PRUNING STRATEGIES

Gather clues from the tree to determine if, when, and how it should be pruned. Think ahead. Develop one dominant leader on large-maturing trees. The sooner you shorten or remove a double leader the better. Do not let low branches grow fast and become large if they are weakly attached or if they will be in the way later. Do not allow branches to get large if they will droop and have to be removed later, because large wounds are created which can lead to trunk decay, and cracks. Remember not to remove more than about 10 to 15 percent of the living canopy at one time on mature trees. Slightly more can be removed on younger trees. As a rule of thumb, a tree that is properly pruned often does not look pruned. Keeping all this in mind, the tree will tell you when, where, and how it needs to be pruned if you train yourself how to "read" trees. This takes practice.

The frequency of pruning required to create good structure depends on many factors, including the climate, tree species, growth rate, tree size at planting, structure of the planted tree, quality of the planting stock, and location and function of the tree. The interval between prunings may need to be shorter in warmer climates than in cooler climates. Some trees require less pruning than others to develop good structure (Appendix 1). Fast-growing trees may need pruning sooner than slow growers because defects develop quicker. Trees that are small at planting need more training than larger trees, provided the large trees are well structured to begin with. Well-developed nursery trees planted in the landscape will need less structural pruning than poor-quality trees. A tree may not need to be structurally pruned if it will not be located close to people or if future limb breakage is of no concern. From this discussion you can see how picking the right trees from the nursery for your situation can have an impact on pruning requirements and maintenance budgets.

Since the main structure of a tree should be developed early in its life, young trees less than 30 years old need more frequent structural pruning than established ones. More live branches can be removed from young trees than from older ones. Fortunately, young trees are relatively easy and inexpensive to prune simply because they are small. Most decurrent trees benefit from annual or every other year pruning until the main structure of the tree is established, but this is usually not practical. At the very least, follow the guidelines in Appendix 6 since this has been shown to result in trees with sound structure. This program will correct potential defects early enough and cuts only small-diameter branches. This minimizes tree injury which results from pruning any live branch and minimizes major structural pruning requirements and remedial action later. However, if you have the opportunity to structurally prune a large-maturing shade tree with a decurrent growth habit only two or three times (provided it was well structured at planting), you can prune about every ten years in the first twenty years after planting. The problem with this program is that you may have to remove significant portions of the live canopy at each pruning. This can cause root problems, reduces vigor, and can create large wounds on the trunk, but this may be better than not pruning at all.

Waiting to structurally prune a tree until ten to fifteen years after planting can cause lasting problems. Although major pruning may be needed, the tree often presents to the arborist only poor pruning options. The required pruning can compromise the tree's defense system, which can shorten its life. For example, major portions of the canopy may be growing from a branch that needs to be removed in order to improve the strength and form of the tree or to provide clearance under the canopy. If this is a street tree, citizens may complain. If this is your tree, you may not prune it because the canopy could be very one-sided and unsightly following pruning. However, an unpruned tree with included bark in the major branch unions can become structurally weak and place a landscape at risk.

If decurrent trees cannot be pruned regularly until the main structure is developed, due to budget constraints, lack of interest, or lack of understanding of the needs of trees, perhaps you cannot afford to plant trees at this time. It may be best to wait until you can afford to care for them properly. This sustainable approach will become more popular as we educate others. Regular pruning can create a strong structure with the potential to last decades and provide a bounty of benefits to the community. Use the successes evident in a community with well-cared-for trees as an example for increasing tree budgets elsewhere.

CHECK YOUR KNOWLEDGE

1) Which is most likely to be accepted across an entire country?
   a. a standard
   b. a specification
   c. neither—they would be equally unacceptable nationwide
   d. both—they would both be acceptable nationwide
2) Which of the following is the safety standard for the arboricultural industry?
   a. A300
   b. Z60.1
   c. Z133.1
   d. Z28
3) The pruning standard ANSI A300 defines terms; well-written pruning specifications:
   a. are an improvement.
   b. essentially duplicate A300 so they are usually not needed.
   c. describe in specific terms what is to be done.
   d. provide further explanation and clarification.
4) Which of the following do not need to be part of a pruning specification?
   a. objectives of pruning
   b. size of parts to be removed
   c. pruning types
   d. reference to A300
   e. All are appropriate.

Answers: a, c, c, e

CHALLENGE QUESTIONS

1) Could many of the statements in the A300 pruning standard be placed in a pruning specification? Why or why not?
2) Explain the difference in the usage of the words "shall" and "should" in an ANSI standard.

SUGGESTED EXERCISES

1) Find a group of trees at a small condominium or apartment complex. Write pruning specifications for the trees on the property.
2) Describe to someone the difference between the A300 pruning standard and a specification.
3) Obtain a copy of the ANSI A300 pruning standard. Attempt to make the standard better by modifying it.
4) Obtain a copy of a pruning specification written by a local municipality. Improve it by making modifications to it.

CHAPTER 16

SHRUB PRUNING

OBJECTIVES

1) Compare shearing with canopy reduction.
2) Describe how to establish and maintain a hedge.
3) Show how to reduce shrub size.
4) Describe how pollarding can be used on shrubs in the garden.

KEY WORDS

Formal hedge
Informal hedge
Pollarding
Renovate
Stooling

INTRODUCTION

Shrub pruning is much simpler than tree pruning. Unlike trees, there are usually no structural concerns to worry about because stem and branch breakage typically places no one at risk. Most shrubs do not have to be pruned to remain prosperous or flower. Shrubs are most often pruned for people reasons rather than health and other reasons. Some shrubs loose vigor as they grow old and benefit from radical size reduction or renovation.

Perhaps the most common reason people prune shrubs is to maintain or reduce size. This can occur because a shrub was placed in the wrong spot too close to a window, walk, entryway, or building. Their size can be maintained by traditionally shearing the plant into a geometric shape, or with reduction techniques that provide a longer lasting, more natural appearance. Shrubs are thinned to reduce a top-heavy appearance or open up a dense canopy. Some shrubs and trees are cut each year in late winter nearly to the ground to provide a wonderful show of lush foliage and a tight canopy during the growing season. Do this following flowering on plants that flower on older growth (Table 16-1).

THINNING

To reinvigorate old shrubs or to reduce plant density, **thin** in late winter or early spring, after danger of frost (Figure 16-1). First remove dead stems and cut back to about 4 to 6 inches long a third of the main stems (cut the oldest third). Although thinning is not

TABLE 16-1. Plants flowering on previous season's wood that should be pruned just following flower display to prevent flower bud removal.

Aesculus (Horsechestnut)	*Ilex* (Holly)
Amelanchier (Serviceberry)	*Illicium* (Anise)
Aronia (Chokecherry)	*Itea* (Sweetspire)
Berberis (Barberry)	*Kalmia* (Mountain laurel)
Calycanthus (Sweet shrub)	*Kolkwitzia* (Kolkwitzia)
Camelia (Camelia)	*Leucothoe* (Drooping leucothoe)
Caragana (Peashrub)	*Lonicera* (Honeysuckle)
Cercis (Redbuds)	*Magnolia* (Magnolia)
Chaenomales (Quince)	*Malus* (Crabapple)
Chionanthus (Fringetree)	*Michelia* (Banana shrub)
Cornus (Dogwood)	*Philadelphus* (Mockorange)
Deutzia (Deutzia)	*Pieris* (Andromeda)
Enkianthus (Enkianthus)	*Prunus* (Cherry)
Exochorda (Pearlbush)	*Pyrus* (Pear)
Forsythia (Forsythia)	*Rhaphiolepis* (Indian hawthorn)
Fothergilla (Fothergilla)	*Rhododendron* (Azalea and Rhododendron)
Fremontodendron (Fremontia)	*Syringa* (Lilac)
Guaiacum (Lignumvitae)	*Viburnum* (Viburnum)
Hamamelis (Witchhazel)	*Weigela* (Weigela)
Hydrangea macrophyllum (Hydrangea)	

meant to reduce plant size, the plant might be slightly smaller following this treatment. You may cut all of the stems to about 4 to 6 inches long to **renovate** the plant, but this usually is not necessary. Cutting all stems might kill some very old or unhealthy plants, but most can tolerate this and recover nicely.

When removing only the oldest third, take care not to damage the nearby younger stems and foliage that will become the new shrub canopy. Remove or cut back stems that extend far beyond the edge of the canopy. When growth begins in the spring, new vigorous, upright stems typically arise from dormant buds near the ground and from the pruning cuts. You may want to remove some of these or cut some back if there are too many. In other cases, these can be retained to create a denser lower canopy. Plants look fresh and rejuvenated following this treatment. Many shrubs can be thinned each year in this manner.

MAINTAINING AND REDUCING SIZE

Maintaining size

Shrubs such as *Berberis, Fatsia, Mahonia, Nandina, Spiraea, Viburnum,* and other cane-type plants can be pruned to maintain their size by cutting to the ground each year one-third of the tallest canes (Figure 16-2). The remaining two-thirds of the canes remain intact until the following year, when again, the tallest one-third are cut to the ground. Cane-type plants, including many common shrubs, have aggressive upright stems, usually in the center of the shrub. These upright stems often produce few branches until they reach the top of the shrub.

The height and spread of most round-headed shrubs can be maintained by cutting only the longest shoots 6 to 24 inches back inside the outer edge of the plant (Figure 16-3, top). Try to "see" a smaller canopy inside of the existing canopy and simply cut back to expose it. *Make no cuts on the edge of the new, smaller canopy.* Instead, make reduction

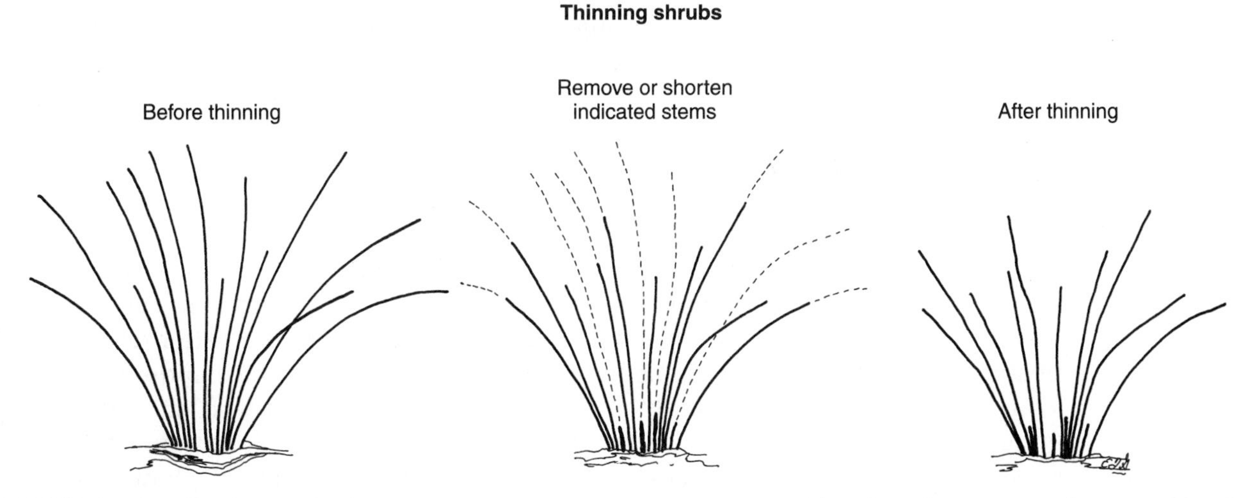

FIGURE 16-1. To thin a shrub, some stems are cut to near ground level and some are shortened back inside the canopy.

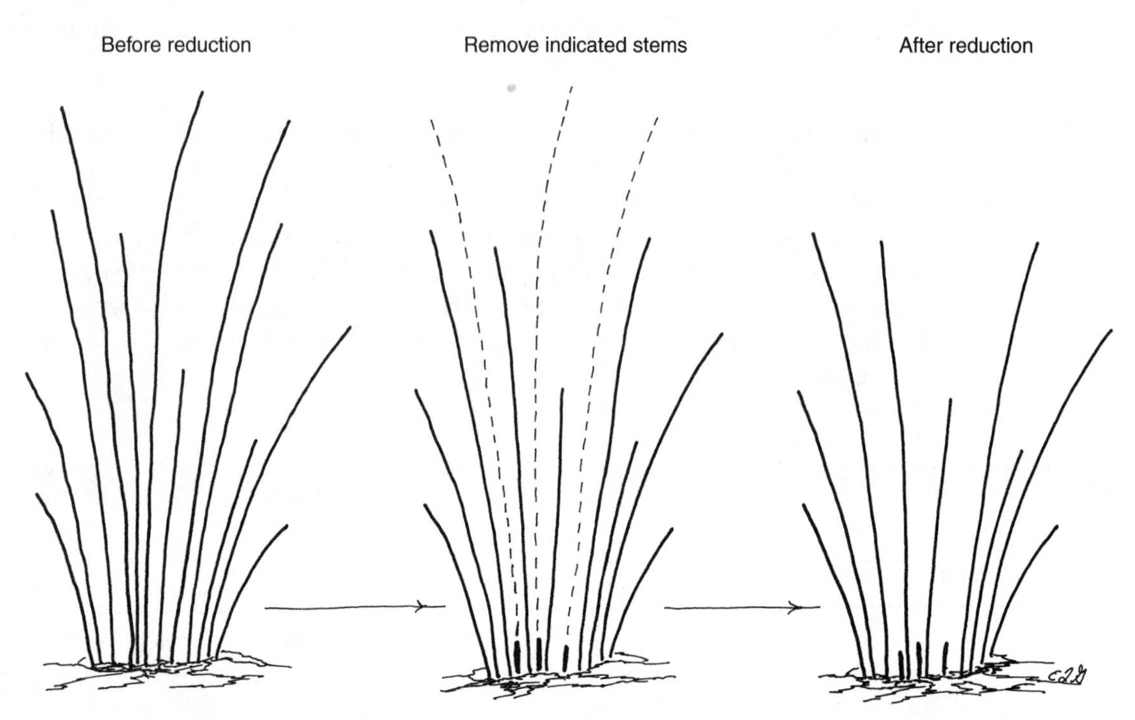

FIGURE 16-2. To reduce cane-type shrubs, cut about one-third of the stems back to near ground level. The shrub is shorter and less dense following this treatment.

cuts back to lateral branches with plenty of foliage. (Instead of reducing all the way back to a lateral, leave a short stub a couple inches long if you want sprouts to form at the location of the cut.) With this technique, you will be removing the larger stems back to a smaller live branch. The base of these smaller branches is located 6 to 24 inches inside the edge of the existing canopy. These smaller branches deeper inside the shrub are not cut, or they are shortened very little, and they form the outside edge of the now smaller plant. This technique allows the shrub to retain its original form and texture without an unnatural sheared appearance (Figure 16-4). Lower foliage is encouraged to grow and remain on the plant because more light reaches these lower leaves. The canopy will be thinner than it was before reduction so you will see more of the trunk and branch structure at least until new growth fills in the voids.

Tip: *In order to hide pruning cuts, make them inside the canopy instead of on the outside edge.*

It is best to begin this treatment long before the shrub reaches the desired size. Be sure to cut shoots at different lengths back inside the canopy in order to hide the cuts. Because cuts are hidden, the shrub will not look like it was pruned after receiving this treatment, at least compared to hedging. While performing this technique, attempt to visualize a smaller shrub inside the existing overgrown shrub, and cut away everything to the outside of this smaller shrub. This canopy reduction technique gives plants a less formal (some say more natural) look compared to hedging, which makes cuts all at the same height at the edge of the canopy. It is easy to perform this pruning with an extension pruner.

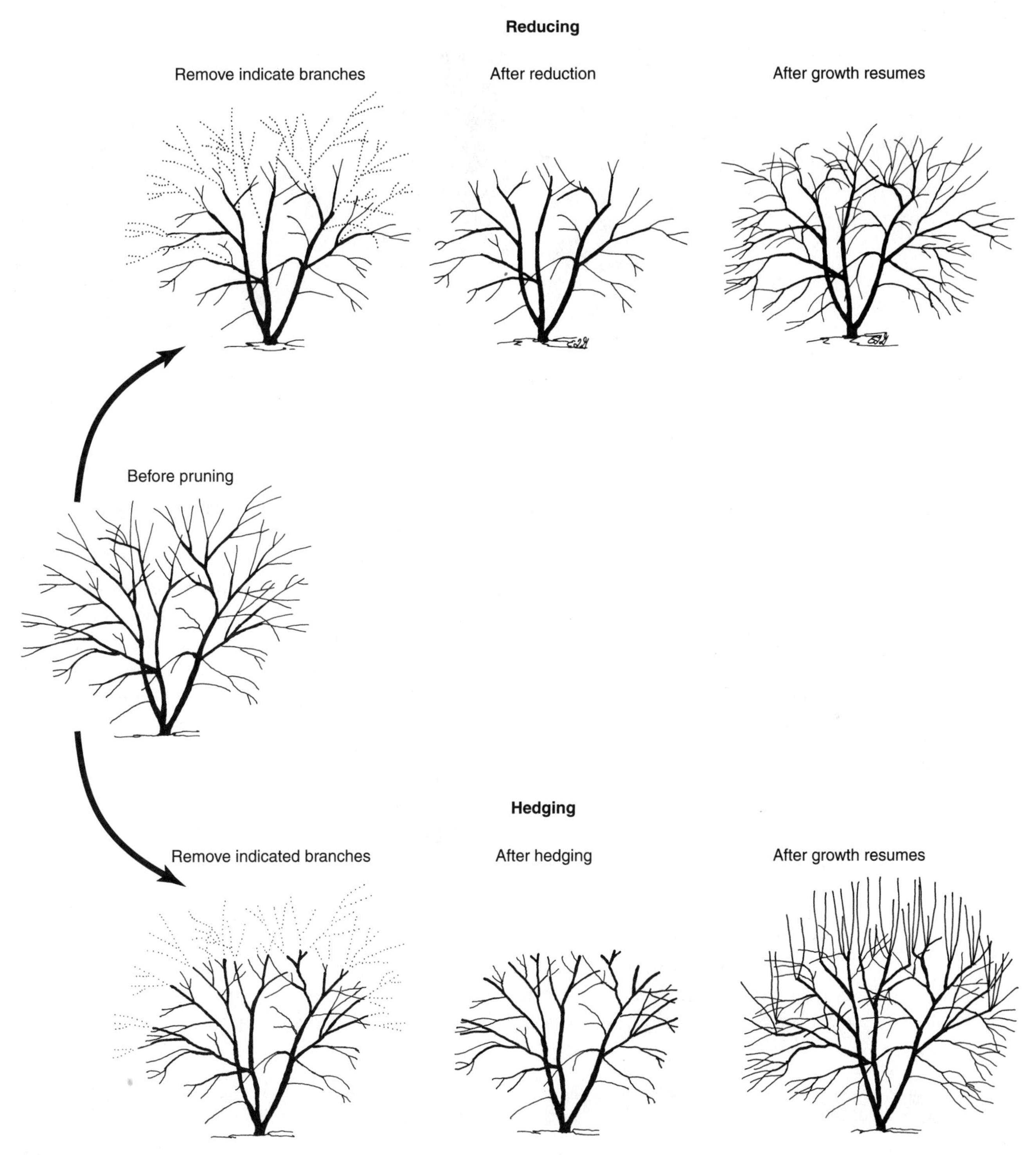

FIGURE 16-3. Shrubs that are reduced re-grow in a more natural habit (top). Those that are hedged look as though they were pruned because many sprouts quickly extend above the canopy (bottom).

FIGURE 16-4. Shrubs can be reduced in size by cutting the longest stems and branches back inside the canopy.

Reducing size

To substantially reduce the size of the plant, cut the longest stems back deep into the shrub, leaving some foliage intact to form the new, smaller canopy. In most respects, this is a more severe version of what was described earlier (Figure 16-3, top). Simply cut deeper inside the shrub canopy. This usually cuts the largest diameter, oldest stems. If there will be little or no foliage left following this treatment, cut the shoots to several different lengths so the new foliage develops in several layers, not all at the same height. Locate the shortest stems more or less in front and the longest ones in back. The plant will look less like it was pruned than if all stems are cut to the same length and the emerging foliage will be pleasingly displayed in layers. On plants that flower on current year's growth, this is best done in late winter/early spring before buds emerge, so the plant can recover before the next dormant season. Perform this immediately after flowering on shrubs that bloom on the previous season's growth in order to preserve the flower display (Table 16-1). Again, attempt to visualize a smaller shrub inside of the overgrown shrub.

Consider performing drastic pruning over a period of two years. Instead of cutting all stems nearly to the ground all at once, cut the oldest half close to the ground (4 to 6 inches) the first spring and allow the sprouts to grow all year. In the following spring, cut the remaining stems close to the ground. In this way, the shrub will be rejuvenated over a two-year period. Many shrubs are remarkably able to withstand severe pruning.

HEDGING

Developing a hedge

Start with the smallest plants you can find. Purchase plants that are as short and wide as possible, not the tall narrow ones. Some shrubs are grown too close in the nursery and develop an upright habit with few leaves on the lower portion of the plant. These are not good candidates for planting as a hedge. Plants can be spaced 2 to 5 feet apart, depending on your budget and how fast you want the hedge to form. It is generally best to space shrubs with large foliage farther apart than shrubs with small leaves.

The primary objective the first year or two is to increase the width of the plants. At planting, or several weeks into the growing season, cut upright shoots on small plants in 1-gallon containers 6 to 12 inches from the ground if there are not enough branches originating from the base of the plants. To encourage growth on the lower portion of the hedge, do not cut many lower branches growing in the horizontal direction unless they are growing far outside the rest of the plant where you do not want branches. Leave branches that are touching the ground. Cutting these lower branches could encourage upright sprouts to form from the cuts, and this could leave gaps in the lower portion of the hedge. If left intact, some of these lower horizontal branches on certain species may form roots if they are allowed to touch the ground. This will hasten the development of the hedge.

As the shrub grows the first three years, cut off about half the length of new upright shoots as they grow about 6 to 12 inches beyond the last cut. Again, refrain from cutting low branches unless they become overextended and leggy. This will further encourage the lower portion of the plant to fill in and create a denser hedge. As the lower portion fills in and becomes wider than the top, allow the plant to grow taller, if you wish. Plants should be maintained so the base of the hedge is wider than the top. Pruned hedges with a narrow base often lose lower foliage because they are shaded from the top. This condition will worsen with age, resulting in sparse growth at ground level and an unattractive hedge that does not provide desired privacy.

Tip: *Hedges should be maintained so the base is wider than the top.*

Maintaining a hedge

The location of pruning cuts to maintain a hedge should be determined by the type of hedge desired. An **informal hedge** is maintained by making heading or reduction cuts only on the longest shoots, 6 to 18 inches back inside the outer edge of the hedge (Figure 16-3, top). The shorter shoots remain intact and form the outside edge of the now smaller hedge. The outer edge of the canopy appears more open and maintains a softer (some people say more natural) appearance than the clipped formal hedge described next. Shape the hedge so the bottom is wider than the top to allow adequate sunlight to reach lower foliage. Pruning on an informal hedge can be performed at any time. Some horticulturists prefer to do it just before new growth emerges in spring so voids fill in quickly. Informal hedges can recover easier than formal hedges from damage caused by vandals and people falling into them. The damage is much less noticeable.

The desired appearance of a **formal hedge** is a sharply defined geometric shape. Globes, boxes and squares are most common. These are maintained with regular shearing (shearing makes heading cuts) using hand operated or power tools. Be sure to shear so the bottom of the hedge is wider than the top to allow adequate sunlight to reach all foliage.

There are advantages and disadvantages to maintaining shrubs in a formal shape with shearing (Table 16-2). One down side of shearing is that most foliage is produced on the outside edge of the canopy (Figure 16-3, bottom). In contrast, informal hedges

TABLE 16-2. Comparing hedging strategies.

Shearing	
Advantages	**Disadvantages**
+ Looks neat (for a short period) + Economical pruning + Easy to train unskilled labor and inexperienced gardeners to perform	– Grows out quickly requiring frequent pruning – Tough to reduce size gracefully – Grows slightly larger each year – All foliage is on outside edge of canopy – Damaged hedge has void for long time – Damage from people very noticeable – Looks neat for a shorter period due to rapid re-growth – May look out of place in a natural garden
Reduction	
Advantages	**Disadvantages**
+ Recovers from damage quicker + Damage from people is less noticeable + Looks neat for a longer period + Provides a more open, natural look + Easier to reduce in size + Easier to maintain at a given height	– Provides a more open, less formal look – More training and time needed to execute – May look out of place in a formal garden

maintained with reduction cuts tend to have some foliage on the interior of the canopy because more sunlight reaches inside. Although shearing provides a neat, clean formal appearance, hedges recover slowly from damage. If people fall into it or vandalize it a lasting hole remains—sometimes for many years. This results because there is usually no foliage on the inside. Informal hedges maintained with reduction cuts are also easier to reduce in size should they become too large.

Formal hedges should be clipped while new growth is green and succulent—normally during the growing season. If new growth is not too long and unkempt, wait to prune until new growth has slowed when the plants are entering a quiescent period. Then the effects of pruning will last longer. Flowering hedges grown formally should only be sheared after flowering to prevent reduction in the number of subsequent blooms. To keep the plant looking sharp during the flowering period, developing shoots without flowers can be carefully cut with a hand pruner until flowers are gone. Then resume normal shearing. If the blooms are of secondary importance, hedge pruning can be conducted at any time.

Shear new growth on mature, formal hedges to within an inch or so of the last shearing. This will leave a nice dense canopy of foliage on the outer edge of the hedge. The hedge will be slightly larger after each shearing than it was following the last shearing. Each year or two, you might choose to cut back into the previous season's growth to keep plants from growing too tall or wide. This usually results in cutting below the original heading cut made to keep the plant at the desired height. Since this treatment often removes a good portion of the living foliage, do it in the spring, as new growth is just about to emerge, so plants are bare for only a short period.

Rejuvenating a hedge

Eventually, most formal and informal hedges become too large, or the bottoms become thin with little foliage. This is especially a problem if the top was allowed to grow wider than the bottom. To correct this on nonconiferous plants, in spring (as new growth is first emerging) cut back the hedge to about half its size. This treatment will leave the plant with little or no foliage. A small, sharp pruning saw or loppers may be the best tools for this job, especially if the hedge is more than a few years old. Cut a formal hedge into the same shape; for example, if it was a box shape, cut it back into a smaller box. Be sure to leave the lower branches longer than the top branches so all new foliage receives sunlight. Allow new shoots to develop several inches long, then resume regular shearing to redevelop the hedge. Cut the stems on an informal hedge back to different lengths to preserve the more open, informal appearance. Prune coniferous plants tightly each time you prune to avoid having to perform this drastic treatment, which most will not survive (see Chapter 14).

ENHANCING CANOPY DENSITY, SLOWING GROWTH, AND INCREASING FLOWER NUMBER

The density of a healthy shrub can be increased by pinching or heading new growth on the outer edge of the canopy when twigs are soft and pliable. If performed when new growth is 6 to 12 inches long, it will be hard to tell the plant was pruned. Pinching may not enhance density on plants lacking vigor or on old plants. These may require rejuvenation. Pinching only new shoots allows the plant to grow larger each year, but the plant will increase in size slower than if shoots were not pinched. This is a nice way to both keep the canopy full and slow the growth rate. Last year's shoots can also be headed, but this may thin out the canopy slightly. If last year's stems are pruned, locate the cuts back inside the canopy so cuts are hidden to present a nicer look.

Increase flower number by pinching at the appropriate time of year. Spring and early summer is a good time to pinch shrubs that flower on current season's wood; pinching just after flowering is most appropriate on shrubs that flower on older wood (Table 16-1). Pinching to create more flowers could reduce flower size.

POLLARDING AND STOOLING

Pollarding and stooling cuts a plant back each year. Cutting back to a position above the ground is usually referred to as **pollarding;** cutting nearly to the ground is usually called **stooling.** Both are good methods of controlling height on plants that would normally grow to a large size. Most plants listed in Table 16-3 can be stooled, but not all pollard gracefully. Those that generate many small stems crowded together are difficult to pollard so they are usually stooled. Some people refer to stooling as coppice.

Pollarding is typically performed on those plants that generate only a few, relatively large trunks or stems such as *Catalpa* and other very woody plants (Table 16-3). Many of these plants would grow as large shrubs or even trees if left unpruned. Pollarding cuts back to the exact same position each year. Eventually a swollen knob of tissue develops referred to as a pollard head (Figure 16-5). All stems are cut back to the head.

TABLE 16-3. Partial list of plants suitable for stooling, pollarding, and rejuvenation pruning.

For foliage effect:	For flower and/or fruit effect:
Acacia (acacia) S, P	*Abelia* x *grandiflora* (abelia) S
Catalpa bignonioides 'Nana' (bean plant) P	*Bubbleia davidii* (butterfly bush) S
Corylus maxima 'Purpurea' S, P	*Calliandra* (powderpuff) S, P
Cotinus coggygria (smokebush) S, P	*Callicarpa* (beautyberry) S
Eucalyptus species S, P	*Callistemon* (bottlebrush) S, P
Fatsia japonica (fatsia) S	*Caryopteris* x *clandonensis* (blue spirea) S
Firmiana simplex (parasol tree) S, P	*Cassia* (shower tree) S, P
Hydrangea macrophylla (plants will flower very little with stooling) S	*Combretum fruiticosum* (flamevine) S
Nandina domestica (heavenly bamboo) S	*Feijoa sellowiana* (guava) S, P
Paulownia tomentosa (princess tree)—will not bloom P	*Hamelia patens* (firebush) S
Prunus x *cistena* and others (plum) S, P	*Hydrangea arborescens* (smooth hydrangea) S
Salix species (willow) S, P	*Hydrangea paniculata* (panicle hydrangea) S
Sambucus canadensis (elderberry) S	*Indigofera decora* (indigo) S
Schefflera (dwarf schefflera) S, P	*Jasminum* (jasmine) S
Ternstroemia gymnanthera (cleyera) S, P	*Justicia* (shrimp plant) S
Weigela (weigela) S, P	*Kerria japonica* S
	Kolkwitzia amabilis (kolkwitzia) S, P
	Lagerstroemia (crape-myrtle) P
	Lespedeza thunbergii S
	Nerium (oleander) S, P
	Perovskia atriplicifolia S
	Plumbago auriculata (cape plumbago) S
	Russelia equisetiformis (firecracker plant) S
	Tamarax ramoissima (tamerisk) S
	Tibochina (princess flower) S
	Vitex agnus-castus (chastetree) S, P

P = plant can be pollarded; S = plant can be stooled; all can be rejuvenation pruned.

The head is never removed. The pollard head can be located at any position desired from ground level to much higher in the plant. Chapter 10 described how to use pollarding to create a small tree form. Pollarding is a good way to keep plants artificially small while preserving health and vigor.

Tip: *Stooling, pollarding, and reduction are good methods of keeping large plants small.*

Stooling is performed on more shrubby plants such as *Abelia* that have numerous small stems. Stems usually do not grow large enough to develop pollard heads so cuts are made an inch or so above last year's cuts. A thick mass of dead stubs sometimes develops at the base of the plant. There may be dozens of stems to cut through each year. Stooling is also a good way to manage plants that are not quite cold hardy in your region. It results in aggressive growth and a very uniform canopy.

Cutting is typically performed in late winter/early spring prior to the flush of new growth. Pruning too early could cause cold injury, especially in warmer climates, should the plant begin growing prior to a late-winter freeze. New shoots grow straight and fast filling the space in the garden with a lush canopy of foliage. Stems grow close together and branch sparingly. Stooling can be used on young shrubs or on older plants that never received this treatment provided they are able to generate sprouts after

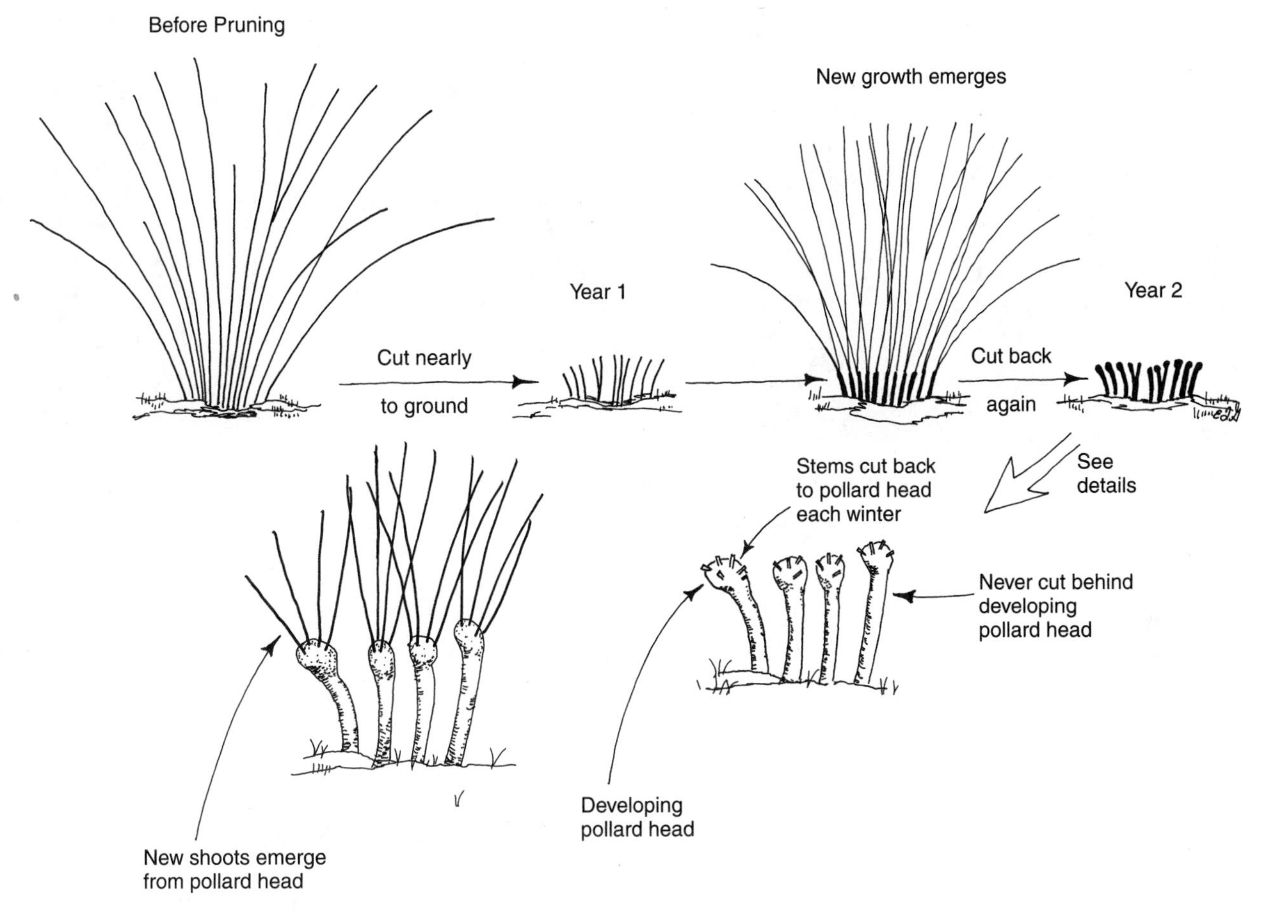

FIGURE 16-5. Pollarding is a good way to maintain large shrubs or trees in a smaller size. Pollarding cuts back to the same point each year, eventually developing a swollen knob of tissue called a pollard head (bottom). Stooling is similar but cuts all stems nearly to the ground each year without regard to location of a knob of tissue.

pruning. Plants not able to sprout could be killed by this treatment. It is best to make heading cuts back to a bud if visible. Buds may not be visible on older stems.

RENOVATING

Most shrubs can be cut nearly to the ground in spring to rejuvenate or remote them. This is done on old plants that lack vigor or those that have become way too big for their location. New growth is often vigorous. Even plants such a *Rhododendron, Weigela,* and *Callistemon* tolerate this radical pruning.

CREATING A SMALL TREE FROM AN OVERGROWN SHRUB

Older shrubs with canopies extending close to the ground can be quickly and easily converted into a small tree (Figure 16-6). This could extend the shrub's utility many years, especially if it has become too wide or tall for its location. First locate the lowest

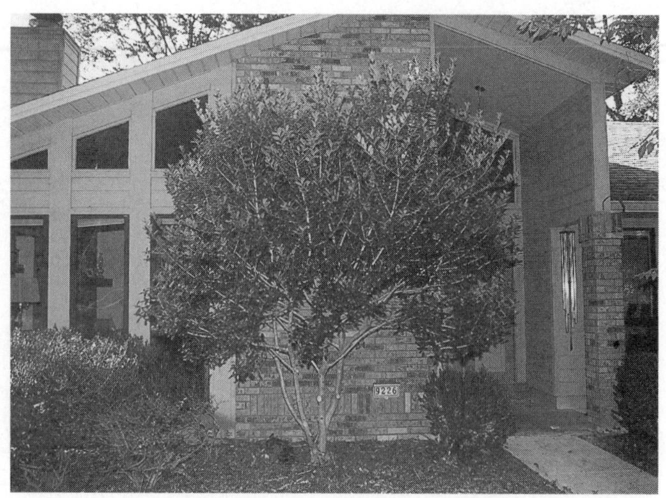

FIGURE 16-6. Large overgrown shrubs can be converted into small trees by removing the lower branches. This exposes the lower branch and stem structure and can add many years to the life of a plant.

branches on the existing shrub that will be saved. Then remove all branches back to the trunks. Some thinning or reduction of the canopy may enhance the appearance of the small tree, especially if the canopy is very dense, or some branches extend way beyond others.

TIME OF YEAR

Many shrubs not grown for their flower display are pruned in late winter/early spring, if pruning is needed only once a year. Shrubs grown for their flower display that flower on last year's stems (Table 16-1) should be pruned immediately following the flower display before next year's flower buds form. There is nothing wrong with pruning these shrubs in summer, fall, or winter, but the flower buds will be removed. Shrubs flowering on current season's stems can be pruned in late winter through spring and early summer up until flower buds begin forming. Heavy pruning like size reduction should be done in late winter/early spring. If this is performed in late summer or fall, growth could be stimulated and new stems injured in the first freeze.

CHECK YOUR KNOWLEDGE

1) Which technique usually leaves the least amount of live foliage on a shrub?
   a. thinning
   b. reducing
   c. hedging
   d. All result in about the same amount.
2) The damage resulting from a person falling into a sheared plant appears greater than on shrubs maintained by other methods because:
   a. sheared plants are usually less healthy.
   b. sheared plants are usually shorter than plants maintained by other methods.

c. the bottom of a sheared plant is usually wider than the top.
d. all foliage of a sheared plant is on the outer edge of the canopy so it is easily damaged.

3) To reduce the size of an overgrown shrub substantially:

a. do not reduce substantially because the shrub could die.
b. use pollarding.
c. perform the needed pruning over a two-year period if possible.
d. shear the canopy.

4) When shearing a hedge, cut:

a. several inches beyond the last cut.
b. several inches shorter than the last cut.
c. at the same position as the last cut.
d. only slightly beyond the last cut.

5) Which technique would you NOT use to keep a plant artificially small for a very long time?

a. pollarding
b. reduction
c. hedging
d. thinning

6) What would be the worst time to drastically reduce the size of a shrub?

a. spring
b. late summer to early fall
c. late fall
d. winter

Answers: b, d, c, d, d, b

CHALLENGE QUESTIONS

1) Construct a list of shrubs in your area that generate flower buds on last year's stems.
2) Construct a list of shrubs that generate flowers on current year stems.

SUGGESTED EXERCISES

1) Find an overgrown group of shrubs. Demonstrate how to reduce their size by half using reduction cuts. Leave as much live foliage intact as possible. Divide the group up into subgroups and give each a unique colored ribbon role. Have each group mark with colored ribbon which stems should be cut and where to make the cuts on each shrub. Gather the group together and begin at the first marked shrub. Critique the decisions made by each subgroup. Reduce each shrub by half.
2) Locate two shrubs that were never pruned or were pruned long ago. Reduce one shrub to a given height with heading cuts and reduce the other to the same height using reduction cuts. Have the group evaluate each technique and discuss the advantages and disadvantages of both techniques.

APPENDIX 1

NONCONIFEROUS TREES REQUIRING ONLY MODERATE PRUNING TO DEVELOP STRONG STRUCTURE

When only the genus or species of tree is listed, most cultivars also require only moderate pruning for good structure. More frequent pruning on trees in this list may be necessary for other reasons. Note that conifers typically require little pruning to develop good structure. An asterisk indicates a tree that may need pruning to balance the canopy.

Acer barbatum—Florida Maple, Southern Sugar Maple
Acer barbatum var. caddo—Caddo Florida Maple, Southern Sugar Maple
Acer buergeranum—Trident Maple
Acer griseum—Paperbark Maple
Acer japonicum 'Acontifolium'—'Acontifolium' Fullmoon Maple, Fernleaf Maple
Acer miyabei—Miyabe Maple
Acer platanoides 'Emerald Queen'—'Emerald Queen' Norway Maple
Acer platanoides 'Superform'—'Superform' Norway Maple
Acer pseudoplatanus—Sycamore Maple, Planetree Maple
Acer rubrum 'Autumn Flame'—'Autumn Flame' Red Maple
Acer rubrum 'Red Sunset'—'Red Sunset' Red Maple
Acer saccharum and cultivars—Sugar Maple
Acer triflorum—Three-Flower Maple
Acer truncatum—Purpleblow Maple

Acer x freemanii 'Armstrong'—'Armstrong' Freeman Maple
Aesculus spp.—Horse Chestnut, Buckeye
Alnus glutinosa 'Pyramidalis'—'Pyramidalis' Black Alder, 'Pyramidalis' European Alder
Alnus rhombifolia—White Alder
Amelanchier spp.—Serviceberry
Araucaria bidwillii—False Monkey-Puzzletree
Araucaria heterophylla—Norfolk Island Pine
Arbutus menzesii—Pacific Madrone
Arbutus texana—Texas Madrone
Asimina triloba—Pawpaw
Betula spp.—Birch
Bumelia lanuginosa—Chittamwood, Gum Bumelia, Gum Elastic Buckthorn
Callistemon citrinus—Red Bottlebrush, Lemon Bottlebrush
Calocedrus decurrens—California Incense Cedar
Calodendron capense—Cape Chestnut
Calophyllum brasiliense—Santa Maria
Camellia oleifera—Tea-Oil Camellia
Carpinus betulus 'Fastigiata'—'Fastigiata' European Hornbeam
Carpinus caroliniana—American Hornbeam, Blue Beech, Ironwood
Carya spp.—Hickory
Catalpa spp.—Catalpa
Cercidiphyllum japonicum—Katsuratree
Chionanthus retusus—Chinese Fringe Tree
Chrysophyllum oliviforme—Satinleaf
Coccoloba diversifolia—Pigeon-Plum
Coccothrinax argentata—Silverpalm, Thatchpalm
Cornus alternifolia—Pagoda Dogwood
Cornus controversa—Giant Dogwood
Cornus drummondii—Roughleaf Dogwood
Cornus florida—Flowering Dogwood
Cornus kousa—Kousa Dogwood, Chinese Dogwood, Japanese Dogwood
Cornus mas—Cornelian Cherry
Cornus walteri—Walter Dogwood
Corylus colurna—Turkish Filbert, Turkish Hazel
Cotinus coggygria—Smoketree, Wig-Tree, Smokebush
Cotinus obovatus—American Smoketree, Chittamwood
Crataegus spp.—Hawthorn
Cydonia sinensis—Chinese Quince
Diospyros texana—Texas Persimmon
Diospyros virginiana—Common Persimmon
Eucalyptus citriodora—Lemon-Scented Gum
Eucalyptus ficifolia—Red-Flowering Gum

Eucommia ulmoides—Hardy Rubber Tree
Eugenia spp.—Stopper, Eugenia
Fagus spp.—Beech
Firmiana simplex—Chinese Parasoltree
Franklinia alatamaha—Franklin-Tree, Franklinia
Fraxinus americana 'Autumn Applause'—'Autumn Applause' White Ash
Fraxinus americana 'Autumn Purple'—'Autumn Purple' White Ash
Fraxinus ornus—Flowering Ash
Fraxinus oxycarpa 'Raywood'—Raywood Ash, Claret Ash
Fraxinus pennsylvanica 'Summit'—'Summit' Green Ash
Fraxinus texensis—Texas Ash
Ginkgo biloba—Maidenhair Tree, Ginkgo*
Gordonia lasianthus—Loblolly Bay, Sweet Bay
Guaiacum sanctum—Lignumvitae
Gymnocladus dioicus—Kentucky Coffee Tree*
Halesia spp.—Silverbell
Hamamelis mollis—Chinese Witch Hazel
Hibiscus syriacus—Rose of Sharon, Shrub-Althea
Hovenia dulcis—Japanese Raisintree
Ilex latifolia—Lusterleaf Holly
Ilex opaca—American Holly
Ilex x attenuata 'East Palatka'—'East Palatka' Holly
Ilex x attenuata 'Fosteri'—Foster's Holly
Ilex x attenuata 'Savannah'—Savannah Holly
Juglans nigra—Black Walnut*
Kalopanax pictus—Castor-Aralia, Prickly Castor-Oil Tree*
Lagerstroemia fauriei—Japanese Crape-Myrtle
Lagerstroemia indica—Crape-Myrtle
Ligustrum lucidum—Glossy Privet, Tree Ligustrum
Liquidambar styraciflua—Sweet Gum
Liriodendron tulipifera—Tuliptree, Tulip Poplar, Yellow Poplar
Lithocarpus densiflora—Tanbark Oak
Magnolia acuminata—Cucumbertree, Cucumber Magnolia
Magnolia grandiflora—Southern Magnolia
Magnolia kobus—Kobus Magnolia, Northern Japanese Magnolia*
Magnolia macrophylla—Bigleaf Magnolia*
Magnolia stellata—Star Magnolia
Magnolia virginiana—Sweetbay Magnolia, Swamp Magnolia
Magnolia x soulangiana—Saucer Magnolia*
Malpighia glabra—Barbados Cherry
Manilkara zapota—Sapodilla
Melaleuca quinquinervia—Punk Tree, Melaleuca

Morus alba—White Mulberry*
Nyssa ogeche—Ogeechee Tupelo, Ogeechee-Lime
Nyssa sinensis—Chinese Tupelo
Nyssa sylvatica—Black Gum, Sour Gum, Black Tupelo
Osmanthus fragrans—Sweet Osmanthus
Ostrya virginiana—American Hophornbeam, Eastern Hophornbeam
Oxydendrum arboreum—Sourwood, Sorrel Tree
Parrotia persica—Persian Parrotia
Photinia serrulata—Chinese Photinia
Photinia villosa—Oriental Photinia
Pinckneya pubens—Pinckneya, Fevertree*
Platanus spp.—Sycamore, Planetree
Platanus orientalis—Oriental Planetree
Platanus racemosa—Western Sycamore
Platanus x acerifolia—London Planetree
Podocarpus falcatus—Podocarpus
Podocarpus gracilior—Weeping Podocarpus, Fern Podocarpus
Podocarpus latifolius—Podocarpus
Podocarpus macrophyllus—Podocarpus, Yew-Pine, Japanese Yew
Podocarpus macrophyllus var. *angustifolius*—Podocarpus
Podocarpus nagi—Nagi Podocarpus, Broadleaf Podocarpus
Populus alba—White Poplar*
Populus deltoides—Eastern Cottonwood*
Populus freemanii—Fremont Poplar
Populus nigra 'Italica'—Lombardy Poplar
Prunus caroliniana—Cherry Laurel, Carolina Laurelcherry
Prunus cerasifera—Purple Leaf Plum
Prunus maackii—Amur Chokecherry, Manchurian Cherry
Prunus mexicana—Mexican Plum
Prunus sargentii—Sargent Cherry
Prunus serotina—Black Cherry
Prunus serrulata 'Kwanzan'—Kwanzan Cherry
Prunus subhirtella 'Autumnalis'—'Autumnalis' Higan Cherry
Prunus subhirtella 'Pendula'—Weeping Higan Cherry*
Prunus umbellata—Flatwoods Plum
Prunus x incamp 'Okame'—'Okame' Cherry
Prunus x yedoensis—Yoshino Cherry
Psidium littorale—Cattley Guava, Strawberry Guava
Quercus acutissima—Sawtooth Oak
Quercus alba—White Oak
Quercus austrina—Bluff Oak
Quercus bicolor—Swamp White Oak

Quercus cerris—Turkey Oak, Moss-Cupped Oak
Quercus coccinea—Scarlet Oak
Quercus falcata—Southern Red Oak, Spanish Oak
Quercus imbricaria—Shingle Oak, Northern Laurel Oak
Quercus lobata—Valley Oak, California White Oak
Quercus lyrata—Overcup Oak
Quercus macrocarpa—Bur Oak
Quercus michauxii—Swamp Chestnut Oak
Quercus muehlenbergii—Chinkapin Oak, Chestnut Oak
Quercus nuttallii—Nuttall Oak
Quercus palustris—Pin Oak
Quercus robur—English Oak
Quercus rubra—Northern Red Oak
Quercus texana—Texas Red Oak, Texas Oak
Robinia pseudoacacia 'Tortuosa'—'Tortuosa' Black Locust
Robinia pseudoacacia 'Umbraculifera'—Umbrella Black Locust
Sambucus mexicana—Mexican Elder
Sapindus drummondii—Western Soapberry
Sapindus saponaria—Florida Soapberry, Wingleaf Soapberry
Sassafras albidum—Sassafras
Sophora secundiflora—Texas Mountain Laurel, Mescalbean
Sorbus alnifolia—Korean Mountain Ash
Stewartia koreana—Korean Stewartia
Stewartia monadelpha—Tall Stewartia
Stewartia pseudocamellia—Japanese Stewartia
Styrax japonicus—Japanese Snowbell
Styrax obassia—Fragrant Snowbell
Syringa pekinensis—Peking Lilac
Syringa reticulata—Japanese Tree Lilac
Tabebuia spp.—Trumpet Tree*
Terminalia catappa—Tropical Almond, India Almond
Terminalia muelleri—Mueller's Terminalia
Tilia americana 'Fastigiata'—'Fastigiata' American Basswood
Tilia americana 'Redmond'—'Redmond' American Basswood
Tilia cordata—Littleaf Linden
Tristania conferta—Brisbane Box*
Tsuga canadensis—Canadian Hemlock, Eastern Hemlock
Tsuga canadensis 'Lewisii'—Lewis Hemlock
Viburnum rufidulum—Rusty Blackhaw, Southern Blackhaw
Viburnum sieboldii—Siebold Viburnum
Xanthoceras sorbifolium—Yellowhorn*

APPENDIX 2

TREES OFTEN GROWN WITH A MULTI-TRUNK HABIT

Some trees could develop included bark in the crotches. Some are trained to one trunk by selected nurseries. Trees marked with an asterisk are difficult to train to one trunk.

Acacia farnesiana—Sweet Acacia, Huisache*
Acacia wrightii—Wright Acacia, Wright Catclaw
Acer buergeranum—Trident Maple
Acer campestre—Hedge Maple
Acer cissifolium—Ivy-Leaf Maple*
Acer ginnala—Amur Maple
Acer griseum—Paperbark Maple
Acer japonicum 'Acontifolium'—'Acontifolium' Fullmoon Maple, Fernleaf Maple*
Acer miyabei—Miyabe Maple
Acer palmatum—Japanese Maple*
Acer tataricum—Tatarian Maple
Acer triflorum—Three-Flower Maple*
Acer truncatum—Purpleblow Maple
Acoelorrhaphe wrightii—Paurotis Palm*
Aesculus californica—California Buckeye*
Aesculus pavia—Red Buckeye
Aesculus x carnea—Red Horse Chestnut
Albizia julibrissin—Mimosa, Silktree*

Amelanchier spp.—Serviceberry
Aralia spinosa—Devil's Walkingstick, Hercules Club
Arbutus texana—Texas Madrone
Arbutus unedo—Strawberry Tree
Asimina triloba—Pawpaw
Bauhinia spp.—Orchid Tree*
Beaucarnea recurvata—Ponytail*
Betula ermanii—Erman Birch
Betula utilis var. *jacquemantii*—Jacquemantii Birch
Betula nigra—River Birch
Betula populifolia—Gray Birch
Bumelia lanuginosa—Chittamwood, Gum Bumelia, Gum Elastic Buckthorn
Caesalpinia pulcherrima—Dwarf Poinciana, Barbados Flowerfence*
Calliandra haematocephala—Powderpuff*
Callistemon citrinus—Red Bottlebrush, Lemon Bottlebrush
Callistemon viminalis—Weeping Bottlebrush
Camellia oleifera—Tea-Oil Camelia*
Carpinus caroliniana—American Hornbeam, Blue Beech, Ironwood
Caryota mitis—Fishtail Palm*
Cassia suratensis—Glaucous Cassia*
Castanea mollissima—Chinese Chestnut*
Cercidiphyllum japonicum—Katsuratree
Cercis canadensis—Eastern Redbud
Cercis mexicana—Mexican Redbud
Cercis occidentalis—Western Redbud, California Redbud
Cercis reniformis 'Oklahoma'—Oklahoma Redbud
Chilopsis linearis—Desert Willow*
Chionanthus retusus—Chinese Fringe Tree
Chionanthus virginicus—Fringe Tree, Old-Man's Beard
x Chitalpa tashkentensis—Chitalpa*
Chrysalidocarpus lutescens—Yellow Butterfly Palm, Bamboo Palm, Areca Palm*
Conocarpus erectus—Buttonwood
Conocarpus erectus var. *sericeus*—Silver Buttonwood
Cordia boissieri—Wild Olive, Anacahuita*
Cordia sebestena—Geiger Tree
Cornus alternifolia—Pagoda Dogwood
Cornus controversa—Giant Dogwood
Cornus drummondii—Roughleaf Dogwood
Cornus kousa—Kousa Dogwood, Chinese Dogwood, Japanese Dogwood
Cornus mas—Cornelian Cherry*

Cotinus coggygria—Smoketree, Wig-Tree, Smokebush*
Cotinus obovatus—American Smoketree, Chittamwood
Crataegus spp.—Hawthorn, Apple Hawthorn
Crescentia cujete—Calabash Tree*
Diosphyros texana—Texas Persimmon
Elaeagnus angustifolia—Russian Olive, Oleaster*
Erythrina crista-galli—Cockspur Coral Tree*
Erythrina variegata var. *orientalis*—Coral Tree*
Eucalyptus ficifolia—Red-Flowering Gum
Eugenia spp.—Stopper, Eugenia
Evodia danielii—Korean Evodia, Bebe Tree*
Feijoa sellowiana—Guava, Feijoa, Pineapple Guava*
Franklinia alatamaha—Franklin Tree, Franklinia
Fraxinus ornus—Flowering Ash
Fraxinus texensis—Texas Ash
Guaiacum sanctum—Lignumvitae*
Halesia spp.—Silver Bell
Hamamelis mollis—Chinese Witch Hazel
Hamamelis virginiana—Witch Hazel*
Hibiscus syriacus—Rose of Sharon. Shrub-Althea*
Hydrangea paniculata—Panicle Hydrangea*
Ilex cornuta 'Burfordii'—Burford Holly*
Ilex decidua—Possumhaw*
Ilex latifolia—Lusterleaf Holly
Ilex rotunda—Round-Leaf Holly
Ilex serrata—Fine-Toothed Holly*
Ilex verticillata—Winterberry*
Ilex vomitoria—Yaupon Holly*
Ilex vomitoria 'Pendula'—Weeping Yaupon Holly
Jatropha integerrima—Peregrina, Fire-Cracker*
Juniperus ashei—Ashe Juniper. Mountain Cedar
Juniperus chinensis 'Torulosa'—'Torulosa' Juniper
Juniperus deppeana 'McFetter'—McFetter Alligator Juniper
Juniperus scopulorum—Rocky Mountain Juniper, Colorado Red Cedar
Juniperus scopulorum 'Tolleson's Green Weeping'—'Tolleson's Green Weeping' Rocky Mountain Juniper*
Laburnum alpinum—Goldenchain Tree
Lagerstroemia fauriei—Japanese Crape-Myrtle
Lagerstroemia indica—Crape-Myrtle
Lagerstroemia speciosa—Queen's Crape-Myrtle

Leptospermum laevigatum—Australian Tea Tree*
Leucaena retusa—Goldenball Leadtree, Littleleaf Leucaena, Littleleaf Leadtree*
Ligustrum japonicum—Japanese Privet, Wax-Leaf Privet*
Ligustrum lucidum—Glossy Privet, Tree Ligustrum
Lithocarpus henryi—Henry Tanbark Oak
Maackia amurensis—Amur Maackia
Maclura pomifera—Osage Orange. Bois D'arc
Magnolia cordata—Yellow Magnolia
Magnolia denudata—Yullan Magnolia
Magnolia kobus—Kobus Magnolia
Magnolia stellata—Star Magnolia*
Magnolia x soulangiana—Saucer Magnolia*
Malpighia glabra—Barbados Cherry*
Malus spp.—Crab Apple
Melaleuca quinquinervia—Punk Tree, Melaleuca
Murraya paniculata—Orange Jessamine*
Musa spp.—Banana*
Myrica cerifera—Southern Waxmyrtle, Southern Bayberry*
Olea europaea—European Olive
Osmanthus americanus—Devilwood, Wild Olive
Osmanthus fragrans—Sweet Osmanthus
Osmanthus x fortunei—Fortune's Osmanthus*
Ostrya virginiana—American Hophornbeam, Eastern Hophornbeam
Parkinsonia aculeata—Jerusalem Thorn*
Parrotia persica—Persian Parrotia
Phoenix reclinata—Senegal Date Palm
Photinia glabra—Red-Leaf Photinia, Red-Top
Photinia serrulata—Chinese Photinia*
Photinia villosa—Oriental Photinia*
Photinia x fraseri—Fraser Photinia
Pinus bungeana—Lacebark Pine
Pinus cembra—'Nana'—Dwarf Swiss Stone Pine
Pinus cembroides—Mexican Pinyon, Pinyon Pine
Pinus densiflora—Japanese Red Pine
Pinus densiflora 'Umbraculifera'—'Umbraculifera' Japanese Red Pine*
Pinus mugo—Mugo Pine, Swiss Mountain Pine
Pinus strobus 'Nana'—'Nana' Eastern White Pine
Pinus sylvestris 'Watereri'—'Watereri' Scotch Pine*
Pinus thunbergiana—Japanese Black Pine*
Pithecellobium flexicaule—Ebony Blackbead, Texas Ebony

Pittosporum spp.—Pittosporum
Plumeria alba—White Frangipani
Plumeria rubra—Frangipani
Podocarpus falcatus—Podocarpus
Podocarpus gracilior—Weeping Podocarpus, Fern Podocarpus
Podocarpus latifolius—Podocarpus
Podocarpus macrophyllus—Podocarpus, Yew Pine, Japanese Yew
Podocarpus nagi—Nagi Podocarpus, Broadleaf Podocarpus
Prosopis glandulosa—Mesquite, Honey Mesquite
Prunus spp.—Cherry
Pseudocydonia sinensis—Chinese Quince
Psidium littorale—Cattley Guava, Strawberry Guava
Quercus glauca—Blue Japanese Oak, Ring-Cupped Oak
Ravenala madagascariensis—Traveler's Tree*
Rhamnus caroliniana—Carolina Buckthorn*
Rhus chinensis—Chinese Sumac
Rhus glabra—Smooth Sumac
Rhus lancea—African Sumac
Robinia pseudoacacia 'Tortuosa'—'Tortuosa' Black Locust
Robinia pseudoacacia 'Umbraculifera'—Umbrella Black Locust
Sambucus canadensis—American Elder, Common Elder*
Sambucus mexicana—Mexican Elder
Sassafras albidum—Sassafras
Schefflera arboricola—Dwarf Schefflera
Sophora secundiflora—Texas Mountain Laurel, Mescalbean
Stewartia koreana—Korean Stewartia
Stewartia monadelpha—Tall Stewartia
Stewartia pseudocamellia—Japanese Stewartia
Strelitzia nicolai—White Bird-of-Paradise, Giant Bird-of-Paradise
Styrax japonicus—Japanese Snowbell
Styrax obassia—Fragrant Snowbell
Syringa pekinensis—Peking Lilac
Syringa reticulata—Japanese Tree Lilac
Taxus baccata—English Yew*
Tecoma stans—Yellow Elder, Yellow Trumpet-Flower*
Tristania conferta—Brisbane Box
Tsuga canadensis 'Sargentii'—Weeping Canadian Hemlock*
Tsuga canadensis 'Lewisii'—Lewis Hemlock*
Viburnum odoratissimum—Sweet Viburnum*
Viburnum odoratissimum var. *awabuki*—Awabuki Sweet Viburnum*

Viburnum plicatum f. 'Tomentosa'—Doublefile Viburnum*
Viburnum rufidulum—Rusty Blackhaw, Southern Blackhaw*
Viburnum sieboldii—Siebold Viburnum
Vitex agnus-castus—Chastetree, Vitex*
Vitex agnus-castus 'Alba'—'Alba' Chastetree, 'Alba'* Vitex
Xanthoceras sorbifolium—Yellowhorn*
Yucca elephantipes—Spineless Yucca, Soft-Tip Yucca*

APPENDIX 3

TREES SUITED FOR TRAINING INTO A STANDARD

Acer campestre—Hedge Maple
Acer japonicum—Fullmoon Maple
Acer palmatum—Japanese Maple
Acer triflorum—Three-Flowered Maple
Acer truncatum—Purpleblow Maple
Brugmansia spp.—Trumpet Tree
Caesalpinia granadillo—Bridalveil Tree
Calliandra haematocephala—Powderpuff
Calliandra surinamensis—Pink Powderpuff
Camellia oleifera—Tea-Oil Camellia
Canella winterana—Winter Cinnamon
Caragana arborescens—Pea-Shrub
Cassia alata—Candlebrush
Chrysobalanus icaco—Cocoplum
Citrus spp.—Citrus
Clerodendron trichotomum—Harlequin Glorybower
Clusia rosea—Pitch-Apple
Cordia boissieri—Wild-Olive
Cotinus coggygria—Smoketree
Cotinus obovatus—American Smoketree
Eriobotrya deflexa—Bronze Loquat

Eriobotrya japonica—Loquat
Eugenia spp.—Stopper
Eugenia rhombea—Red Stopper
Eugenia uniflora—Surinam Cherry
Ficus benjamina—Weeping Fig
Ficus elastica—Rubber Tree
Hibiscus syriacus—Rose of Sharon
Hydrangea paniculata—Panicle Hydrangea
Jatropha integerrima—Peregrina
Lagerstroemia—Crape-Myrtle
Ligustrum japonicum—Japanese Privet
Ligustrum lucidum—Tree Ligustrum
Malus—Crabapple
Myrcianthes fragrans—Simpson's Stopper
Myrica cerifera—Southern Waxmyrtle
Myrica pennsylvanica—Waxmyrtle
Myrsine guianensis—Myrsine
Nerium oleander—Oleander
Photinia spp.—Photinia
Podocarpus gracilior—Weeping Podocarpus
Podocarpus macrophyllus—Podocarpus
Prunus caroliniana—Cherry-Laurel
Prunus serrulata 'Kwanzan'—Kwanzan Cherry
Prunus triloba var. *multiplex*—Flowering Almond
Raphiolepis umbellata—Round-Leaf Hawthorn
Rhamnus caroliniana—Carolina Buckthorn
Sambucus canadensis—American Elder
Schefflera arboricola—Dwarf Schefflera
Senna spectabilis—Cassion
Styrax japonicus—Japanese Snowbell
Syringa reticulata—Japanese Tree Lilac
Taxus baccata—English Yew
Tibouchina granulosa—Purple Glory Tree
Tibouchina urvilleana—Princess-Flower
Vitex agnus-castus—Chastetree

APPENDIX 4

PLANTS SUITED FOR ESPALIER TRAINING

Arbutus unedo—Strawberry Tree
Bougainvillea spp.—Bougainvillea
Callianda haematocephala—Powderpuff
Calliandra surinamensis—Pink Powderpuff
Callistemon citrinus—Bottlebrush
Calophyllum brasiliense—Santa Maria
Calophyllum inophyllum—Beautyleaf
Camellia spp.—
Canella winterana—Winter Cinnamon
Carpinus betulus 'Fastigiata'—Upright European Hornbeam
Carpinus caroliniana spp.—Hornbeam
Cedrus spp.—Cedar
Citrus spp.—Citrus
Clerodendron trichotomum—Harlequin Glorybower
Clusia rosea—Pitch-Apple
Crataegus spp.—Hawthorn
Dodonaea viscosa—Varnish-Leaf
Duranta erecta—Golden Dewdrop
Eriobotrya deflexa—Bronze Loquat
Eriobotrya japonica—Loquat
Ficus spp.—Fig

Forestiera segregata—Florida Privet
Forsythis spp.—Forsythia
Gordonia lasianthus—Loblolly-Bay
Ilex x attenuata—Holly
Ilex cornuta 'Burfordii'—Burford Holly
Ilex latifolia—Lusterleaf Holly
Ilex vomitoria—Yaupon Holly
Juniperus chinensis 'Torulosa'—Torulosa Juniper
Laburnum spp.—Goldenchain Tree
Magnolia denudata—Yulan Magnolia
Magnolia grandiflora—Southern Magnolia
Magnolia kobus var. *loebneri* 'Merrill'—Merrill Magnolia
Magnolia virginiana—Sweetbay Magnolia
Magnolia hybrids—Magnolia
Magnolia x soulangiana—Japanese Magnolia
Malus spp.—Crabapple
Noronhia emarginata—Madagascar Olive
Ochrosia elliptica—Ochrosia
Podocarpus gracilior—Weeping Podocarpus
Podocarpus macrophyllus—Podocarpus
Prunus spp.—Cherry, Plum
Psidium littorale—Guava
Pyracantha spp.—Pyracantha
Pyrus spp.—Pear
Schefflera arboricola—Dwarf Schefflera
Stewartia koreana—Korean Stewartia
Stewartia malacodendron—Silky-Camellia
Stewartia monadelpha—Tall Stewartia
Stewartia pseudocamellia—Japanese Stewartia
Tecoma stans—Yellow-Elder
Tibouchina urvilleana—Princess-Flower
Viburnum obovatum—Blackhaw

APPENDIX 5

SAMPLE PRUNING SPECIFICATIONS

SAMPLE PRUNING SPECIFICATIONS

Note: The word *shall* indicates a practice that is mandatory. The word *should* refers to a practice that is highly recommended.

Objectives

Mature trees shall be thinned to allow more light to reach the ground under the tree. The foliage removed shall be taken primarily from the outer edge of the canopy, not from the interior. No more than 20 percent of the foliage shall be removed from any tree without authorization from our office.

Some water sprouts shall be removed from selected trees in order to space remaining sprouts along main limbs.

Dead branches greater than 1.5 inches in diameter (measured at the base of the branch) shall be removed from the canopy of all trees that are pruned.

Branches considered at risk or potentially hazardous shall be reduced, removed, or cabled.

Tree canopies that were cold damaged or topped several years ago shall be restoration pruned in order to improve their structure and form.

Young trees shall be pruned primarily with reduction cuts, with the intention of developing a dominant trunk.

Pruning Techniques

When removing a live branch at its point of origin on the trunk or from a parent branch, the final pruning cut shall be made in branch tissue just outside the branch bark ridge and collar. No stubs shall be left. (A stub is the remaining branch tissue to the outside of the collar and branch bark ridge.)

When removing a dead branch, the final cut shall be made outside the collar of the living woundwood tissue. If the collar has grown out along the dead branch stub, only the dead stub shall be removed; the living collar shall remain intact and uninjured regardless of its length.

When reducing the length of a branch or the height of a leader (making a reduction cut), the final cut shall be made just beyond the branch bark ridge. The cut shall approximately bisect the angle formed by the branch bark ridge and an imaginary line perpendicular to the trunk or branch being cut. Reduction cuts shall be made back to a branch with a diameter at least ⅓ the diameter of the branch or leader cut.

To prevent damage to the parent limb when removing a branch with a narrow branch attachment, the final cut shall begin at the bottom of the branch.

Pruning cuts shall be smooth and clean, leaving the bark at the edge of the cut firmly attached to the wood. Final pruning cuts should be approximately circular in cross section on small branches.

Thinning shall be conducted by removing branches from the parent branch, especially toward the end of the parent branch. "Lions-tailing" shall not be performed. (Lions-tailing is the practice of removing only the inner branches closest to the trunk on a parent branch and leaving the branches located toward the end of the parent branch.)

Removal and reduction cuts shall be used, and not heading cuts. A removal cut shall consist of the removal of a lateral branch at its point of origin. A reduction cut is a cut that shortens a branch or stem by cutting it back to a lateral branch at least ⅓ the size of the portion being cut. A heading cut consists of cutting a branch back to a stub or lateral branch less than ⅓ the size of the branch cut.

Limbs to be cut that are too heavy to support with one hand shall be precut to avoid splitting or tearing of the bark on the trunk or parent branch.

Large branches that cannot be thrown clear should be lowered via ropes to prevent injury to the tree, other plants, and property.

Cut limbs shall either be removed from the canopy upon completion of the pruning or periodically at times when the tree would be left unattended or at the end of the workday.

Not more than 10 percent of the live foliage of mature trees shall be removed.

Neither wound dressings nor tree paints should be applied to any pruning cuts.

Tools and Equipment

Climbing spurs shall not be used when climbing trees except to climb a tree to be removed or to perform an aerial rescue of an injured worker.

Equipment and work practices that damage bark or cambium should be avoided.

Rope injury to thin-bark trees from loading out heavy limbs should be avoided by installing rigging in the tree to carry the load. Avoid injury to the crotches from the climber's line.

General

Preference will be given to firms that have at least one certified arborist on their staff. Certification is obtained through the International Society of Arboriculture (Champaign, IL).

Violation of these specifications can result in termination of the contract without payment.

All debris shall be removed from the site at the completion of the job. Lawns shall be raked clean, handscape shall be swept clean.

All work shall be performed in accordance with the latest edition of the ANSI A300 pruning standard and the ANSI Z133.1 safety standard.

SAMPLE PRUNING SPECIFICATIONS FOR MEDIUM-AGED AND MATURE TREES

(Following some minor wording changes, you can use this on residential and commercial jobs to help ensure that the customer knows what work will be performed on the trees.)

The word "shall" indicates a practice that is mandatory. The word "should" refers to a practice that is highly recommended. Please indicate in writting to the owner why any "should" practices will not be carried out. We will notify the pruning contractor in writing if these recommendations can be waived.

Objectives

Reduce potential hazardous or "at risk" conditions in trees and improve tree structure by:

1) removing dead branches.
2) reducing the weight toward the ends of branches or stems with included bark, cracks, decay, or other serious defects.
3) reducing the weight toward the ends of all but one codominant stem.
4) thinning portions of the canopy and restoring topped trees.

General Procedures

Live branches less than 1.5 inch diameter should not be removed.

Dead branches greater than 1.5 inch diameter measured at the base of the branch shall be removed from the canopy of all trees.

No live branches or stems greater than 3 inches diameter should be removed from the tree without authorization from owner or owner's agent. In other words, these should be no pruning cuts greater than 3 inches diameter.

Remove no more than 20 percent (use your own number here, depending on species and age) of live foliage from the tree unless indicated in the following specific procedures.

Live crown ratio should be at least 60 percent when pruning is completed meaning that no more than the lower 40 percent of the tree shall be clear of branches.

Specific Procedures

1) Weight on main scaffold limbs with included bark or other defects shall be reduced by approximately one-third by removing some tertiary and perhaps secondary branches toward the ends of the limbs and/or by removing the end of the branch using a reduction cut.
2) If a medium-aged large-maturing tree (less than 24 inches trunk diameter) divides into two or more codominant leaders of about equal size in the bottom two-thirds of the tree, reduce the end weight by approximately one-third (you can adjust this percentage depending on the size hole in the canopy the customer is willing to tolerate) using reduction and removal cuts on all stems except the one that you believe could become the strongest and most dominant leader. To accomplish this, remove secondary and tertiary branches growing upright and toward the center and some of those at the edge of the canopy, and reduce stem length with reduction cuts. Use mostly removal cuts, not reduction cuts, on larger trees. (Note: On some trees, you may not be able to perform all of this because you cannot remove more than 20 percent of the foliage. Make a note of this on the site map.)
3) Evaluate trees for possible cabling or large branch removal. Cabling or large branch removal may be performed under separate contract or as a modification of this contract. Identify subordinated limbs and trunks with included bark, vertical cracks, cankers, or cavities. These may be cabled later.
4) If less than 20 percent of the foliage was removed on a mature tree following procedures 1 and 2, thin the denser portions of the canopy to allow more light to reach the ground under the tree and to reduce damage from wind storms. The foliage removed shall be taken primarily from the outer edge of the canopy, not from the interior. This should be accomplished by removing tertiary, quadernary or quinary branches. Interior branches shall be left on the tree. Do not remove water sprouts from the interior of the tree.
5) Crowns of trees that were cold-damaged or topped will be pruned to improve structure and form. Remove half of the sprouts and shorten the remaining half except one, which will become the dominant stem at that point. You may remove up to 25 percent of the foliage when performing this work.

Pruning techniques

Pruning cuts shall be in accordance with ANSI A300 pruning standards (latest edition).

Tools and equipment

Climbing spurs shall not be used when climbing trees, except to climb a tree to be removed or to perform an aerial rescue of an injured worker.

Equipment and work practices that damage bark or cambium should be avoided.

Rope injury from loading out heavy limbs should be avoided.

General

Preference will be given to firms who have at least one certified arborist on their staff. Certification is through the International Society of Arboriculture, Champaign, IL. An ISA certified arborist shall be on site at least once each day.

Violation of these procedures and techniques could result in termination of your contract without payment.

All debris shall be removed from the site at the completion of the job.

Safety

All work shall be performed by workers trained in accordance with ANSI Z133.1 safety guidelines as required by OSHA.

Areas of inclusion

(Provide a brief description of the trees to be pruned so there can be no confusion. A map is often very helpful.)

Exclusions

(Provide a brief description of the trees and large shrubs that are not included in the bid.)

Additional Requirements

All debris and equipment shall be removed from the site by the end of each workday. Lawn areas will be raked and landscape swept at the end of the job.

The selected contractor shall be required to furnish a certificate of insurance to include liability, automotive, and worker's compensation before commencing work.

Note to user of this sample specification: With these detailed specifications, all firms will be bidding on the same work, and with time, bids will be more competitive and closer to each other in price. These specifications will also scare away the underqualified firms that should not be working on these trees anyway. Sentences within parentheses are notes to you, the reader. You will need to construct the text to be included here.

SAMPLE PRUNING SPECIFICATIONS FOR YOUNG SHADE TREES

C. Way Hoyt, Tree Trimmers and Associates, Ft. Lauderdale, Florida
Edward F. Gilman, Professor, Environmental Horticulture Department, University of Florida

The word "shall" indicates a practice that is mandatory. The word "should" refers to a practice that is highly recommended.

General

The primary purpose of pruning our trees is to improve trunk and branch structure, remove or shorten low limbs for underclearance, thin the canopy to allow better air flow, and to maintain health. Although we are concerned with aesthetics, the appearance of the trees will be secondary to health and structural concerns. The single greatest structural concern is the large number of codominant trunks or main leaders.

The specified pruning may require the removal of up to 30 percent of the foliage in many instances. Under no circumstances shall more than 40 percent of the foliage from an individual tree be removed.

Other than when shortening limbs for clearance over roadways or sidewalks, removal of live limbs smaller than 1 inch in diameter on the interior of the canopy will not be required. In fact, this is discouraged and is considered unnecessary. Removal of dead limbs or stubs 1 inch in diameter or larger will be required.

All pruning is to be done within the scope of the approved techniques as described in the most recent edition of ANSI A300. Work is to be done by workers trained in compliance with ANSI Z133.1 safety guidelines, as required by OSHA.

Codominant leaders and stems

Due to the recognized potential hazards associated with codominant leaders, the subordination (shortening using a reduction cut) or removal of one side of a codominant leader is the primary objective. Branches, trunks, or leaders not considered the main leader, 2 inches (note: use your own number here) diameter or larger should be subordinated or removed (see Figure 8-9). The main leader shall not be subordinated, headed, or removed. Codominant leaders are considered to be two or more branches, trunks, or leaders of approximately the same size, originating in close proximity to one another. If there is no stem considerably larger than others, subordinate all but one stem. Where there is included bark as part of the condition, preference should be given to the removal of one side, but only if such removal will not destroy the aesthetic value of the canopy or remove more than 40 percent of the foliage.

Canopy raising or lifting

Branches over paved areas should be shortened (with a reduction cut back to a living side branch at least one-third the diameter of the removed portion) or removed to allow approximately 7 or 8 feet (8′) of clearance for cars and delivery vans as practical. Preference shall be given to shortening instead of removing. Over landscape areas and sidewalks they should also be shortened or removed to allow 8 feet for pedestrian traffic and utility use as practical. Shortening of branches is the preferred method for attaining adequate clearance. When pruning is completed, approximately one-half of the foliage should originate from branches on the lower two-thirds of each tree. Live crown ratio should be at least 60 percent when pruning is completed meaning that no more than the lower 40 percent of the tree shall be clear of branches.

Canopy cleaning

Although small-diameter limbs may occasionally be pruned to gain access into the tree, it will not be necessary to make cuts smaller than 1 inch in diameter, other than where branches may be shortened to accommodate clearance beneath the canopy. Do not strip out the interior foliage leaving only live branches at the ends of branches.

Canopy cleaning is to include the following: (1) remove dead or broken limbs 1 inch in diameter or larger; (2) if two limbs are crossing or touch each other, shorten or remove one of them; (3) if two limbs (1 inch diameter or larger) originate within 12 inches of each other on the trunk, shorten or remove one of them. Clearance from buildings, lights, or other structures should be a minimum of 3 feet or as practical. Use directional pruning where possible so future growth is directed away from buildings and lights.

Palms

All large-growing palms, should be pruned to remove dead fronds, and fronds with a petiole that droops below horizontal. Dead fronds are those with less than 50 percent green tissue. Only those live fronds with petioles drooping below horizontal (9:00–3:00) should be removed. All seedpods should also be removed including those originating among remaining fronds. When removing fronds and seedpods, care shall be taken so those fronds that are to remain are not nicked or wounded.

APPENDIX 6

A PLAN FOR TRAINING SHADE TREES IN THE URBAN LANDSCAPE

Trees are like children; they require about 25 years of training to create good, solid structure that will last them a lifetime.

FIRST THINGS FIRST

Establish a pruning cycle and objectives. Pruning cycle depends on quality of nursery stock, growth rate, climate, and species. Then decide on strategies to meet objectives.

Objectives:

(25yrs)

1) Establish and maintain a dominant leader—subordinate all but one codominant stem.
2) Space main scaffold limbs apart—remove or shorten nearby branches.
3) Anticipate future form and function—train and prune early to avoid cutting large branches later; removing large branches can initiate decay in the trunk (i.e., instead of allowing a low branch to get large then removing it when it droops, anticipate this by shortening it earlier).
4) Position the lowest main scaffold limb high enough so it will not droop and have to be removed later.
5) Keep all branches less than half the trunk diameter.

Strategies:

At planting

- If structure is good with one dominant leader, do not prune.
- All branches will eventually be removed on trees less than 4 inches caliper.
- Do not remove more than about 25 percent of live foliage.
- Shorten or remove leaders and branches competing with the main leader.
- If there is no dominant leader, create one by cutting back all leaders except one.

Two or three years after planting

- All branches will eventually be removed on trees less than 4 inches caliper.
- Do not remove more than 35 percent of live foliage.
- Shorten or remove all competing leaders (may have to do in two stages if >3).
- Shorten or remove large, low branches to improve clearance.
- Shorten or remove branches within 12 inches of largest diameter branches in top half of trees greater than about 4 inches caliper.

Five years after planting

- Most branches are still temporary and will eventually be removed from the tree.
- Do not remove more than 35 percent of live foliage.
- Shorten or remove competing leaders.
- Shorten or remove large, low branches to improve clearance.
- Shorten or remove branches within 12 inches of largest diameter branches in top half of tree.
- Shorten aggressive branches growing from the bottom half of the canopy that have reached into the top third of the tree.
- There should be only one large branch per node (no clustered branches); shorten those nearby so only one is present.

Ten years after planting

- Shorten or remove competing leaders.
- Do not remove more than 25 to 35 percent of foliage.
- Determine where you want the lowest permanent scaffold limb and shorten all aggressive branches lower than this limb.
- Shorten branches within 12 to 18 inches of largest diameter branches (there should be only one large branch per node (no clustered branches).
- Shorten aggressive branches growing from the bottom half of the canopy that have reached into the top third of the tree.
- Shorten low branches that will have to be removed later.

Fifteen years after planting

- Shorten or remove competing leaders.
- Identify several permanent scaffold limbs.

- Shorten aggressive branches within 18 to 36 inches of permanent scaffold limbs.
- There should be only one large branch per node (no clustered branches).
- Shorten or remove large branches lower (on the trunk) than the first permanent scaffold branch.
- Shorten aggressive branches growing from the bottom half of the canopy that have reached into the top third of the tree.
- Shorten low branches that will have to be removed later.

Twenty years after planting
- Shorten or remove competing leaders.
- Identify five to ten permanent scaffold limbs.
- Shorten aggressive branches within 18 to 36 inches of permanent scaffold limbs.
- There should be only one large branch per node (no clustered branches).
- Shorten or remove large branches lower (on the trunk) than the first permanent scaffold branch.
- Shorten aggressive branches growing from the bottom half of the canopy that have reached into the top third of the tree.
- Shorten low branches that will have to be removed later.

Twenty-five years after planting
- Shorten or remove competing leaders.
- Continue to develop and space permanent scaffold limbs.
- Shorten branches within 36 inches of permanent scaffold limbs.
- There should be only one large branch per node (no clustered branches).
- Shorten or remove large branches lower (on the trunk) than the first permanent branch.
- Shorten aggressive branches growing from the bottom half of the canopy that have reached into the top third of the tree unless they are permanent scaffold limbs.
- Shorten low branches that will have to be removed later.

With seven prunings in the first twenty-five years after planting, a good structure can be developed that can place the tree on the road to becoming a permanent fixture in the landscape. Less-frequent pruning may be needed if good-quality nursery trees were planted with a dominant leader, and trees were irrigated appropriately until established.

APPENDIX 7

NURSERY PRODUCTION PROTOCOL FOR UPRIGHT TREES

APPENDIX FIGURE 7A (Opposite). In the first year, the trunks on liners can be staked when planted or sometime in the first year to make them grow straight. Stakes are not shown. Prune as little from the tree as is needed to meet objectives for the most rapid growth. Do not shorten or remove *any* branches or stems early in the year unless they compete with the leader. All branches on the tree now are temporary and will be removed by the time the tree leaves the nursery. Shorten temporary branches if their tips are above the lowest part of the permanent nursery canopy.

In year two, many stems grow upright parallel to the leader. The lower portion of the permanent nursery canopy begins to appear. Later in the year, begin to shape the permanent nursery canopy into a tear drop or rounded cone using heading or reduction cuts on branches. The lowest branches in the permanent nursery canopy will be left the longest. Maintain low temporary branches to increase caliper and to moderate canopy growth rate. Cut back any that grow up into the permanent nursery canopy (eight temporary branches were cut back in summer because they were up in the permanent canopy—see Appendix 7A, arrow center left). A small number of temporary branches may be removed at the end of the year if the tree will be marketed at the end of year three. Remove those that grow larger than half the trunk diameter or larger than half an inch diameter. Make reduction cuts in the permanent nursery canopy on main branches back to more horizontal branches to increase canopy spread if desired.

In the third year, use reduction cuts to reduce length of stems and branches shading the sides of and competing with the leader. Shape the tree by making reduction cuts and perhaps some heading cuts (heading cuts are less desirable but some are acceptable) toward the edge of the canopy. Lower temporary branches should be removed in time for most wounds to close over before trees are delivered to the customer.

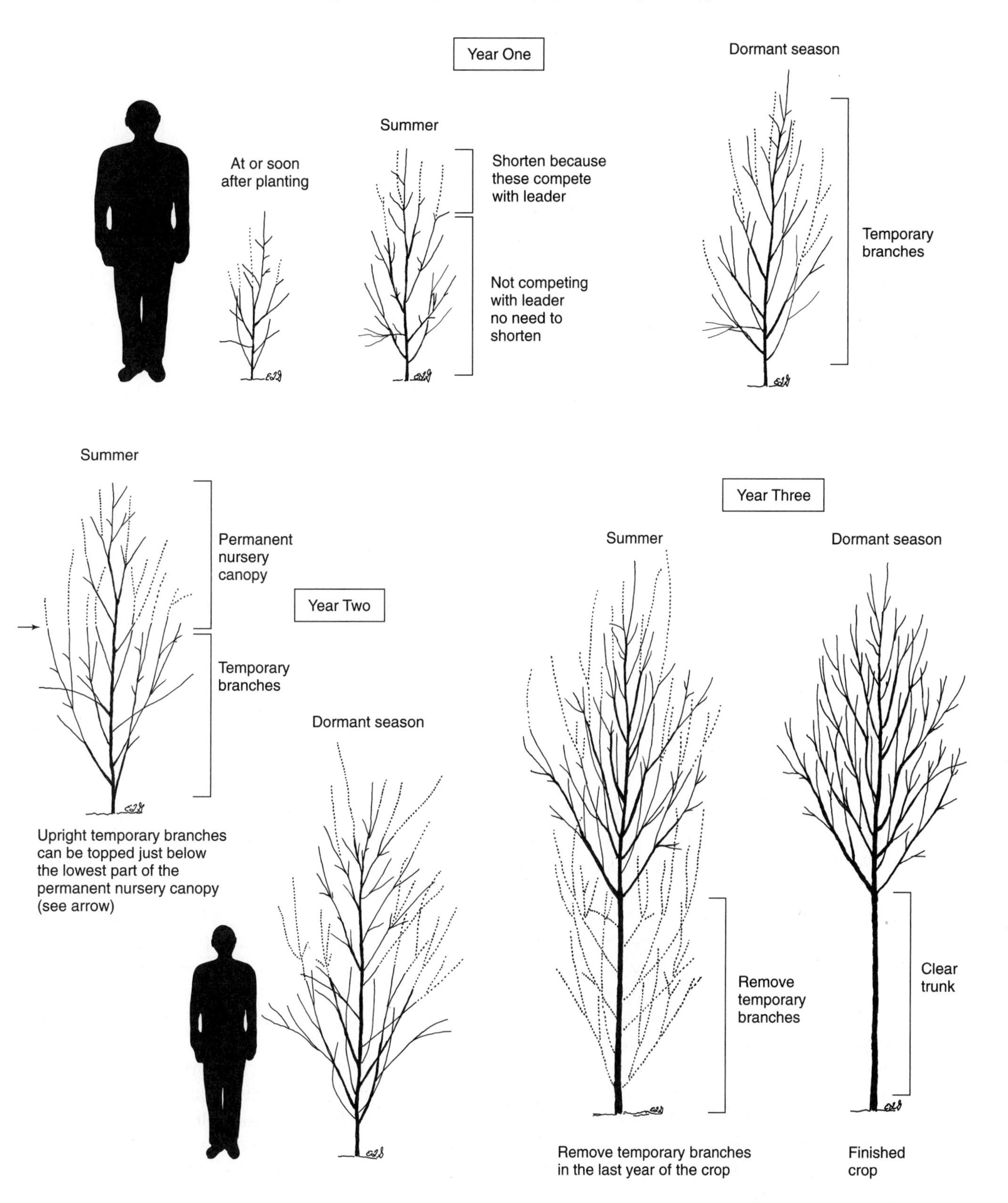
Nursery production protocol for upright trees (3-year crop)
Year One
At or soon after planting
Summer
Shorten because these compete with leader
Not competing with leader no need to shorten
Dormant season
Temporary branches
Summer
Permanent nursery canopy
Temporary branches
Year Two
Upright temporary branches can be topped just below the lowest part of the permanent nursery canopy (see arrow)
Dormant season
Year Three
Summer
Dormant season
Remove temporary branches
Clear trunk
Remove temporary branches in the last year of the crop
Finished crop

APPENDIX FIGURE 7B (Opposite). Protocol for the third and fourth years for producing a 4-year crop for an upright tree form. The first two years of pruning are identical for producing a 3-inch or 4-inch crop (Appendix 7A). In the third year of a 4-year crop, cut back upright temporary branches substantially. If they are allowed to grow, they will shade out portions of the permanent canopy spoiling its shape when temporary branches are removed. They could also slow the rate of canopy spread if allowed to grow too aggressively. Reduce the length of stems and branches in the permanent nursery canopy competing with the leader using reduction cuts. If allowed to grow, they will shade the sides of the leader forcing it to grow too tall with few lateral branches. Competing stems will also develop into codominant stems that are undesirable. The largest diameter temporary branches can be removed back to the trunk in the dormant season. Splint the leader if needed (splint not shown).

In the fourth year of a 4–year crop, the leader should be straight enough so it does not need to be splinted any more. Reduce the length of competing leaders early in the season. Shape the canopy to the desired form with heading and reduction cuts through one year old twigs—this is not the time to make major changes in canopy shape. Remove half of the temporary branches early in the season and the rest in time so most wounds are closed when the tree is delivery to the customer. Adjust the timing of temporary branch removal so wounds close in time for delivery.

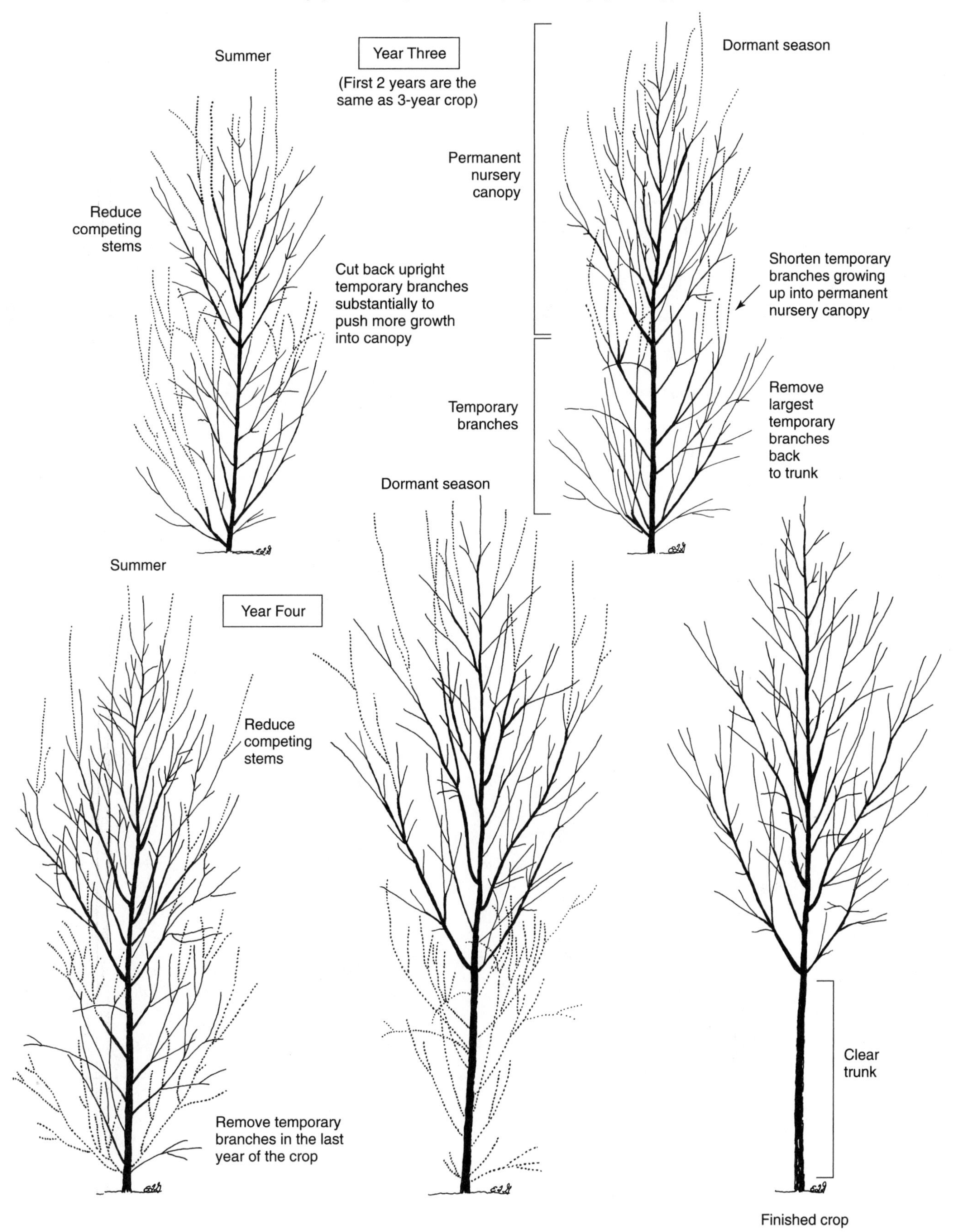
Nursery production protocol for upright trees (4-year crop)
Year Three
(First 2 years are the same as 3-year crop)
Summer
Reduce competing stems
Cut back upright temporary branches substantially to push more growth into canopy
Permanent nursery canopy
Temporary branches
Dormant season
Shorten temporary branches growing up into permanent nursery canopy
Remove largest temporary branches back to trunk
Year Four
Summer
Reduce competing stems
Remove temporary branches in the last year of the crop
Dormant season
Clear trunk
Finished crop

APPENDIX 8

TREES THAT OFTEN FORM MULTIPLE TRUNKS

Many medium- and large-maturing trees form multiple upright trunks and codominant stems in an open landscape unless pruned correctly. Some are listed below. Many tend to form included bark where branches meet the trunk. It is especially important to prune these trees regularly in the first 25 years after planting.

African Tulip Tree *(Spathodea)*
Ash *(Fraxinus)*
Black Olive *(Bucida)*
'Bradford' Callery Pear *(Pyrus)*
Castor-Aralia *(Kalopanax)*
Coral Tree *(Erythrina)*
Elm *(Ulmus)*
Eucalyptus *(Eucalyptus)*
Fig *(Ficus)*
Floss-Silk Tree *(Chorisia)*
Hackberry *(Celtis)*
Honeylocust *(Gleditsia)* (some)
Jacaranda *(Jacaranda)*
Lindens *(Tilia)*
Mahogany *(Sweitenia)*
Maple *(Acer)*
Mesquite *(Prosopis)*
Oak *(Quercus)*
Pongam *(Pongamia)*
Royal Poinciana *(Delonix)*
Sea Grape *(Coccoloba uvifera)*
Sugarberrys *(Celtis)*
Tamarind *(Lysiloma)*
Weeping Willow *(Salix)*
Yellow Poinciana *(Caesalpinia)*
Yellowwood *(Cladrastis)*
Zelkova *(Zelkova)*

GLOSSARY

Aerial root: A root originating from a branch or trunk that eventually touches the ground and takes root in the soil.

Aggressive branches (limbs): Fast-growing stems.

ANSI A300: The American National Standards Institute standard for pruning trees and shrubs in landscapes.

ANSI Z60.1: The American National Standards Institute standard for nursery stock.

ANSI Z133.1: The American National Standards Institute standard for safe working practices in and near trees.

Anvil pruner: A tool designed to cut by passing a sharpened blade through a twig against a metal anvil.

Apoplast: Network of open, dead conducting elements in xylem.

Arborist: A person with technical knowledge of tree care practices gained through experience and training.

Architectural pruning: Shapes and maintains trees to a specific form and size with regular pruning.

At risk: See *hazardous condition.*

Auxin: A plant growth regulator that inhibits shoot formation in high concentrations and initiates roots.

Balancing: Removes branches to redistribute weight.

Bark inclusion: See *included bark.*

Best management practice: The best available treatment, considering the benefits and drawbacks, based on current knowledge.

Bonsai: A technique combining root and shoot pruning designed to keep a plant very small.

Border tree: A tree jointly owned by adjacent property owners due to its location on the property line.

Bracing: Placing a stiff support under a low limb to prevent it from falling from the tree; securing stems together.

Branch: A stem arising from a larger stem; a subdominant or subordinate stem; the pith in true branches has no connection to the parent stem.

Branch angle: The angle formed in the union between stem and branch.

Branch arrangement: Orientation and distribution of branches along a trunk.

Branch bark ridge: A more or less commonly occurring raised area of bark tissue in the union of two branches or two stems or in the union of branch and stem.

Branch collar: A swelling at the base of a branch where it joins the trunk or larger branch resulting from overlapping trunk and branch tissue.

Branch protection zone: A thin zone of starch-rich tissue at the base of a branch where chemicals are deposited to retard the spread of discoloration and decay.

Branch stub: The part of the branch beyond the collar inadvertently left following branch removal.

Branch union: The place where two branches or stems join or where a branch meets a trunk. See *crotch.*

Bypass pruner: A tool that pushes a sharp blade through a twig past a hooked or curved metal anvil.

Caliper: Trunk diameter measured 6 inches from the ground; if caliper is greater than 4 inches, the measurement is taken at 12 inches from the ground.

Callus: Undifferentiated, meristematic tissue with little lignin formed by the cambium layer; callus can form sprouts.

Cambium: The layer of dividing meristematic cells beneath bark giving rise to xylem, phloem, and more cambium.

Canker: A depression or opening in the bark usually caused by a fungus or bacterium.

Canopy: The portion of the tree with foliage from the lowest branch to the topmost part of the tree; synonymous with crown.

Carpenter's saw: A saw designed to cut through dried lumber, not fresh wood.

Central leader: A dominant stem located more or less in the center of the canopy.

Certified arborists: An arborist who has passed an exam and receives, on a regular basis, continuing education administered by the International Society of Arboriculture (Champaign, IL).

Chain saw: A power tool designed to cut through large branches and stems.

Christmas tree: Spruces, pines, firs, Douglasfirs, cedars, Junipers, and other evergreens marketed as decorative trees for the holidays.

Clean (cleaning): Removes dead, broken, rubbing, or diseased branches and foreign objects; could also include removing or subordinating weakly-attached branches.

Clear trunk: The lower portion of a trunk lacking lateral branches on a nursery tree.

Clustered branches: Branches that are closely spaced, originating from nearly the same position on the trunk.

Codominant stem: A stem growing at about the same rate, and with nearly the same diameter, as another stem originating from the same union; often the piths are connected in the union.

Collar: See *branch collar.*

Collar cut: See *removal cut.*

Compartmentalization: The boundary-setting process that resists loss of normal wood function and resists the spread of discoloration and decay; a process that separates injured or decayed tissue from healthy tissue.

Crotch: See *branch union.*

Crotch spreading: Increasing the angle of a branch union by placing a wooden dowel or stick in the union.

Crown: See *canopy.*

Cultivar: A cultivated variety of a species typically propagated by cuttings, tissue culture, grafting, or budding.

Cultural problems: Too little or too much sunlight, water, fertilizer, air, pest infestations, or other factors resulting in poor growth.

Curtain: Creates a flat wall-like surface of foliage and twigs with regular shearing.

Decay: Degradation of tissue caused by biological organisms; the orderly breakdown of tissue resulting in strength loss.

Decurrent: Round-headed tree form; no leader to the top of the canopy in an open landscape without pruning.

Defects: Cracks, poor branch or trunk structure, included bark, and other conditions that can reduce a plant's utility or value.

Directional pruning: Guides the tree to grow in a certain direction by removing live branches from another portion of the tree.

Discoloration: Darkened tissue in the xylem resulting from the orderly response of the tree to microorganisms.

Dogleg: Typically, an S-shaped bend in a tree trunk.

Dominant leader/trunk: The one stem that grows much larger than all other stems and branches; at least ⅓ bigger than lateral branches located nearby.

Double leader: Two codominant stems originating more or less in the center of a tree and jointly assuming the role of the leader.

Drop-crotch cut: See *reduction cut.*

Drop cut: Making three cuts, beginning with an undercut, to remove a branch to prevent bark tearing.

Edge trees: Trees with access to sunlight from only one side that grow more on that side.

Eradication: Removes branches with pest infestations or disease.

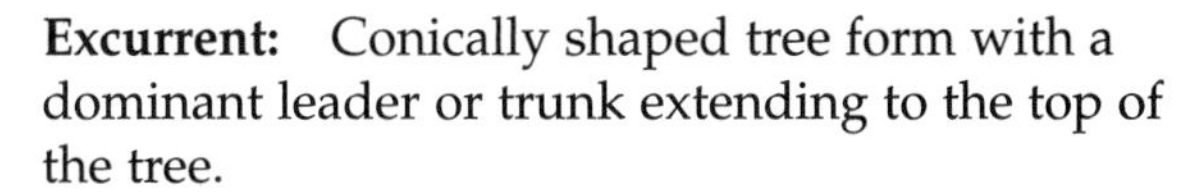

Excurrent: Conically shaped tree form with a dominant leader or trunk extending to the top of the tree.

Extension pruners: A bypass pruner with integrated extension pole.

Fail: To break or fall.

Faults: See *defects.*

Feature trees: Trees located by themselves with few other trees nearby surrounded by turf, ground cover, or shrubs.

Flush cut: A destructive pruning cut made on the trunk side of the branch bark ridge or through the collar.

Formal hedge: A shrub maintained as a sharply defined geometric shape by shearing regularly.

Frost crack: A visible vertical crack in a trunk or branch usually originating from a ring crack inside the tree; a split or crack in wood that extends out through the bark.

Good compartmentalizer: A plant with good resistance to movement of discoloration and decay.

Good-quality trees: Trees with a good trunk, branch, and root structure.

Good structure (architecture or form): Branch and trunk architecture resulting in a canopy form that resists failure.

Growth ring: A new layer of xylem produced by the secondary growth system (cambium) typically demarcated by a visible change in color.

Hand pruners: Mechanical, single-handed pruners designed to cut twigs up to about ½ inch diameter.

Hazard reduction: Reduces potentially hazardous conditions.

Hazardous condition: A condition in a tree that could result in injury to people or damage to property.

Heading cut: A type of pruning cut that prunes a shoot no more than 2 years old back to a bud; cutting an older stem back to a lateral branch less than ⅓ the diameter of the cut stem; cutting a stem to an indiscriminate length.

Healing: A physiological, regenerative process not known to occur in plants.

Heartwood: Xylem in the center of a trunk or branch that receives deposits from other portions of the tree; wood lacking living cells.

Hedging shear: A two-handed mechanical or power tool designed to cut many shoots at once.

Included bark: Bark pinched or embedded between two stems or between a branch and trunk preventing formation of a branch bark ridge; an indication of a weak union; a crack in the union.

Informal hedge: A shrub maintained by making heading or reduction cuts only on the longest shoots, 6 to 18 inches back inside the outer edge of the hedge.

Large caliper trees: Nursery trees greater than about 4 inches trunk diameter.

Large-maturing trees: Trees that grow to a height or spread greater than about 40 feet.

Large wound: A wound that can lead to defects.

Latent bud: A suppressed bud lying just beneath bark, capable of forming a shoot, that grows enough each year to stay even with the bark.

Lateral branch: A stem arising from a larger stem.

Lateral pruning (cut): See *reduction.*

Leader: A stem that dominates a portion of the canopy by suppressing lateral branches.

Leader training process: The technique that leads to development of one leader.

Limb: A large branch that is among the biggest on a tree.

Liners: Young seedlings planted in a container or field nursery for growing on to landscape sized trees.

Lions-tailing: The improper practice of removing all of most secondary and tertiary branches from the interior portion of the canopy leaving most live foliage at the edge of the canopy.

Live crown ratio: The ratio of the top portion of the tree baring live foliage to the cleared lower portion, that includes the trunk, without live foliage.

Lopper: A tool best suited for cutting branches once they have been removed from a tree; a tool with two long handles used to cut stems on shrubs up to an inch diameter.

Lopping: A term used to describe topping.

Lowest permanent limb: The lowest large branch or scaffold limb that will remain on the tree for a long time.

Main branches: Those that are the largest several on the tree. See also *scaffold limbs.*

Major limbs: See *scaffold limbs.*

Matching trees: A set of trees of the same species or cultivar with like sizes and shapes.

Mature trees: Trees that have reached at least 75 percent of their final height and spread.

Maximum critical diameter: The largest diameter pruning cut you are willing to make on a certain species.

Medium-aged shade trees: Trees more than about 15 to 20 years old that are not yet mature.

Modified central leader: A system of training small maturing trees to a single, short trunk with five to eight scaffold limbs.

Mop top: Trees that will grow as a ground cover or sprawling or mounded shrub if not pruned initially to an upright trunk; plants trained with many weeping branches on top of one straight trunk.

Multiple leaders (trunks): A group of two or more leaders or trunks with a similar diameter.

Natural tree form: The form that develops in the tree's native habitat without disturbance from human activities.

Neglected tree: A poorly formed tree that has not been pruned for some time, or that has never been structurally pruned.

Node: The point on a stem where a leaf and bud emerge. Branches emerge from nodes.

Open-center system (open-vase): A training technique used on fruit trees that allows sunlight to reach developing fruit from above.

Open landscape: An area with few trees within a few dozen feet.

Ornamental tree: Those that never reach a large size.

Over-mature trees: Trees that have reached their final height and spread that are declining in vigor.

Overthinning: Removal of too much foliage typically from the interior portion or lower portion of the canopy.

Parent branch (or parent stem): A main branch or stem from which smaller lateral branches arise.

Permanent branches (permanent limbs): Those that will remain on the tree for many years, perhaps until maturity.

Permanent canopy: That portion of the tree that will remain for a long time.

Permanent nursery canopy: The portion of a nursery tree canopy that will be present when the tree is sold.

Phloem: Living cells located just outside the cambium that move sugars and other components about the plant.

Photosynthesis: The process that turns light energy into chemical energy in green plants.

Pinching: The equivalent of heading performed on a soft young stem.

Pith: The center, typically soft portion of a branch or stem that forms the first year that lacks living tissue.

Pleaching: Intertwines branches and trunks to form a hedge, archway, or tunnel.

Pole saw: A saw with a long handle several feet long.

Pole lopper: A device on a pole that passes a sharp blade past an anvil operated from the ground or in a tree used to remove small diameter branches.

Pollard head: The starch-rich swollen living tissue comprised of callus, collars, and buds at the end of a branch or stem resulting from many years of removing shoots back to the same point.

Pollarding: The specialized training technique used to maintain a tree at a specified height with regular heading to the exact same position; not the same as topping.

Pollarding cut: A pruning cut that removes sprouts back to the same location annually or every other year.

Poor compartmentalizer: A plant with poor resistance to movement of discoloration and decay; decay could move rapidly through the plant following mechanical injury.

Poor structure (form): Branch and trunk architecture resulting in a canopy form that could lead to premature failure of a tree part.

Potentially hazardous: See *hazardous condition.*

Pot-in-pot: Arrangement of nursery containers so one touches the others.

Preventive arboriculture: Tree care practices and techniques incorporating strategies designed to prevent problems from occurring on trees in urban and suburban landscapes.

Preventive tree care: See *preventive arboriculture.*

Primary branches: Branches attached directly to the trunk.

Production protocol: A written plan for nursery trees detailing what is to be performed and when.

Pruning: Removal of plant parts.

Pruning cycle: The interval or time between each pruning.

Pruning dose: The amount of live tissue removed at one pruning; can be used in a whole-tree sense, or on one stem only.

Pruning objectives: What is to be accomplished by pruning, for example, create and maintain strong structure by guiding a tree's architecture.

Pruning types: Includes clean, thin, reduce, raise, balance, risk reduction, restore, directional prune, vista, root prune, eradicate and structural pruning.

Quality nursery tree: Trees with good root and branch structure, typically with no circling or kinked roots and a dominant leader (for shade trees).

Quaternary branches: Branches growing from tertiary branches.

Radial crack: A crack beginning from a ring crack that forms along a ray and may extend to the bark.

Raise (raising): Provides vertical clearance under canopy.

Rays: Long groups or plates of living cells that extend from the phloem into the xylem toward the center of the trunk.

Reduce (reduction): Decreases height or spread on entire tree, or one section only, using reduction cuts; also referred to as reduction or reduction pruning.

Reduction cut (drop-crotch cut): Reduces the length of a branch or stem back to a live lateral branch large enough to assume the apical dominance—this is typically at least one-third the diameter of the cut stem.

Regular pruning: Pruning at a more-or-less set interval, such as yearly or every five years.

Removal cut: Removes a branch from the trunk or parent branch.

Renovate: Cutting a shrub back to the ground or nearly so in order to increase vigor.

Restore (restoring, restoration): The process of improving the structure of a tree that was previously topped, damaged, vandalized, or overthinned.

Ring crack: A crack that forms along a wall 4.

Root problems (defects): Conditions in the root system that could lead to poor health, or plants falling over such as circling roots, cut roots, decayed roots, no trunk flare, and deep planting.

Root pruning: Removes circling and girdling roots around trunk base; a technique of cutting many roots on a tree growing in a field nursery or landscape to prepare it for digging; cutting roots regularly to help keep a plant small.

Rounding over (roundover): Reducing the size of a tree by pruning the outer edge of a canopy with small-diameter (typically less than 2 inches) heading cuts; diameter of the cuts are typically small compared to a tree that was topped.

Sapling: A young tree about 1 to 3 years old.

Scaffold limb: A branch that is among the largest in diameter on the tree.

Secondary branches: Branches growing from primary branches.

Seedling: A young tree less than about 1½ years old.

Shade tree: Those that grow to be more than about 35 feet at maturity.

Shears: A tool used to cut many small diameter stems at once.

Single-leadered tree: A tree with a dominant trunk.

Small-maturing tree: A tree that reaches about 25 to 30 feet in height at maturity.

Species selection: The process of choosing a plant for a specific location.

Specifications: Describe what pruning types should be performed and provide other details of the pruning required for the job.

Splints: Stakes that have no contact with the ground used in a nursery to form a straight trunk.

Standard: A tree or large shrub trained into one short, straight trunk with a small dense, round canopy.

Standards: Industry accepted definitions and principles.

Starch: A chain of sugars linked together that stores chemical energy for later use by the plant.

Stem: A slender woody structure bearing foliage and buds that gives rise to other stems.

Stem bark ridge: Raised bark in the union of two stems.

Stomata: Pores or openings in leaves, typically on the undersides, through which gases such as CO_2, O_2 and water vapor pass.

Stooling: Cutting a plant nearly to the ground each year.

Structural pruning: Pruning that influences the orientation, spacing, growth rate, strength of attachment, and ultimate size of branches and stems resulting in a strong tree.

Structure: The spacing, orientation, and size of branches relative to the trunk; the arrangement of trunk and branches; the tree's architecture.

Stub: The piece of branch left beyond the collar after a removal cut.

Subordination (subordination pruning, suppression): Removing the terminal, typically upright or end portion of a parent branch or stem to slow growth rate so other portions of the tree grow faster.

Sunscald: A flattened, dried, or sunken area of the bark resulting from overexposure to the sun.

Suppression: See *subordination.*

Symplast: Network of living cell contents.

Taper: The thickening of a stem or branch toward its base.

Temporary branch: A branch that will remain on the tree for only a short period; not a permanent limb.

Terminal bud cluster pruning: Removing all buds except for the terminal from the end portion of a dormant twig.

Tertiary branches: Branches growing from secondary branches.

Thin (thinning): Removes lateral branches from the edge of the canopy; increases light and air penetration, or reduces weight by removing branches primarily from the outer edge of the canopy.

Thinning cut: See *removal cut.*

Tipping: Similar to topping except heading cuts are made through smaller diameter branches toward the outer edge of the canopy; may be called pencil pruning on some small-maturing trees when cuts are made through pencil-diameter branches.

Topiary: A training system that creates an animal, column, ball, or other shape with regular heading or shearing.

Topping: An inappropriate technique to reduce tree size that makes heading cuts through a stem more than 2 years old; a type of pruning cut that destroys tree architecture and serves to initiate discoloration and perhaps decay in the cut stem.

Transpiration: Evaporation of water vapor from foliage.

Tree evaluation: The process of determining what actions should be taken to improve plant health and reduce risk.

Tree habit: The form or shape taken on by the canopy.

Tree inspection: The careful process of checking for defects that could lead to tree failure while climbing.

Tree shelters: Tubes 2 to 4 inches diameter about 3 to 4 feet long made from various materials that are placed around the trunk of young saplings.

Trimming: Clipping the ends of young branches using heading cuts; see *tipping.*

Trunk: The main woody part of a tree beginning at the ground and extending up into the canopy from which primary branches grow.

Union (crotch): The junction between stem and branch or between stems.

Utility arborist: An arborist with specialized training who prunes trees near energized wires and other utility equipment.

Utility tree care: Tree practices near overhead wires and other potentially hazardous structures.

Vigorous branches: Those that grow at a fast rate compared to most other branches on the tree; aggressive branches.

Vista pruning: A combination of pruning types including thinning, raising, and others designed to enhance a view.

Wall 1: A term used to describe the boundary formed by plugging of xylem vessels, sometimes in response to injury; the weakest boundary that resists movement of discoloration and decay up and down the xylem.

Wall 2: A term used to describe the boundary between one growth ring and the next; a boundary that resists movement of discoloration and decay in toward the pith usually considered stronger than wall 1.

Wall 3: A term used to describe the boundary formed by rays; a boundary that resists movement of discoloration and decay around the trunk along growth rings usually considered stronger than wall 2.

Wall 4: A term used to describe the boundary formed by a reaction zone along the cambium in response to injury; the strongest boundary that resists movement of discoloration and decay into wood formed after injury.

Water sprouts: Stems arising from interior branches often growing upright and vigorously, often as a result of a stress such as overpruning, drought, or root damage.

Weak crotch (union): A union with included bark; a union that is relatively weak compared to other unions.

Wind throw: When a tree falls over due to a strong wind.

Wound closure: The process of forming callus and woundwood over a wound such as a pruning cut.

Wound dressing: A substance, solution or formulation developed for application over a recent pruning cut.

Woundwood: Differentiated woody tissue forming around a wound, such as a pruning cut. See *callus* for comparison.

Xylem: The woody part of the tree that begins on the inside of the cambium.

REFERENCES

American National Standards Institute. 1994. *Tree care operations—Pruning, trimming, repairing, maintaining, and removing trees, and cutting brush—Safety requirements.* ANSI Z133.1 New York: American National Standards Institute.

American National Standards Institute. 1996. *American standard for nursery stock.* ANSI Z60. New York: American National Standards Institute.

American National Standards Institute. 2000. *American National Standard for tree care operations—Tree, Shrub, and Other Woody Plant Maintenance—Standards practices (Support Systems a. Cabling, Bracing, and Guying).* ANSI A300 (part 3). New York: American National Standards Institute.

American National Standards Institute. 2001. *American National Standard for tree care operations—Tree, Shrub, and Other Woody Plant Maintenance—Standards practices (Pruning).* ANSI A300 (part 1). New York: American National Standards Institute.

Eisner, N. and E. F. Gilman. 2002. Branch characteristics impact formation of branch protection zones. *J Arboriculture.* In press.

Florida Department of Agriculture. 1998. Grades and standards for nursery plants. Gainesville, Fl.

Gilman, E. F. 1997. *Trees for urban and suburban landscapes.* Delmar Publishers, Albany, NY. 674 pgs.

Gilman, E. F. 2000. *Horticopia Professional CDROM.* Horticopia, Inc. Purcelleville, VA.

Green, T. L. and G. W. Watson. 1989. Effects of turfgrass and mulch on the establishment and growth of bare root sugar maple. *J. Arboriculture,* 15:268–272.

Karlovich, D. A., J. W. Groninger, D. D. Close. 2000. Tree condition associated with topping in Southern Illinois Communities. *J. Arboriculture,* 26:87–91.

Lonsdale, D. 1999. *Principles of tree hazard assessment and management.* Dept. of Environment, Transport and the Regions, London. 388 pgs.

Matheny, N. P. and J. R. Clark. 1996. *Evaluation of hazard trees in urban areas.* Champaign, IL. International Society of Arboriculture.

Miller, R. W. 1981. Economic evaluation of pruning. *J. Arboriculture,* 7:109.

Research practical speaking, video. 1994. The seasonal differences in the effects of wounds on hardwoods. Series II, Vol. 5. International Society of Arboriculture, Champaign, IL.

Shigo, A. 1991. *Modern arboriculture: A systems approach to the care of trees and their associates.* Shigo and Trees, Associates, Durham, NH.

Struve, D. 1996. Terminal bud cluster pruning. Presented at International Society of Arboriculture Ann. Conf. Cleveland, OH.

INDEX

T